PHYSICS OF THERMAL GASEOUS NEBULAE
(Physical Processes in Gaseous Nebulae)

ASTROPHYSICS AND SPACE SCIENCE LIBRARY

A SERIES OF BOOKS ON THE RECENT DEVELOPMENTS
OF SPACE SCIENCE AND OF GENERAL GEOPHYSICS AND ASTROPHYSICS
PUBLISHED IN CONNECTION WITH THE JOURNAL
SPACE SCIENCE REVIEWS

VOLUME 112

PHYSICS OF THERMAL GASEOUS NEBULAE

(Physical Processes in Gaseous Nebulae)

by

LAWRENCE H. ALLER

University of California, Los Angeles, U.S.A.

D. REIDEL PUBLISHING COMPANY

A MEMBER OF THE KLUWER ACADEMIC PUBLISHERS GROUP

DORDRECHT / BOSTON / LANCASTER

Library of Congress Cataloging in Publication Data

CIP

Aller, Lawrence H. (Lawrence Hugh), 1913–
Physics of thermal gaseous nebulae.

(Astrophysics and space science library; v. 112)
Includes bibliographies and index.
1. Nebulae. I. Title. II. Series.
QB853.A37 1984 523.1′135 84-13403
ISBN 90-277-1814-8

Published by D. Reidel Publishing Company,
P.O. Box 17, 3300 AA Dordrecht, Holland.

Sold and distributed in the U.S.A. and Canada
by Kluwer Academic Publishers,
190 Old Derby Street, Hingham, MA 02043, U.S.A.

In all other countries, sold and distributed
by Kluwer Academic Publishers Group,
P.O. Box 322, 3300 AH Dordrecht, Holland.

Printed in The Netherlands

TABLE OF CONTENTS

PREFACE

Gaseous nebulae offer outstanding opportunities to atomic physicists, spectroscopists, plasma experts, and to observers and theoreticians alike for the study of attenuated ionized gases. These nebulae are often dusty, heated by radiation fields and by shocks. They are short-lived phenomena on the scale of a stellar lifetime, but their chemical compositions and internal kinematics may give important clues to advanced stages of stellar evolution.

The material herein presented is based on lectures given at the University of Michigan, University of Queensland, University of California, Los Angeles, and in more abbreviated form at the Raman Institute, at the Scuola Internazionale di Trieste, and elsewhere. Much of it is derived origionally from the series "Physical Processes in Gaseous Nebulae" initiated at the Harvard College Observatory in the late 1930s. I have tried to emphasize the basic physics of the mechanisms involved and mention some of the uncertainties that underlie calculations of many basic parameters. Emphasis is placed on ionized plasmas with electron temperatures typically in the neighborhood of 10,000°K. Dust and other ingredients of the cold component of the interstellar medium are treated briefly from the point of view of their relation to hot plasmas of H II regions and planetaries. Chemical composition determinations for nebulae are discussed in some detail while the last section deals with interpretations of elemental abundances in the framework of stellar evolution and nucleogenesis. Gaseous nebulae offer some particularly engaging opportunities for studies of stellar evolution.

This volume could not have been prepared without the loyal help of many colleagues and associates here and abroad. Illustrations have been supplied by N. Walborn, G. Jacoby, and P. Kupferman. Line drawings or reprints have been sent by John Raymond, N.K. Reay, R.W. Russell, Martin Cohen, Paul Scott, M.J. Seaton, and H. Dinerstein. Stimulating discussions (particularly at IAU meetings and at the Ithaca and London symposia on planetary nebulae) with the Peimberts, Karen Kwitter, M.J. Seaton, D.E. Osterbrock, T. Barker, H. Ford, R. Dufour, R. Costero, H.B. French, F.D. Kahn, J. Lutz, C. Mendoza, M. Felli, G. Tofani, D.G. Hummer, J. Köppen, M. Rosa, R.P. Kirshner, H. Nussbaumer, Y. Terzian, M. Perinotto, A. Renzini, I. Iben, V. Weidemann, D. Schonberner, G. Shields and many others have been extremely useful. I am heavily indebted to close associates such as M.F. Walker, D.K. Milne, and particularly S.J. Czyzak.

I am substantially indebted to those who have read one or more chapters and have supplied critical comments and suggestions. Donald P. Cox rendered invaluable help on Chapter 9; J.P. Harrington and R. Tylenda offered important suggestions on Chapter 7, as did Bruce Balick for Chapter 10. S.R. Pottasch read the entire manuscript, made many helpful suggestions, and supplied data on the He continuum (Table 9, Chapter 4). J.B. Kaler reviewed the entire manuscript with great care; he made numerous helpful suggestions and supplied illustrative material. To C.D. Keyes I am indebted specifically for most of Table 9 in Chapter 4, Table 5, Chapter 5, and the diagostic digrams, Figures 4 through 12 of Chapter 5, but even more importantly for much indirect help and encouragement, particularly with regards to the nebular modeling described in Chapters 7 and 11. Walter Feibelman has made many useful comments and suggestions. I am grateful to the Royal Astronomical Society and to Astronomy and Astrophysics for permission to reproduce illustrations and to University of Chicago Press for permission to use copyrighted material from the Astrophysical Journal and Supplements. The appendix on atomic parameters is reproduced from C. Mendoza, IAU Symposium No. 103, with his kind permission. Finally, for the patience and perseverence of Mr. Robert L. O'Daniel, who prepared the photo-ready copy, I am most grateful.

Chapter 1

TYPES OF GASEOUS NEBULAE

A gaseous nebula is an ionized low-density plasma, very large in size compared with laboratory or even stellar dimensions. There exist a great variety of these objects. Densities range from a few ions and electrons/cm^3 to perhaps $10^{12}/cm^3$, beyond which point the plasma resembles the photosphere or chromosphere of a stellar atmosphere. We restrict our attention to ionized plasmas that emit observable spectral lines and continua. The gas kinetic temperatures of these objects, as measured by speeds of ions and electrons, range from a few thousand degrees in galactic diffuse nebulae to several hundred thousands or even millions in the highly attenuated bubbles associated with supernova events. Our main attention, however, will be devoted to objects with gas kinetic temperatures between 5000°K and 30,000°K.

Many gaseous nebulae are excited by ionizing radiation emitted by hot stars, mostly in the range ($30,000°K < T < 150,000$) or possibly by nonthermal sources as seem to occur in galaxies with active nuclei such as Seyferts or quasars. In other instances, energy may be provided by collisions of gas clouds with one another or with the interstellar medium. The energy of the shocked gas is gradually dissipated as radiation.

Our interest in gaseous nebulae is inspired by several engaging aspects. First, there are interpretations of their observable properties, particularly their spectra in terms of well-defined physical processes. It is this problem that first attracted the attention of physicists and astronomers in the twenties and thirties of this century. The relevance of gaseous nebulae to far broader contexts of stellar birth, demise, element-building scenarios, and the structure and evolution of galaxies was to come later. In fact, most of the gaseous nebulae we shall study can be assigned to some epoch of the life of a typical star. Great diffuse clouds of glowing gas such as the Orion nebula called H II regions are but clumps of the primordial chaos from which stars are formed. These cosmic embers glow because massive, blue-hot stars have been suddenly "turned on" in a region containing great quantities of gas and dust. The copious ultraviolet radiation ionizes the gas. As H or He ions and electrons recombine, they emit the familiar spectra of these elements. Unfamiliar spectral lines of familiar elements such as oxygen and nitrogen, the so-called forbidden lines, are produced by collisional excitation of ions from their ground levels to nearby levels where they cascade back with emission of radiation.

Other types of nebulae, the so-called planetaries, the ejecta of novae, and supernova remnants involve material thrown out from stars in advanced stages of their evolution.

One of the engaging facets which we will emphasize is the chemical composition of these tenuous plasmas. Determination of elemental abundances is beset with many difficulties which we will enumerate later, but it is possible to find ionic concentrations for the more abundant elements of the first two rows of the Periodic Table.

Analyses of the H II regions or diffuse nebulae give us the chemical composition of the interstellar medium. If the temperature is high enough and the solid grains are all evaporated, we may get a reasonably accurate assay. Otherwise, some of the nonrefractory elements such as carbon or magnesium may be tied up in solid particles. Although we may be able to see radiation from only a limited number of ions of a given element, e.g., we normally observe nitrogen only in its neutral and first ionized state, it is possible to make reasonable estimates of the distribution of atoms among various ionization stages, even though the gas is far from equilibrium. Thus, analyses of the spectra of diffuse nebulae offer opportunities for probing the composition of the interstellar medium whenever high-temperature stars cause it to fluoresce and shine.

This circumstance enables us to analyze gas clouds in distant galaxies where it is no longer possible to see, much less obtain, spectroscopic observations for individual stars. Further, one can study the change in chemical composition with distance from the nucleus in such a galaxy and assess various proposed element-building scenarios. In our own galaxy, of course, we can compare the chemical composition of the gas with that of stars recently formed from it, e.g., the Orion nebula with the stars contained therein. Aside from the usual sources of error in stellar or nebular analysis, we must always be mindful of the possibility that some elements may be held in refractory grains.

The chemical compositions of shells ejected from dying stars, e.g., planetary nebulae and supernova remnants (SNR), have been extensively studied. Supernova remnants may reflect the composition of the stellar material which has been processed via nuclear transformations in the deep interior, or a mixture of stuff from the defunct star and the neighboring interstellar medium on which it has impinged. Sometimes, as in Casseopia A, different ejected blobs show different chemical compositions, suggesting that they may have come from different depths in the element-building furnace.

Among planetary nebulae, the situation is less dramatic. The chemical compositions of some may correspond to that of the ambient interstellar medium at the place and remote time of the formation of the parent star. Such a situation pertains to stars of relatively low mass such as the sun. Envelopes ejected by more massive stars may contain a mixture of the initial interstellar medium material and products of nuclear processing in the stellar interior. Nitrogen and carbon are often enhanced. Both planetary nebulae and supernovae are important as sources of new supplies of carbon and heavier elements to the interstellar medium. Shells of ordinary novae can also be regarded as temporary or evanescent nebulae. Some Wolf Rayet stars have expanding envelopes and may be important suppliers of material to the interstellar medium. Combination variables or symbiotic stars show spectral characteristics somewhat similar to those of gaseous nebulae.

Figure 1-1

EXAMPLES OF GASEOUS NEBULAE

H II Regions or Diffuse Nebulae

We start our discussion with some of the most familiar objects where an illuminating star or stars serve as a copious source of ultraviolet radiation. Some relatively simple examples of H II regions are found in our galaxy or in the Magellanic clouds where often single blue stars or a star cluster may be seen at the center of a glowing nebulosity. See Figure 1-1.

The best known diffuse nebulae, such as M 8, the Trifid Nebula, the η Carinae Nebula (see Fig. 1-2) or the great Orion Nebula (Messier 42 or NGC 1976), are far from being simple ionized gas plasmas. The bright portion of NGC 1976 has a diameter of 4' which amounts to half a parsec at a distance of 470 parsecs. Actually, what we observe as the Orion nebula is an ionized zone in a vast complex dominated by cool molecular clouds and irregular dark filaments containing solid grains. This luminous volume is excited by hot O-type stars, the celebrated Trapezium. Behind it lies a very dense cloud containing molecules of CN, H_2CO and HCN in which the average density of molecular hydrogen is $n(H_2) \sim 10^5$ cm^{-3} over a volume of diameter approximately 6'. At the center of this cloud is a condensation called the Kleinman-Low nebula which is observed in the infrared. The mass of the main body of this dusty molecular cloud is of 1000 $M_\odot$.

The luminous Orion nebula is one of the brightest and densest H II regions known, although its density, $N(H) \sim 10^4$ cm^{-3}, is less than that of the background cloud. The gas kinetic temperature is of the order of 10,000°K to 12,000°K. We seem to be observing a time-dependent phenomenon. The Trapezium stars appear to have been turned on "just yesterday," i.e., perhaps a few tens of thousands of years ago, and their ultraviolet radiation is eating into the dense molecular clouds. As H_2 becomes dissociated and ionized, the gas pressure rises precipitously and the material flows into the H II region. Such a model seems to be required to explain the observed motions of the gases. The Orion complex is regarded as an active region of star formation. This mixture of young stars, incandescent and relatively cool gas (much of it in molecular form) and dust, is enclosed in an expanding shell of cool, largely atomic hydrogen with a radius of 8.5° or about 66 parsecs and a total mass of about 100,000 $M_\odot$. Our present picture of this very complicated melange is based on optical, infrared, radio-frequency (including mm waves), and ultraviolet observations which reveal a vast range in physical conditions of the gas. See Chapter 10.

Planetary Nebulae

While an object such as M42 is closely associated with the formation and birth of stars, another type -- called planetary nebulae -- is associated with their demise. It was once though that these extended, low-density envelopes of white dwarf-like stars were relatively simple objects, but more recent work has shown them to be

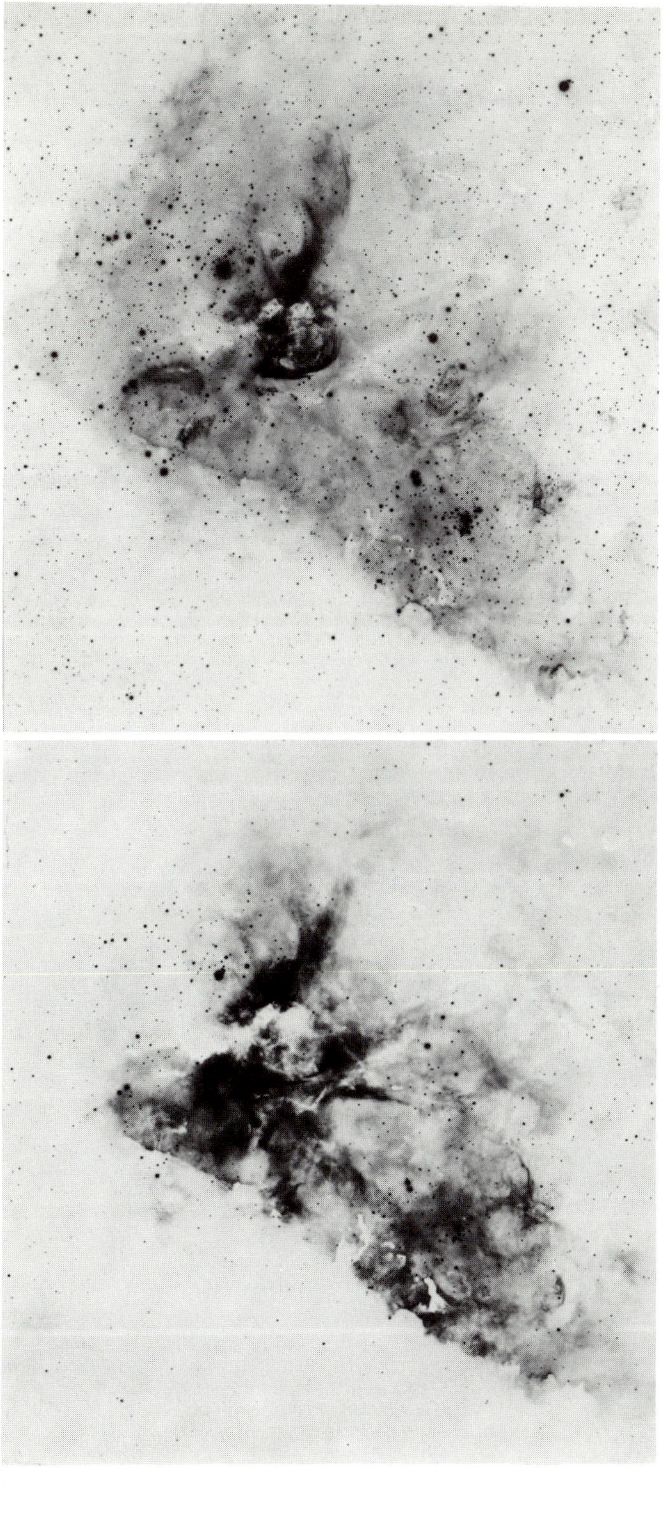

Figure 1-2

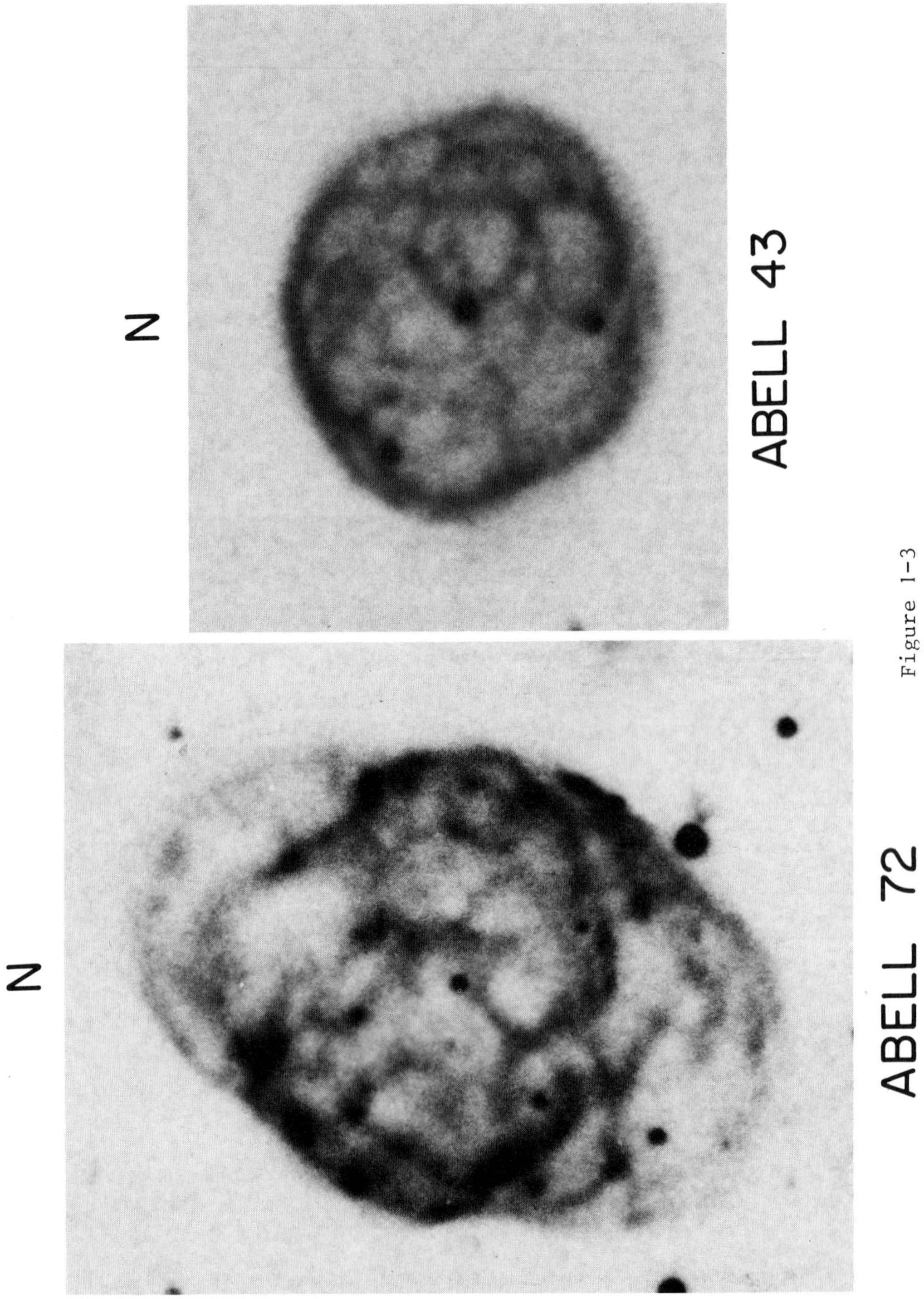

Figure 1-3

frequently rather inhomogeneous objects often containing quantities of dust and molecules (see Fig. 1-3).

The faint green disks of some of the brighter objects of this class reminded early visual observers of Uranus and Neptune, hence the name planetary nebula. A typical object of this class has a radius of about 0.5 pc or 10,000 astronomical units, an ionic density in the neighborhood of 1000 to 10,000 ions/cm^3 and a gas kinetic temperature of 8000°K to 15,000°K. Their spectra display recombination lines of hydrogen and helium, and collisionally excited (often so-called forbidden) ionic lines of abundant elements such as C, N, O, Ne, Mg, Si, S, Cl, and Ar.

Unlike the ionized portions of the interstellar medium (H II regions) which are confined to spiral arms in the plane of the galaxy, planetary nebulae show a spatial distribution akin to that of a stellar population whose ages extend over many billions of years. They are found at high galactic latitudes and in great numbers towards the center bulge of our galaxy. Some are associated with central galactic bulge and halo populations, others with the ancient disk population, and yet others may be assigned to what is called the Type I populations. Thus, although the lifetime of an observable planetary nebula may be limited to ten to thirty thousand years, the age of the star from which it was ejected may vary from the order of tens to millions of years to virtually the age of the galaxy.

They are not few in number; perhaps between 10,000 and 20,000 are to be found in systems like our galaxy or M31. Available data are consistent with the suggestion that many stars in the mass range $1 \rightarrow 8\ M_\odot$ may evolve into planetary nebulae.

Nebulae of high luminosity that are heavily enriched in products of nucleogenesis, e.g., NGC 6302 or NGC 7027, must have come from the more massive stars.

Progress is hampered by the circumstance that nebular distances are poorly known. Nevertheless, although many important details remain to be supplied, the broad scenario of the origin, development, and disappearance of planetary nebulae appears to be clear. At an advanced stage in its life, a star evolves into a red giant and may become a long-period variable (LPV). Eventually, the core of the red giant becomes very hot and compressed while its outer envelope, continuing to expand outwards, becomes unstable. Some material may escape as a wind before ejection of the large chunk that is destined to become the luminous shell of a planetary nebula.

Specific details of the ejection process are far from settled and need not concern us here. At the outset, the ejected material is dense (compared to usual nebular standards), inhomogeneous, and sufficiently cool for solid particles of carbon or silicates to form. The shrinking hot core of the dying star is initially hidden by the gas of the inner envelope, but gradually, this material becomes more and more rarefied and can no longer block the intensely blue flux of the core from the expanding shell. The inner portion of the shell becomes ionized. At this stage we have a dense, compact planetary nebulae such as IC 4997. The shell continues to expand with a velocity of the order of 10 to 30 km/sec. As time goes on, the rich ultraviolet radiation of the central star penetrates more and more of the nebular shell and causes the gas to fluoresce copiously. The nebulae may show a variety of shapes and forms dependent on details of the ejection process, and how much of the material is trapped as neutral gas in cool, dense blobs. The nebular spectrum will depend on these factors and on the energy distribution in the radiation from the hot core.

In the outer parts of planetary nebulae not only neutral atoms but also molecules may exist, particularly H_2. Carbon monoxide is found throughout a cloud five times larger than the visible nebula NGC 7027. Solid (evidently mostly carbon) particles that had been formed by condensation in the atmosphere of the red giant precursor are heated by the radiation field of the star and nebula to temperatures of the order of 100°K.

Finally, the nebula starts to fade as the gas density drops substantially. The surface brightness becomes so low that the nebula would be hidden by even a small amount of interstellar extinction. The low surface brightness planetaries catalogued by Abell (1966) are examples of objects of this type.

Planetary nebulae have long been favorite objects for detailed observational studies, especially spectroscopic ones. The combination of high surface brigntness, the ready accessibility of structural details, a great range in excitation, and apparently well-understood mechanisms of spectral line excitation has inspired intensive investigation for more than half a century.

In contrast, H II regions are more difficult to handle both observationally and theoretically. In both H II regions and planetaries, energy is supplied through photoionization of atoms and ions by radiation from high-temperature stars whose ultraviolet energy distribution sometimes can be estimated with the aid of theory of stellar atmospheres. Other sources of excitation, e.g., the direct contribution of stellar winds usually appears to be relatively unimportant. Ionized plasmas that are excited by photoionization display spectroscopic features different from those such as supernova remnants that are excited by shock wave phenomena.

SUPERNOVA REMNANTS

The Crab Nebula (M1, NGC 1952) is the best known and most intensively studied of all supernova remnants (SNRs). It originated from the 1064 supernova which was observed by the Chinese and Japanese. It appears to be of a type that is relatively infrequent.

There are believed to be two types of SNRs. The most frequently observed type has the appearance of a hollow shell seen in projection on the sky (i.e., it appears as a bright ring with an empty center). This is the kind associated with historical supernovae of 1006, 1572 (Tycho), 1604 (Kepler), and Cass A which is believed to have exploded in 1667. The best known examples are the Network nebula in Cygnus, which has a diameter of about 40 pcs (see Fig. 9-1), and IC 443. After the detonation of the supernova, these structures are formed when the expanding shock front hits the interstellar medium. They fade slowly and may last 50,000 years or so. Many are detected from the characteristic radio frequency emission. Numerous examples are known, not only in the galaxy, but also in the Magellanic Clouds and the spiral galaxy, M 33. SNRs of this type appear to be associated with what are called "Type I" supernovae.

The type of SNR illustrated by the Crab nebula has a very different type of structure from the Network nebula. These objects are brighter in their central regions and do not show a shell structure. In the radio frequency range they show a flat spectrum with linear polarization. These objects (which are often called plerions) emit synchrotron radiation. They appear to be relatively short-lived, their energy being derived from the spindown of a pulsar which came from a Type II supernova.

Except for the Crab nebula, most attention has been paid to hollow shell objects such as the Network nebula. The spectra of the filaments themselves superficially resemble those of conventional gaseous nebulae. The red forbidden lines of ionized sulfur [S II] $\lambda 6717$, 6730 tend to be particularly strong; [O II] $\lambda 3727$ also appears, both ions are indicative of a low level of ionization. These cool filaments, however, are interlaced with plasma whose temperature is of the order of 3,000,000°K and which emits soft X-rays and the coronal line [Fe XIV] $\lambda 5303$.

The excitation of SNRs is derived from shock phenomena and involves much more complicated physical processes than we find in plasmas excited by photoionization. Phenomena are strongly time-dependent, involving nonequilibrium situations. Further complications are imposed by the inhomogeneity of the material involved. All theoretical attempts to explain the observed motions of the material, gas kinetic temperatures, etc., in terms of shock phenomena in a uniform medium have failed. The interstellar medium is markedly heterogeneous.

Thus, intensive studies of the kinematics and spectra of SNRs can teach us about the very involved physics involved and some of the properties of the interstellar medium. Behind the shock fronts, large volumes of space are filled with rarified plasmas at gas kinetic temperatures typically in excess of 1,000,000°K. These hot, low-density regions are often called bubbles. When a supernova explosion occurs, the ejecta moving at velocities of about 10,000 km/sec bulldoze everything in their way. After a while, they have swept up an amount of interstellar material greater than their own total mass. The remnant gradually cools, but the hot gas in the swept-up volume or bubble, now at a temperature of 2,000,000 or 3,000,000°K, emits X-rays and coronal lines. A typical SNR bubble has a diameter of about 50 pcs. Although their sizes are smaller, bubbles can be produced also by O-stars (Chapter 10).

Yet, larger structures called superbubbles have been identified. One example is a cloud of X-ray-emitting gas with a diameter of about 30° (200 pcs) $T_\varepsilon \sim 2,000,000$°K in the constellations of Eridanus and Orion. The faint filamentary structure known as Barnard's Loop (which is very bright in the ultraviolet) appears to define an edge of this giant loop. An even more impressive superbubble has been identified in Cygnus. It lies at a distance of about

Figure 1-4

1,800 pcs and has a diameter of 400 pcs. The X-ray-emitting gas is at a temperature of ~ 2,000,000°K, with a total energy content of about 10^{52} ergs; more than an order of magnitude more energy than is contained in a typical SNR. It can be shown that such a superbubble cannot be created in a single supernova detonation. The origin of these structures seems to lie in a sequence of supernova events. The shock wave from the first supernova compresses the interstellar medium, permitting the formation of massive stars, some of which eventually produce supernovae. The round robin continues and ultimately a superbubble of vast volume appears. It has been estimated than ten percent of the volume of the galaxy in the solar neighborhood is occupied by bubbles of X-ray-emitting gas. The fraction of the energy content of the interstellar medium found in supernova bubbles may be even higher.

THE 30 DORADUS NEBULAR COMPLEX

The most dramatic of gaseous nebulae is certainly the 30 Doradus complex, sometimes regarded as the nucleus of the Large Magellanic Cloud (see Fig. I-4). The 30 Doradus (Tarantula) nebula itself seems to be an enormous H II region that presides as the centerpiece or hot spot of a much larger structure that spreads over a domain of 2 kpc diameter. This extended complex includes far-flung wisps and filaments, bright infrared sources (identified by the IRAS satellite as probable regions of star formation), an IR protostar, CO emission, SNR's and an X-ray source, LMC X-1, associated with large H II regions (N 158, 159, and 160 in Henize's 1966 catalogue). The diameter of the Tarantula nebula itself is about 15' which corresponds to about 240 pcs. Note that at the distance of the LMC, the bright 4' core of the Orion nebula would subtend an angle of only 2"! In the core of the nebula is a remarkable star cluster that contains many hot stars, some of which are certainly multiple systems resembling the Trapezium in Orion. They appear to ionize a cloud of gas amounting to 500,000 solar masses, spread out over a volume of nearly half a kiloparsec diameter (see Chapter 10). Local electron densities ranging from $80/cm^3$ to $800/cm^3$ have been measured, but regions of the same size and density as the Orion nebula cannot be excluded. This H II region is intimately associated with the densest clouds of neutral H in the LMC. Radio frequency observations show that in a manner analogous to the nucleus of our galaxy, the 30 Doradus nebula complex contains a number of distinct thermal and nonthermal sources.

Assessment of optical direct photographs, radio-frequency observations, and spectroscopic data show the 30 Doradus complex to be the site of marvelously complex developments in a dynamic situation. Akin to Orion, but on a much grander scale, the H II region is growing as ionization zones penetrate clouds of neutral material. Bright rims observed in forbidden lines of ionized nitrogen appear to define fronts along which ionization is proceeding. It is unlikely, though, that all the intricate features of 30 Doradus, such as the large, graceful arcs, are to be interpreted as ionization fronts. A substantial number of SNRs have been identified in the LMC and it is probable that expanding supernova shells or shells blown out by O-stars and Wolf-Rayet stars may contribute to the structural complexity of 30 Doradus. Spectroscopic data from [S II] and [O II] lines demonstrate a wide range in density throughout the central region, much of it occurring on a small scale. The [S II] data in particular suggest that shocks from supernovae may have played an important role in this object. In addition, they may have triggered star formation. Much of the nonthermal radio frequency may arise from these SNRs.

Thus, 30 Doradus is perhaps the most fascinating gaseous nebula known. It may be an example of a nucleus of a galaxy, favorably placed so that it can be studied optically as well as at other frequencies. Urgently needed are high-resolution studies at radio frequencies to separate thermal and nonthermal sources and to correct for extinction effects. Analyses of the optical spectra of bright regions are consistent with a photoionization model, i.e., the energy comes from ultraviolet spectra of very luminous imbedded stars, not from the dissipation of shock-wave energy. It is possible that small filamentary wisps with strong forbidden lines of ionized sulfur [S II] are actually SNRs and owe their luminosity at least in part to the dissipation of mechanical energy.

NONSPATIALLY RESOLVABLE OBJECTS, GALAXIES AND STARS

There are many objects for which satisfactory spatial resolution is impossible and for which spectroscopic data clearly indicate a source akin to a nebular plasma. These include objects such as η Carinae, certain old novae, combination variables on the one hand and the nuclei of certain galaxies on the other. Of particular interest in this context are radio galaxies, Seyferts and N galaxies, and finally quasars which are often interpreted as very remote galaxies with extremely bright nuclei, such that the outer portions are unobservable. In this group are also placed the BL Lacertae objects which differ from the others in that they show no bright line spectra.

Radio galaxies, N galaxies, Seyferts and quasars seem to display a continuum of properties grading from one into the other. Some radio galaxies and Seyferts are spatially resolvable, although most of them and all quasars appear as essentially point sources.

Cygnus A (3C 405) and Centaurus A are prototypes of radio galaxies. The spectrum of Cyg A shows bright emission lines ranging from transitions found in the earth's aurora [O I] to those found in the solar corona: [Fe X] nine times ionized iron! The lines are broad; FWHM (full width half maximum) amounts to about 500 km/sec. The spectrum resembles that of a planetary nebula remarkably closely. The forbidden lines of ionized nitrogen [N II] and doubly ionized oxygen [O III] suggest gas kinetic temperatures of 10,000°K and 15,000°K. The forbidden ionized sulfur lines indicate a rather low density $N_\varepsilon \sim 10^3/cm^3$. Densities may be higher in other zones, but the radiating plasma seems to be distributed in a very nonuniform manner. The total mass of the radiating gas has been estimated as about 20 million suns. Most radio galaxies are similar to Cyg A in having so-called "narrow" lines, i.e., FWHM = 500 km/sec. These are the more copious emitters of radio frequency radiation. About one-third have very broad lines, FWHM ~ 10,000 km/sec of H, He, etc., and relatively narrow forbidden lines [N II], [O III]. In the doubly ionized oxygen, O^{++} zone, the plasma density is of the order of 10^6 to approximately $10^7/cm^3$, but in the broad line zone the density must exceed $10^8/cm^3$. A relatively small amount of mass (~ 10 $M_\odot$) compressed in a volume smaller than that of a typical planetary nebula could produce the observed emission, and permit the short time-scale variations observed in this class. Many broad-lined radio galaxies are observed optically as "N" galaxies. The "narrow"-line radio galaxies are often giant ellipticals; some are classed as cD or DE galaxies.

Radio galaxies show strong background continuous spectra in which the energy distribution follows a power law, $F_\nu \sim \nu^{-n}$ where n is typically of the order of 1.5. The source of energy for the bright-line spectrum is photoionization followed by recombination, but the continuum is due mostly not to a mixture of hot stars but to synchrotron emission. The line spectrum of the filaments in the Crab nebula is also excited by synchrotron radiation and in fact it resembles that of Cyg A.

Radio galaxies resemble Seyferts in some regards, but there are important differences. A Seyfert galaxy has a bright compact nucleus, whereas that of an ordinary galaxy tends to be diffuse. The outer parts of a Seyfert galaxy tend to be relatively faint and often difficult to detect. When the background galaxy can be classified, it is found that most of them are spirals, and of spiral galaxies, perhaps 1% are Seyferts. Spectroscopically, most of them have strong, broad emission lines and also intense infrared continua. Most Seyferts are not strong radio emitters. Seyfert galaxies are generally classified into two groups, Sy 1 with very broad lines of H, He and possible Fe II blends up to 10,000 km/sec, and narrow forbidden lines, and Sy 2 where H and forbidden lines have the same width, up to about 1000 km/sec.

The Sy 1 nuclei are very inhomogeneous. The Balmer lines arise in plasmas with densities of ~ 10^8 ions/cm^3, while the forbidden lines arise in low density regions perhaps 1,000,000 times larger in size. The continuum in the ultraviolet, optical and infrared regions follows a power law and appears to be of synchrotron origin. The power supplied to the optical spectrum comes via photoionization from this continuum. These features of Sy I nuclei recall broad line radio galaxies and quasars, but as noted above Seyferts sometimes show spiral structure. The broad spectral lines of all Seyferts are attributed to Doppler motions, although scattering by electrons may play a role. Because of similarities in profiles of different lines in Sy 2 objects, different spectral features, H, He and [O II], etc., are believed to arise in the same physical space. The background continuum in these objects may arise partly from a larger number of hot stars, but a synchrotron source may be present, also. The infrared radiation appears to come mostly from hot dust. In many Seyferts, the largest contribution to the total luminosity comes from the infrared. Radio emission is observed only from narrow-line Sy 2 objects which seem to have some resemblance to the narrow-line radio galaxies. Some Seyferts show spectral line variations.

Radio galaxies, N galaxies, Seyferts, and BL Lacertae objects are all examples of galaxies with active nuclei. There is considerable overlap and whereas a Seyfert can be distinguished from an ordinary galaxy because of its bright nucleus, the brightest Sy 1 objects could qualify as quasars. In fact, many of the properties of quasars suggest scaled-up Seyferts. They differ, however, in one important respect: they do not show underlying galaxies. Although the first examples of quasars were found through their radio emissions, many are radio quiet. Most of them are blue and radiate from the infrared through the visual to the ultraviolet and even the X-ray region.

One factor that sets quasars apart is that they all appear to be remote, particularly if you attribute the red shifts in their spectra to cosmological Doppler shifts. Some have small red shifts; others are so distant that the Ly α-λ3000 spectral region is shifted into the visible, thus astronomers had a chance to view the "satellite" ultraviolet spectrum before they could study effectively this region in nearby galactic objects.

If we adopt the cosmological red shift interpretation, and we have a reasonably good idea of the energy distribution in their spectra, we can estimate their luminosities from their apparent magnitudes. The energy output of a bright galaxy is $10^{43} \rightarrow 10^{44}$ ergs/sec^{-1}; for comparison, L(sun) = 4×10^{33} ergs/sec. Yet, most quasars fall in the range 10^{46} to 10^{48} ergs/sec^{-1}, i.e., 100 to 10,000 times as bright as a "bright" galaxy. Telescopically, quasars are point sources and are generally interpreted as compact nuclei of galaxies. Although most quasars are steady over time scales of the order of 50 years, some show light variability on a time interval of less than one year, thus requiring that the radiating volume be relatively small.

Astrophysicists have paid a great deal of attention to quasar spectra since these appear to offer the greatest hope for understanding the physical conditions in their intensely luminous cores, and perhaps ultimately in unraveling the source of their energy output. Three components are often observed in the spectra of quasars: a) there is always present a continuous spectrum which seems to follow very roughly a power law: $F(\nu) \sim \nu^{-n}$ (where $n \sim 1$). It extends from the optical into the ultraviolet and X-ray regions and is almost certainly of nonthermal, perhaps synchrotron origion. There may be some component of thermal origin; b) emission lines which suggest a plasma of "nebular" densities, i.e., less than 10^{12} ions/cm^3. Permitted lines of hydrogen, carbon, etc., are invariably broad; perhaps because of larger scale mass motions, while the "forbidden lines" characteristic of typical low-density gaseous nebulae are much narrower; c) many high red shift quasars show systems of narrow absorption lines with different red shifts. The red shifts of these absorption lines are always less than that defined by the emission lines. It is not known if these arise from clouds that are intimately associated with quasars or if they just happen to fall in the line of sight. Either interpretation presents difficulties, for if we suppose that they are associated with the quasar itself, they indicate clouds that are ejected with velocities ranging from several thousand km/sec to even a substantial fraction of the velocity of light. If they are intergalactic clouds that just happen by chance to lie in front of quasars, a large number of very remarkable plasma clouds must be postulated.

As with radio galaxies and most Seyferts, a total lack of spatial resolution imposes difficult problems. The real challenge is to see how much information can be extracted from the intensities and shapes of the emission lines. The absorption spectra pertain to regions far removed from quasar cores where the action is, while the background continuous spectrum, although fundamental to the excitation mechanism, tells us very little about the underlying physical conditions in the source.

Hence, we have to fall back on the emission lines and see what clue they can offer us. Details will be explained later; here only some general comments will be offered. Quasar spectra resemble those of nebulae and not stars. A consistent picture is obtained if we assume that the broad permitted lines are produced in a plasma with a density between 10^9 and 10^{10} particles/cm^3 at $T_\varepsilon \sim 15{,}000$ to 25,000°K. These dense regions may contain 100 to 1000 solar masses. The forbidden lines are produced in a vast region of perhaps a thousand times as much mass with a density between 10^3 and 10^6 ions/cm^3. The source of the energy in the gas is photoionization followed by recombination. The chemical composition of the material appears to be similar to that of the sun.

COMBINATION VARIABLES AND OTHER EXOTIC OBJECTS

Lack of spatial resolution also hinders our analyses of extended stellar envelopes which show radiations characteristic of gaseous nebulae. A combination variable such as AX Persei or Z Andromedae may show a background continuum of a cool M star and a bright line spectrum, including not only the usual H and He lines, but also forbidden lines characteristic of gaseous nebulae. At times, the cool M-type spectrum is overwhelmed by a blue continuum and there occurs an overall brightening of the source.

Such objects are usually interpreted as binary systems. An M-type giant at a late evolutionary stage is accompanied by a white dwarf-like companion of high surface temperature. The giant component of the binary swells up to its Roche limit, beyond which material is stripped off by the gravitational intervention of the companion star. Some of this emerging gas may find its way to an accretion ring from which it precipitates onto the hot dwarf companion, producing periodic outbursts. The rest constitutes a tenuous envelope surrounding the entire system and is radiatively excited by copious ultraviolet radiation of the hot companion. In the resultant composite spectrum, the H and He recombination lines and the blue continuum (when visible) must arise from the compact B star and its immediate envelope, and the tenuous envelope produces the nebular spectrum (which is variable in BF Cygni). Underlying all is the characteristic spectrum of the M star which can be smothered at times by outbursts of the blue companion.

Another remarkable object is η Carinae, which had a nova-like outburst in 1843. At the present time, it appears as a red fuzzy nucleus with a diameter of about 2" surrounded by a small bright disk (often called the humonculus) with a diameter of about 10". Both the core and the surrounding shell have a rich spectrum of permitted and forbidden lines of ionized iron with hydrogen and helium lines that show complex profiles involving absorption features. Forbidden lines of singly ionized nitrogen and doubly ionized neon are present. Many of the lines in the nucleus show sharp cores with broad underlying wings. The emission in the shell spectrum appear to be derived from the central nucleus but the sharp cores are gone; only the diffuse features remain. With respect to the background continuum, the absorption lines appear to be strengthened. It appears that the light from the shell is actually stellar radiation scattered by electrons, since the entire character of the shell spectrum appears to be what you would get with the lines from the nucleus broadened by electron scattering and displaced as a consequence of rapid Doppler motions in an expanding shell. The total luminosity of this object is about 10^7 L(sun), which makes it one of the intrinsically brightest objects known. The central star or core may have a temperature of 30,000°K and is probably the source of a massive wind that produces a dusty halo which is visible in the telescope as a fuzzy red nucleus. The bright disk of the humonculus with its complex pattern of broadened lines contains a number of blobs that were ejected from the nucleus with velocities of about 500 km/sec at the 1843 outburst. Beyond the bright disk are a few wisps that were thrown out before the main outburst, during previous variations of this remarkable object.

Gaseous nebulae and nebular envelopes fall into a distinct category with respect to the underlying physics. Perhaps the simplest objects from this point of view are the planetaries where geometrical symmetry and relatively low densities prevail. The most difficult to handle are probably the unresolved nuclei of active galaxies.

The excitation mechanism in many nebulae is photoionization. This category includes H II regions, planetaries, nuclei of active galaxies and, most likely, combination variables and the core lines in η Carinae. Shock wave excitation dominates in supernova remnants.

A Concise List of Selected References. Additional Bibliographies May Be Found in the Articles and Books Cited.

Orion Nebula: Goudis, C. 1982, The Orion Complex, A Case Study of Interstellar Matter, Dordrecht, Reidel Publ. Co.
Symposium on the Orion Nebula, 1982, Annals of the New York Academy of Sciences, 395, ed. Glassgold, A., Huggins, P., and Schucking, E.L.
30 Doradus: I.A.U. Symposium No. 108 (1983) Reidel Publishers and Elliot, K.H., Goudis, C., Meaburn, J., and Tebbutt, N.J., 1977, Astron. Astrophys., 55, 187.
Planetary Nebulae: International Astronomical Union Symposium No. 76, 1978. Edited by Y. Terzian (Dordrecht: Reidel Publ. Co.).
See also: I.A.U. Symposium No. 103, 1983, Reidel Publ. Co., and references cited in these articles.
Pottasch, S.R., Planetary Nebulae, 1983, Dordrecht, Reidel Publ. Co.

A Concise List of References (Continued)

Crab Nebula: "The Crab Nebula," Mitton, S. 1978 (New York: Scribner's). This popular account provides a useful introduction. See also: "The Crab Nebula," International Astronomical Union Symposium No. 46, 1971. Edited by R.D. Davies and F.G. Smith.

Supernova Remnants: Chevalier, R.A. 1977, Ann. Rev. Astron. Astrophys., **15**, 174.

Radio Galaxies: Osterbrock, D.E. in International Astronomical Union Symposium No. 74, 1978, p. 183, "Radio Astronomy and Cosmology," edited by D.L. Jauncey (Dordrecht, Reidel Publ. Co.).

Seyfert Galaxies: Weedman, D.W. 1977, Ann. Rev. Astron. Astrophys., **15**, 69.

Quasars - Emission Line Spectra: Davidson, K., and Netzer, H. 1979, Rev. Modern Phys., **51**, 715.

List of Illustrations

Figure 1-1. NGC 346 in the Small Magellanic Cloud. This is the largest H II region in the Small Magellanic Cloud. It is associated with a star cluster, which overwhelms the nebulosity on broad bandpass, blue-sensitive plates. We have used here a narrow bandpass, Hα filter with a 60-minute exposure with an image tube (Cerro Tololo Interamerican Observatory).

Figure 1-2. The Northern Part of the η Carina Nebula as photographed with the Cerro Tololo 4-meter telescope. Left [S II] λ6717, 6731; Right [O III] λ5007. North is at the top, east is at the left. Notice that many of the features prominent in low excitation [S II] are missing in high excitation [O III]. Courtesy: Nolan R. Walborn, Ap. J., **202**, L129, 1975.

Figure 1-3. The Planetary Nebulae, Abell 72 and Abell 43. (Photographed by George Jacoby at Kitt Peak National Observatory.)

Figure 1-4. The Inner Region of 30 Doradus. NW is at the top. Notice the intricate filamentary structure throughout. Photographed in the light of [O III] with a narrow bandpass filter. Cerro Tololo Observatory.

Chapter 2

SPECTRA OF GASEOUS NEBULAE

In the 19th Century there raged an argument as to whether fuzzy telescopic objects loosely called nebulae were really stellar aggregates or a gas. Although he had resolved some diffuse objects as star clusters, Herschel, who was an astute observer, commented that the light of others appear so "soft" that it might arise from "a luminous fluid."

The spectroscope solved the problem. The small "white nebulae" that avoided the Milky Way yielded spectra akin to those of stars. On the other hand, nebulae, such as Orion and the planetaries, showed bright line spectra, sometimes with an underlying continuum, usually of nonstellar origin.

The familiar Balmer lines of hydrogen were quickly recognized in the spectra of gaseous nebula. Much later the Paschen lines were found, while Brackett series lines have been measured by infrared observers. After it had been discovered in the sun, helium was also noted, but the very strongest radiations in many nebulae, a pair of green lines at λ4959 and λ5007 (as well as other strong features), defied identification. Some workers attributed them to a hypothetical "nebulium," but advances in the knowledge of atomic structure showed no such element could exist. Eventually, the green lines were identified with the so-called forbidden transitions of doubly-ionized oxygen and other lines with similar-type transitions in other familiar elements, such as nitrogen, neon and argon. Nebulium had literally vanished into thin air.

Figure 2-1 shows the blue region of the spectrum of the bright ring of NGC 7009 from λ3440 to λ4782. It includes strong permitted recombination lines of H, He I, He II, weak permitted lines of C II, N II and other ions plus a number of collisionally excited forbidden lines. Figure 2-2 (lower) shows the strong [O III] λ4959 and λ5007 lines, λ4686 He II, [N I], and a number of weaker features. Figure 2-3 shows some of the Paschen lines in the near infrared.

Since oxygen, nitrogen, neon, etc., have abundances 10^{-3}, or less than that of H, why are their forbidden lines so strong? To answer this question, we must consider physical mechanisms whereby the nebular spectrum is produced.

2A. The Primary Mechanism

First, we must realize that in a typical low-density nebula an individual volume element is subjected to dilute radiation, that is, to a low-energy density. At the same time, the quality of the radiation, namely, its energy distribution, may correspond to a high temperature. If R is the stellar radius, and r the distance from the star to the volume element, then since r >> R, the fraction of the celestial sphere filled by the star will be: $W_g = \pi R^2/4\pi r^2 = R^2/4r^2$. This is called the geometrical dilution factor. In a nebular shell of radius 0.1 parsec = 3.08×10^{17} cm, illuminated by a star of solar radius i.e.,

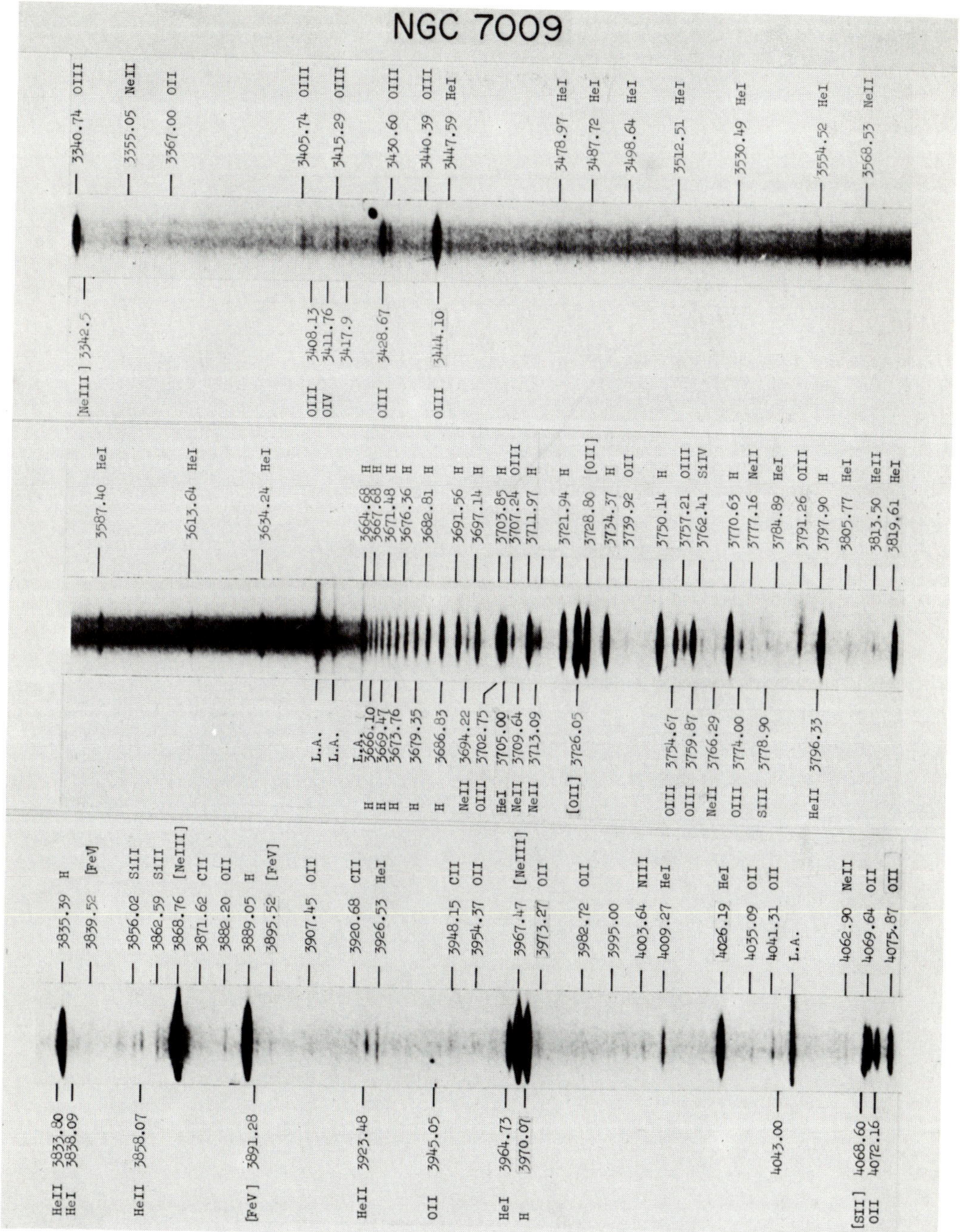

Figure 2.1.a

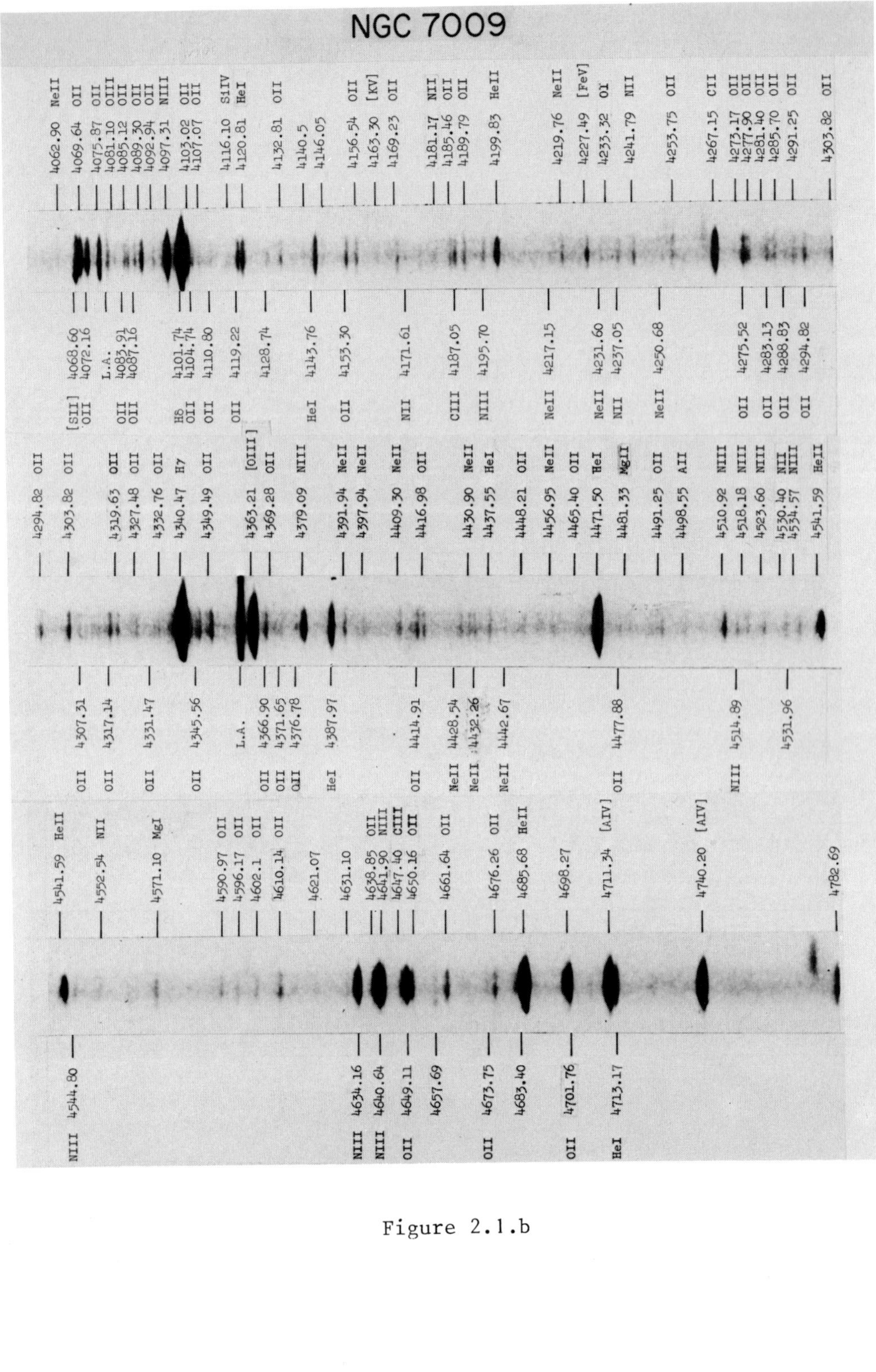
NGC 7009
4062.90 NeII
4069.64 OII
4075.87 OII
4081.10 OIII
4085.12 OII
4089.30 OII
4092.94 OII
4097.31 NIII
4103.02 OII
4107.07 OII
4116.10 SiIV
4120.81 HeI
4132.81 OII
4140.5
4146.05
4156.54 OII
4163.30 [KV]
4169.23 OII
4181.17 NII
4185.46 OII
4189.79 OII
4199.83 HeII
4219.76 NeII
4227.49 [FeV]
4233.32 OI
4241.79 NII
4253.75 OII
4267.15 CII
4273.17 OII
4277.90 OII
4281.40 OII
4285.70 OII
4291.25 OII
4303.82 OII
[SII] 4068.60
OII 4072.16
L.A.
OII 4083.91
OII 4087.16
Hδ 4101.74
OII 4104.74
OII 4110.80
OII 4119.22
4128.74
HeI 4143.76
OII 4153.30
NII 4171.61
CIII 4187.05
NIII 4195.70
NeII 4217.15
NeII 4231.60
NII 4237.05
NeII 4250.68
OII 4275.52
OII 4283.13
OII 4288.83
OII 4294.82
4294.82 OII
4303.82 OII
4319.63 OII
4327.48 OII
4332.76 OII
4340.47 Hγ
4349.49 OII
4363.21 [OIII]
4369.28 OII
4379.09 NIII
4391.94 NeII
4397.94 NeII
4409.30 NeII
4416.98 OII
4430.90 NeII
4437.55 HeI
4448.21 OII
4456.95 NeII
4465.40 OII
4471.50 HeI
4481.33 MgII
4491.25 OII
4498.55 AlI
4510.92 NIII
4518.18 NIII
4523.60 NIII
4530.40 NII
4534.57 NIII
4541.59 HeII
OII 4307.31
OII 4317.14
OII 4331.47
OII 4345.56
L.A.
OII 4366.90
OII 4371.65
OII 4376.78
HeI 4387.97
OII 4414.91
NeII 4428.54
NeII 4432.26
NeII 4442.67
OII 4477.88
NIII 4514.89
4531.96
4541.59 HeII
4552.54 NII
4571.10 MgI
4590.97 OII
4596.17 OII
4602.1 OII
4610.14 OII
4621.07
4631.10
4638.85 OII
4641.90 NIII
4647.40 CIII
4650.16 OII
4661.64 OII
4676.26 OII
4685.68 HeII
4698.27
4711.34 [AIV]
4740.20 [AIV]
4782.69
NIII 4544.80
NIII 4634.16
NIII 4640.64
OII 4649.11
4657.69
OII 4673.75
4683.40
OII 4701.76
HeI 4713.17

Figure 2.1.b

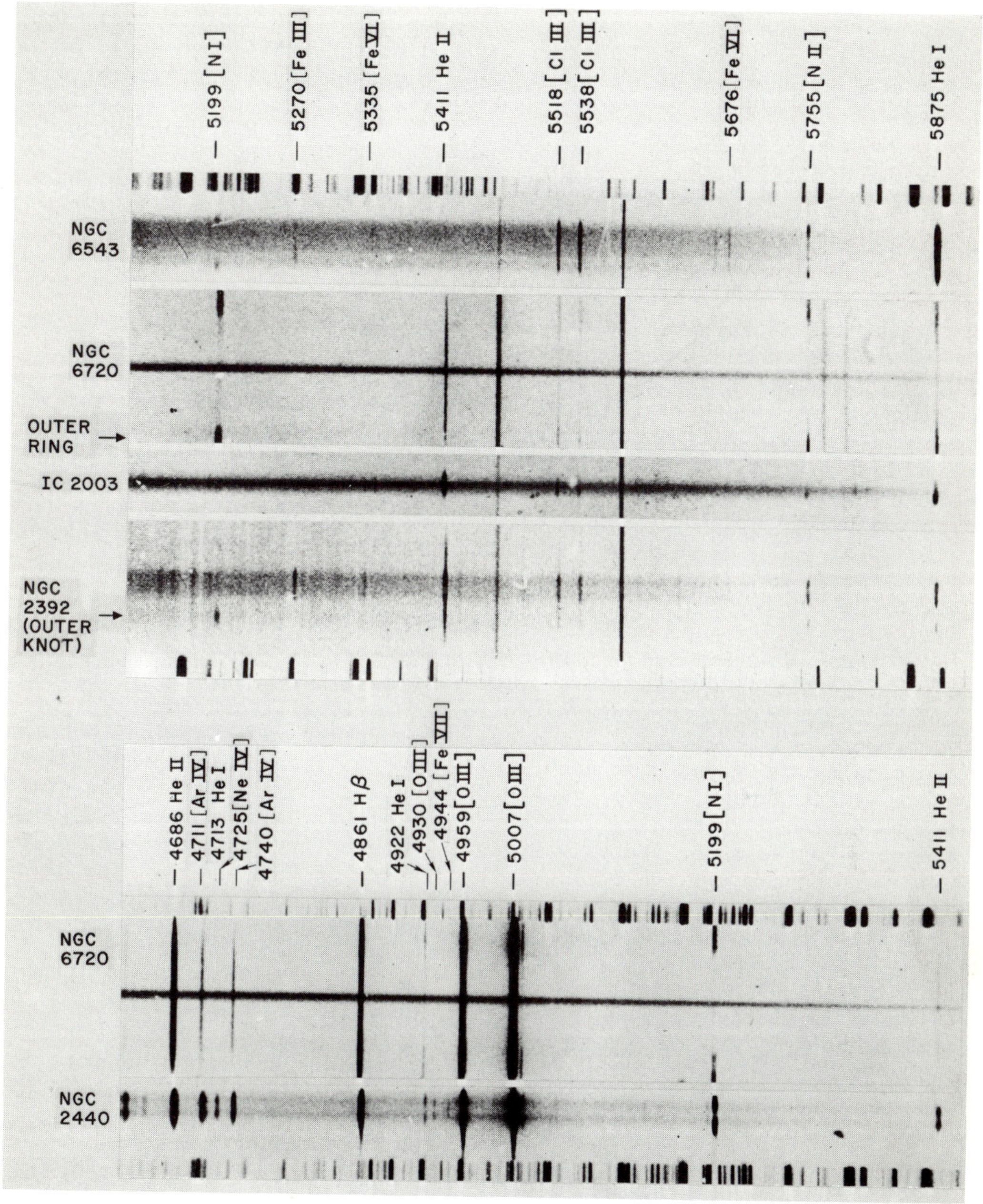
5199 [N I]
5270 [Fe III]
5335 [Fe VI]
5411 He II
5518 [Cl III]
5538 [Cl III]
5676 [Fe VI]
5755 [N II]
5875 He I
NGC 6543
NGC 6720
OUTER RING
IC 2003
NGC 2392 (OUTER KNOT)
4686 He II
4711 [Ar IV]
4713 He I
4725 [Ne IV]
4740 [Ar IV]
4861 Hβ
4922 He I
4930 [O III]
4944 [Fe VII]
4959 [O III]
5007 [O III]
5199 [N I]
5411 He II
NGC 6720
NGC 2440

Figure 2-2

6.96×10^{10} cm), W_g would be 1.28×10^{-14}. Thus, although the emission from the central star might resemble that of a black body with a temperature between 30,000°K and 150,000°K, this radiation would be attenuated by a factor of 10^{14}. Atomic processes involving absorption of radiation, such as excitation to discrete levels or photoionization, will proceed at rates cut down roughly by a factor of W_g compared to corresponding rates in thermal equilibrium. On the other hand, processes such as recombinations will occur at rates dependent on the local ionic and electron density and the gas kinetic temperature. For example, a hydrogen atom in any excited level n will almost certainly cascade downwards to some lower level before it can absorb a light quantum capable of exciting it to a higher level or ionizing it. Consequently, at any instant of time, virtually all nonionized hydrogen atoms will be concentrated in the ground level.

Take first the most elementary view of the hydrogen atom in which we group together all levels of the same n-value and essentially the same energy. All photoionizations occur from the ground level, but recombinations can take place to any level. If recapture occurs on the ground level, a Lyman continuum quantum is restored. If it is captured on the second level, a Balmer continuum quantum is created, followed by a Lyα quantum as the atom jumps to ground level. Capture on the third level results in creation of a Paschen continuum quantum. From the third level it may jump to the second level emitting Hα and then Lyα or it may jump directly to the ground level with emission of Lyβ. Other scenarios can be visualized readily.

Note that except for Lyα, every time a Lyman line or continuum quantum is absorbed, there is a finite chance it will be converted into two or more quanta of lower frequency. Zanstra showed that if a nebula is optically thick, so that each Lyman quantum gets absorbed many times before it can escape, every Lyman continuum quantum emitted by the star will be degraded eventually into a quantum of Lyα, plus a Balmer quantum. For example, there is a finite possibility that absorption of a Lyman γ quantum will be followed by emission of Hβ and Lyα. The Hβ quantum escapes from the nebula and Lyα is scattered. Escape from the n = 4 level may entail emission of a Paschen α quantum followed by Lyβ which, after repeated absorptions, is degraded into Hα + Lyα. Due consideration of all possible scenarios shows that the Lyman continuum radiation is continually degraded until there is no Lyman radiation left except Lyα, and for every Lyα there exists some Balmer continuum or line quantum. Thus, although Paschen, Brackett and other quanta will be produced, the important fact is that for an optically thick nebula

Number of stellar Lyman continuum quanta absorbed	=	Number of recombinations	=

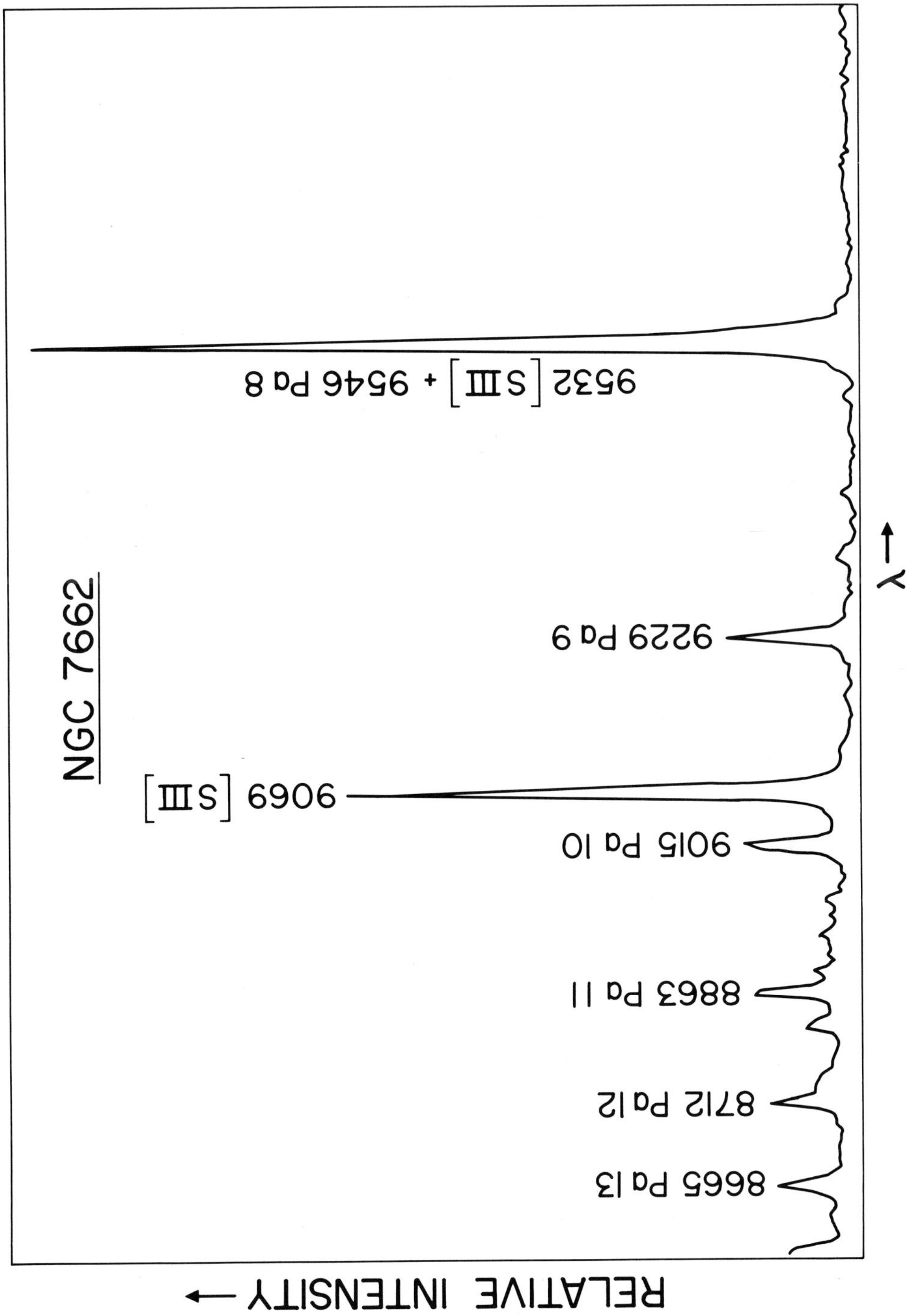
NGC 7662
9532 [S III] + 9546 Pa 8
9229 Pa 9
9069 [S III]
9015 Pa 10
8863 Pa 11
8712 Pa 12
8665 Pa 13
RELATIVE INTENSITY →
λ →

Figure 2-3

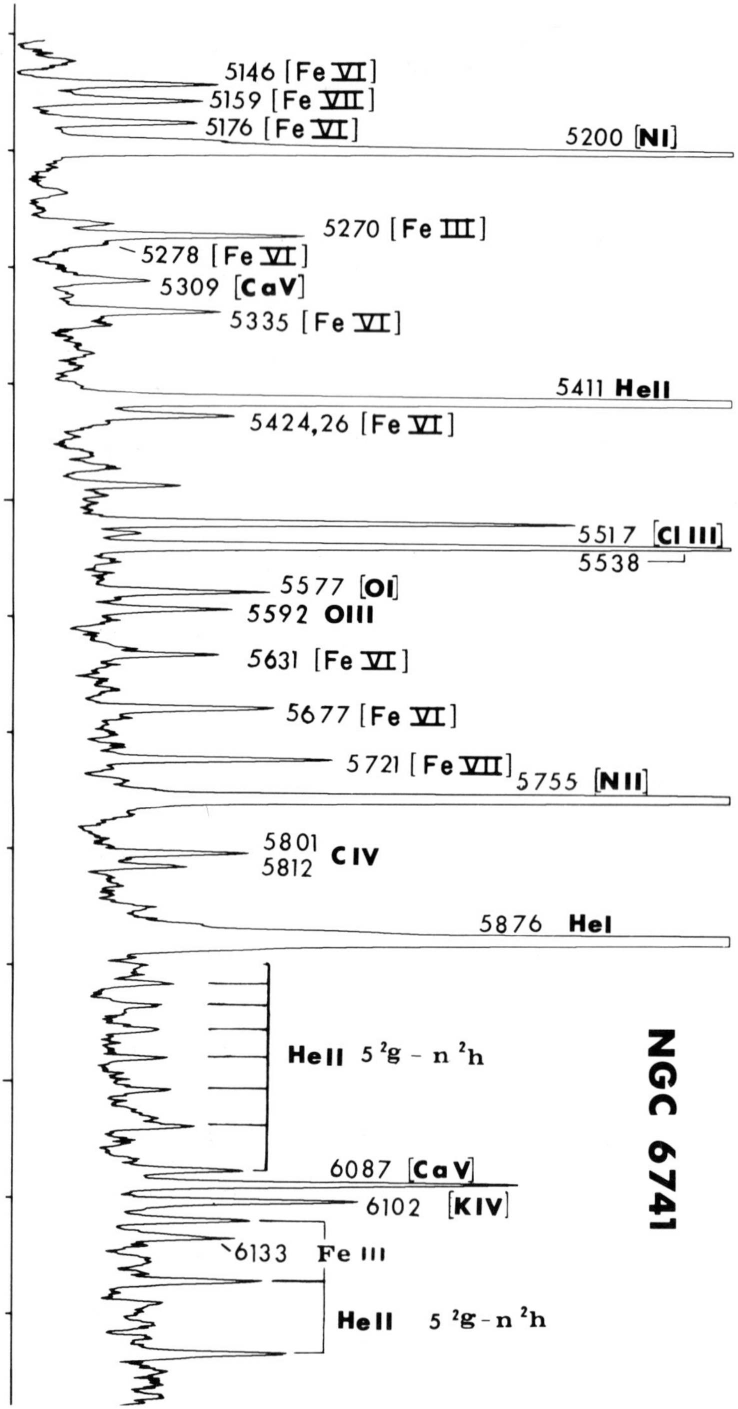
5146 [Fe VI]
5159 [Fe VII]
5176 [Fe VI]
5200 [NI]
5270 [Fe III]
5278 [Fe VI]
5309 [CaV]
5335 [Fe VI]
5411 HeII
5424,26 [Fe VI]
5517 [Cl III]
5538
5577 [OI]
5592 OIII
5631 [Fe VI]
5677 [Fe VI]
5721 [Fe VII]
5755 [NII]
5801
5812
C IV
5876 HeI
HeII 5 ²g – n ²h
6087 [CaV]
6102 [KIV]
6133 Fe III
HeII 5 ²g-n ²h
NGC 6741

Figure 2-4

Number of Balmer line and continuum quanta emitted by the nebula	=	Number of Lyα quanta emitted by the nebula

$$P_{UV}(*) = N_{recomb}(neb) = Q(Balmer) = Q(Ly\alpha)$$

Similar arguments may be applied to the recombination spectrum produced when doubly-ionized helium ions capture electrons. Since the ionization potential of He^+ is 54.4 eV, or four times that of H, the corresponding series is shifted to shorter wavelength: Hα of He^+ falls at λ1640, Paα is λ4686, alternate members of the Brackett series coincide with the Balmer series and Pfund series lines (5 - n) are seen in high-excitation planetaries (see Fig. 2-4).

Zanstra's principle can be applied to show that the temperatures of central stars of planetaries must be very high (see Chapter 3). Actually, the hydrogen problem is somewhat more complicated than depicted here. Each level is characterized by a value of the azimuthal quantum number, ℓ, as well as by a principal quantum number, n. Because of the coloumbic potential field, the energy levels are degenerate in ℓ; that is, the energy depends only on n. The actual transitions are still governed by the Laporte parity rule, $\Delta\ell = \pm 1$, however. The Lyman series consists of the single sequence 1s → 2p, 1s → 3p ... 1s → np, but the Balmer series comprises transitions of the type 2p → ns, 2p → nd, 2s → np, while the Paschen and higher series offer many more possibilities (see Fig. 2-5). Electrons are recaptured in levels of all (n, ℓ) configurations, and cascading occurs to lower levels. Roughly two-thirds of all recaptures lead to np levels whence quick decay to 1s can take place, but about one-third of all recombinations lead to the 2s level.

The 2s level is metastable. A transition to the ground level would violate the Laporte rule. A radiative transition to the $2p\,^2P_{1/2}$ level which lies 0.0354 cm^{-1} lower than the $2s\,^2S_{1/2}$ level would occur only after an interval of about thirty million years. Atoms may escape from the 2s level by one of several processes:

a) Photoionization to the continuum, which can become important if the Balmer continuum radiation field is intense, as in quasars. This process is not important in planetary and diffuse nebulae;

b) Collisional transitions to 2p, whence decay to 1s can occur. The efficiency of this process depends strongly on the density;

c) The two-photon emission. An atom in a 2s level jumps to a "conjured" level that may lie anywhere between the n = 1 and n = 2 levels.

Thus, each transition produces two quanta subject only to the condition that $\nu_1 + \nu_2 = \nu_{Ly\alpha}$. This is the two-photon emission that appears often to contribute substantially to the underlying continua of many planetaries. The Einstein A coefficient for this process is 8.227 sec^{-1}. The distribution function of the radiated photons is symmetric about $\nu_{2p} = 1/2\nu(Ly\alpha)$, but the radiated energy (which is proportional to hν) is not.

In an optically thick nebula, many Lyα quanta are created by degradation of Lyman continuum quanta. What happens to them? Some diffuse out of the nebula (see § 4F). Others may be destroyed by ionizing atoms of C, Si, and metals. If grains are present, many Lyα quanta will be absorbed to heat these particles, thereby raising their temperatures sometimes sufficiently to permit detection by infrared techniques.

If the flux in Lyα is strong, and N_ε is large, the population of 2p levels may be built up enough for collisional transitions to 2s to occur often. Thus, the two-photon emission could be enhanced. We can note that under such circumstances, Balmer line absorptions can become significant.

Similarly, when the photoionizing source is sufficiently rich in far UV quanta to twice ionize helium (as in high-excitation planetaries), Lyα (He^+), λ303.780, can attain a moderately high energy density. There is an important practical difference from Lyα (H), however. There is a transition from the low-lying $2p^2\ {}^3P_2$ level in O III to a higher level, 2p 3d ${}^3P_2^o$, the wavelength of which is λ303.799. Hence, when Doppler motions are taken into account, many O III atoms can be excited to this level whence they can cascade to lower levels emitting the strong O III permitted transitions of the Bowen fluorescent mechanisms (see Chap. 3); also, Lyα (He^+) quanta can ionize H both from ground and 2s levels, He atoms, and ions of many elements.

Gaseous nebulae also show lines of He I λ5876, 4471, 6678, etc. (see Fig. 2-6). There are two distinct types of spectroscopic terms, singlets and triplets. Helium atoms are photoionized from 1^1S, but recapture may occur in either singlets or triplets in the ratios of 1 to 3 (since the statistical weight of a triplet term is thrice that of a singlet). Because helium is in practically perfect LS coupling, radiative transitions between the two kinds of terms are excluded.

Although both 2^3S and 2^1S are metastable, the radiative lifetime of 2^3S exceeds two hours. In his pioneering study, Ambarzumian regarded the 2^3S level as rigorously metastable; i.e., he neglected the role of depopulating events. Then 2^3S could serve as a trap (like a pseudo-highly-elevated ground level), a large population could be built up in this level and the 2^3S to 2^3P transition could play a role analogous to Lyα. These quanta would be scattered in the nebula. The 2^3S - 3^3P λ3889 line (which is blended with H7) would also be enhanced. In the Orion nebula, O.C. Wilson observed the nebular λ 3889 line in absorption in the spectra of background stars. Although λ3889 is usually simply scattered, transitions from 3^3P - 3^3S at 4.3 μm can occur with subsequent decay to 2^3P with the emission of λ7065. Analogous transitions can occur following the absorption of λ3188. Hence, as shown, e.g., by Robbins (see Chapter 4B), the 7065/4471 and 3188/4471 intensity ratios could be used to assess the optical depth in λ10830 or λ3889. In H II regions and planetaries, the intensities of λ5876 and λ4471 will not be much affected. In dense nebulae such as QSOs, radiative transfer effects can be important and the prediction of helium line intensities can become complicated.

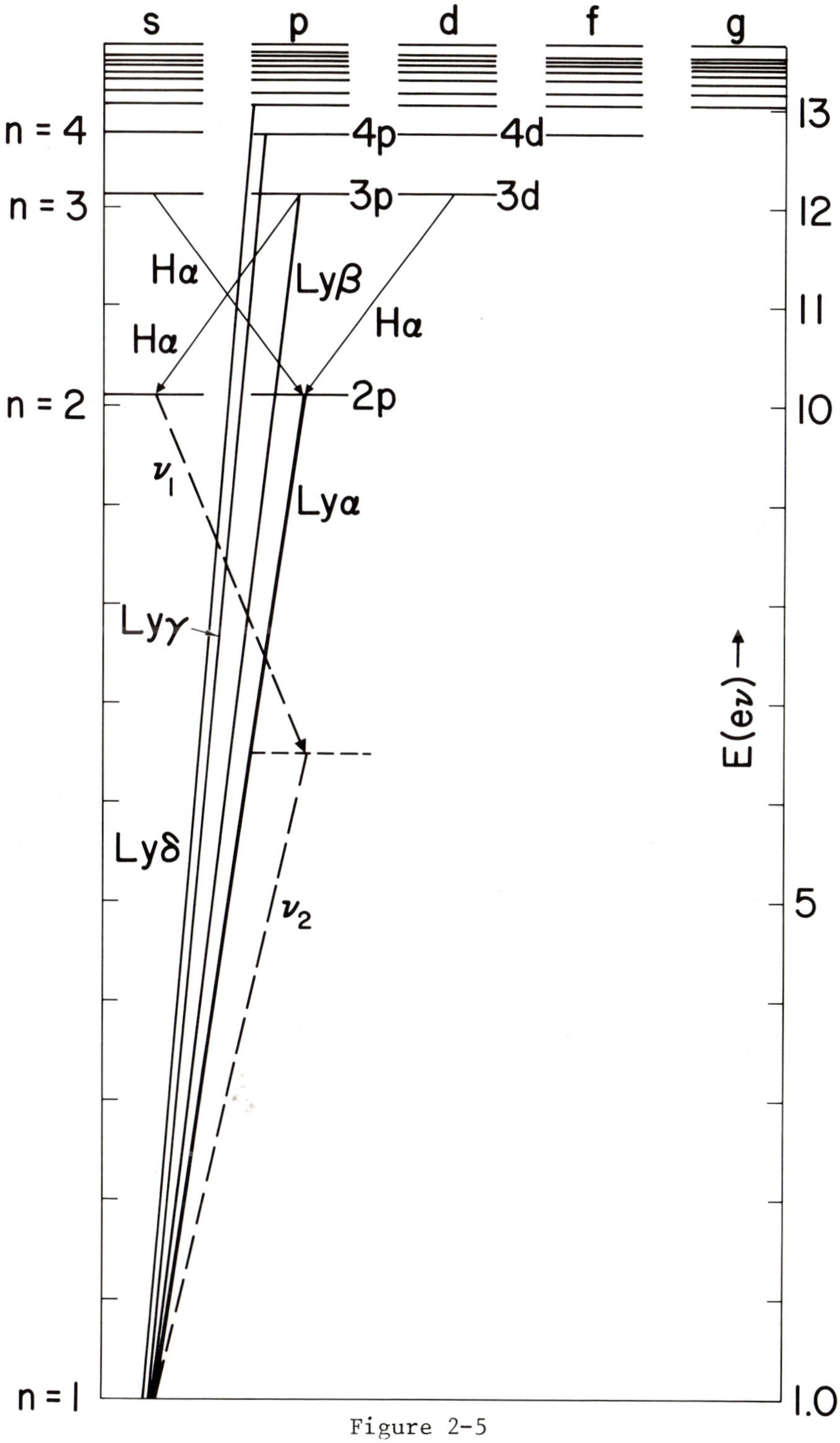
s
p
d
f
g
n = 4
n = 3
n = 2
n = 1
4p
4d
3p
3d
2p
Hα
Hα
Hα
Lyβ
Lyα
Lyγ
Lyδ
ν_1
ν_2
13
12
11
10
5
1.0
E(eν) →

Figure 2-5

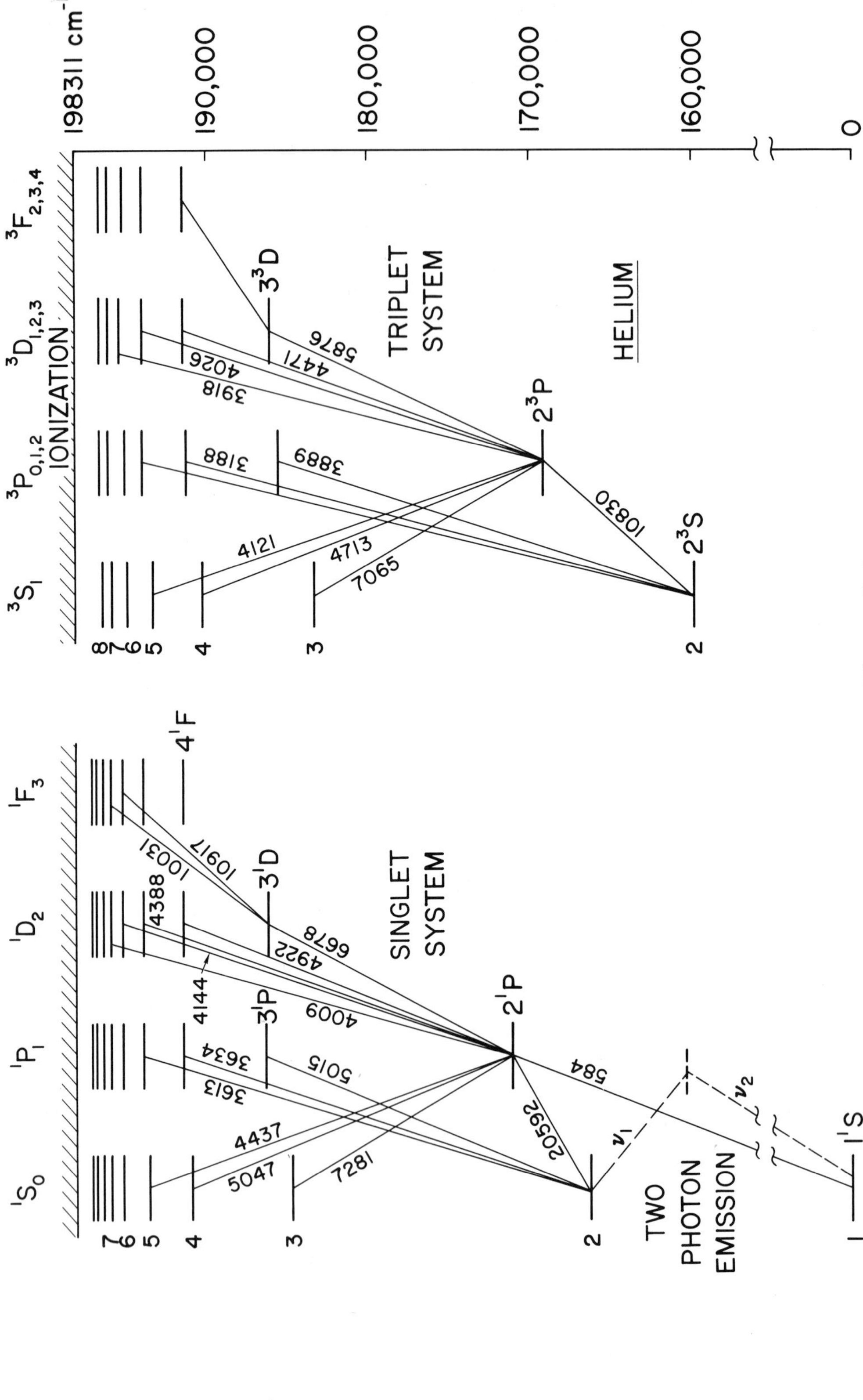

198311 cm⁻¹
190,000
180,000
170,000
160,000
0
IONIZATION
$^3S_1$
$^3P_{0,1,2}$
$^3D_{1,2,3}$
$^3F_{2,3,4}$
TRIPLET SYSTEM
HELIUM
3^3D
2^3P
2^3S
5876
4471
4026
3918
3188
3889
10830
4121
4713
7065
8 7 6 5 4 3 2
$^1S_0$
$^1P_1$
$^1D_2$
$^1F_3$
4^1F
SINGLET SYSTEM
3^1D
3^1P
2^1P
1^1S
10917
10031
4388
6678
4922
4144
4009
3634
3613
5015
584
20592
4437
5047
7281
ν_1
ν_2
TWO PHOTON EMISSION
7 6 5 4 3 2 1

Figure 2-6

2B. Collisionally-Excited Lines

The strongest lines in the visible region of many planetary nebulae are the "forbidden" transitions between levels in the ground $2p^2$ configuration of O III, i.e., between the $^1D_2$ level and the $^3P_1$ and $^3P_2$ levels in the ground ^3P term. The excitation potential of the $^1D_2$ term is about 2.5 eV. The highest term in this configuration, ^1S (see Fig. 2-7), lies at 5.35 eV above the ground level. Other energy levels in O III have substantially higher excitation potentials. Similar situations exist in many other ions of $2p^n$ and $3p^n$ configurations. Examples of term structures for various ions that produce important forbidden lines in gaseous nebulae are shown in Fig. 2-7 (note the strong [S III] transition in NGC 7662, Fig. 2-3).

Involved here are transitions between levels belonging to the same configuration. They are called forbidden lines because they violate the Laporte parity rule, $\Delta\ell = \pm 1$. Ordinary, permitted transitions which satisfy the Laporte rule are of the electric dipole type. Also, $\Delta J = 0, \pm 1$ but $J = 0 \rightarrow J = 0$ is excluded. Forbidden lines are indicated by square brackets, as [N II], [O III], and [S III].

In pure LS coupling, as in helium, transitions occur only between terms of the same multiplicity, from singlets to singlets and triplets to triplets. Ions of light atoms of the lithium and sodium rows of the periodic table, such as C III, O III and N IV are frequently fairly close to LS coupling. Intercombination transitions of the type $2s^2\ ^1S - 2s\ 2p\ ^3P$ at λ1906, 1908 in C III, N IV $2s^2\ ^1S_0 - 2s\ 2p\ ^3P_1^0$ λ1487, which are prominent in the satellite ultraviolet spectra of planetaries and other high-excitation gaseous nebulae, are sometimes called semiforbidden lines and are indicated by a bracket on the right-hand side, as N III].

For p^n configurations ($n = 1 \rightarrow 5$), which contribute all prominent nebular forbidden lines, there are three patterns of energy levels. Those of p and p^5 are simply ^2P terms; for example, S IV, Ne II. Transitions between the two fine-structure levels normally fall in the infrared, although such transitions for [Fe X] and [Fe XIV] are found in optical regions of the solar corona. Note that p^2 and p^4 configurations consist of a ground ^3P term, which is "inverted" for p^4 (where the $^3P_2$ level falls lowest), a $^1D_2$ and a $^1S_0$ (which lies highest). A p^3 configuration also contains five distinct levels, the lowest of which is $^4S_{3/2}$. Above this lie the ^2D and ^2P terms.

For historical reasons, transitions between the middle and lowest terms are called nebular-type transitions and those between the highest and middle terms are called auroral-type transitions. The strong green line in the Earth's aurora, λ5755 [O I] is from $^1S_0$ to $^1D_2$, while the strongest lines in most planetary nebulae are [O III] λ4959, 5007 ($^1D_2 - {}^3P$). Transitions from the top to the lowest term, such as λ4068, 4076 in [S II] are called transauroral transitions. Jumps between individual levels of a ground configuration, such as $^3P_2 - {}^3P_1$ or $^2P_{3/2} - {}^2P_{1/2}$, are called "fine structure" transitions.

Forbidden transitions are of the magnetic dipole or electric quadrupole type. Magnetic quadrupole transitions also have been observed. Magnetic dipole transitions obey the rules:

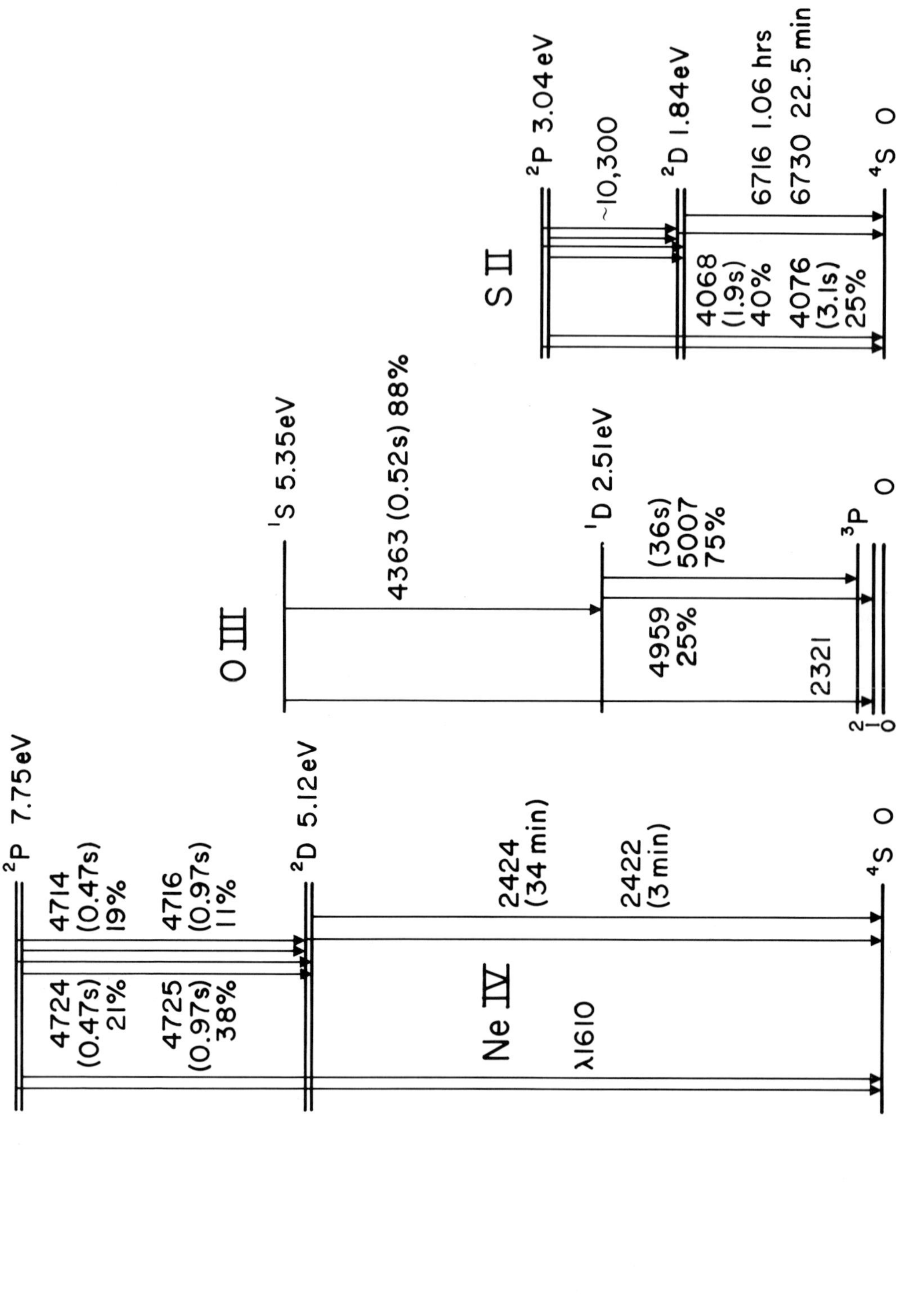
Ne IV
²P 7.75eV
²D 5.12eV
⁴S 0
4714 (0.47s) 19%
4716 (0.97s) 11%
4724 (0.47s) 21%
4725 (0.97s) 38%
2424 (34 min)
2422 (3 min)
λ1610
O III
¹S 5.35eV
¹D 2.51eV
³P 0
2 1 0
4363 (0.52s) 88%
(36s) 5007 75%
4959 25%
2321
S II
²P 3.04eV
²D 1.84eV
⁴S 0
~10,300
6716 1.06 hrs
6730 22.5 min
4068 (1.9s) 40%
4076 (3.1s) 25%

Figure 2-7

a) Spectroscopic parity cannot change; the jumps are even-to-even or odd-to-odd;
b) The principal quantum number, n, cannot change, $\Delta n = 0$ and likewise $\Delta \ell = 0$;
c) $\Delta J = \pm 1, 0$, $J = 0 \to J = 0$ is excluded.

Electric quadrupole transitions obey rule a) above; however, Δn is arbitrary and $\Delta \ell$ can be ± 2 as well as 0. $\Delta J = 0, 1, \pm 2$, but $0 \to 0$, $1/2 \to 1/2$ and $1 \leftrightarrows 0$ are excluded. In pure LS coupling $\Delta L = 0, \pm 1, \pm 2$ ($0 \to 0$ is excluded) and no magnetic dipole transitions can occur except between two levels of the same term, like fine-structure lines such as $^3P_2 - {}^3P_1$, [Ar III], 9.0 μm. The green nebular lines of [O III] and the corresponding red lines of [N II] are magnetic dipole transitions between ^3P and ^1D. They actually occur only because of deviations from strict LS coupling. The so-called auroral-type transitions of the $^1S_0 - {}^1D_2$, and transauroral transitions of the $^1S_0 - {}^3P_2$ kinds are clearly of the electric quadrupole types, since $\Delta J = 2$. The Einstein A-value (transition probabilities) for forbidden transitions are often very low, e.g., 0.014 sec^{-1} and 0.007 sec^{-1} for the λ5007 Å and λ4959 Å lines, respectively. Hence, the lifetime in the $^1D_2$ level of O III is 1/(0.021) = 36 seconds. Even lower transition probabilities are found for many astrophysically important lines, e.g., the red ones of [S II] (see Fig. 2-7).

An immediate question is why forbidden lines attain such great strength in gaseous nebulae as compared with permitted lines, particularly those of H and He. The answer lies partly in the vast extent of a nebular plasma and partly in the structures of energy level diagrams. First, we must ask how the lines are produced. Recombination and cascade certainly will not suffice as then we would observe lines arising from high levels as well as these low-level transitions. Except for certain O III transitions excited by fluorescence (see Chapter 3), all O III lines involving high levels are very weak. Note that the metastable levels lie but a few eV above the ground level, whereas all other levels involved in optical region lines lie at very high energies. The energy of the average electron in the plasma is of the order of an electron volt. Some electrons will have energies of 2 to 4 eV, a few may have energies to 5 or 7 eV, but none will have energies of, say, 20 eV. Hence, there will be a supply of electrons capable of exciting atoms to the metastable levels. Once in an excited level, they may escape by superelastic collisions or by the emission of a forbidden line. The total number of emissions per second will be N_nA, where N_n is the number in the upper level and $A = \Sigma A\ nn'$ is the sum of Einstein coefficients to lower levels.

We can now grasp why the forbidden lines dominate the optical region of the spectra of gaseous nebulae. The emission per unit volume is governed by the number of atoms in the upper level. Under laboratory conditions this number would be given by Boltzmann's equation; in the gaseous nebula, the proportion in the metastable level would fall somewhat because atoms could now escape by radiation as well as by collision. On the other hand, a radiating column of 10^{15} meters vastly exceeds anything one could produce in the laboratory. Thus, the emission per unit volume in the nebula may be within one or two orders of magnitude below that which exists in the laboratory, but the

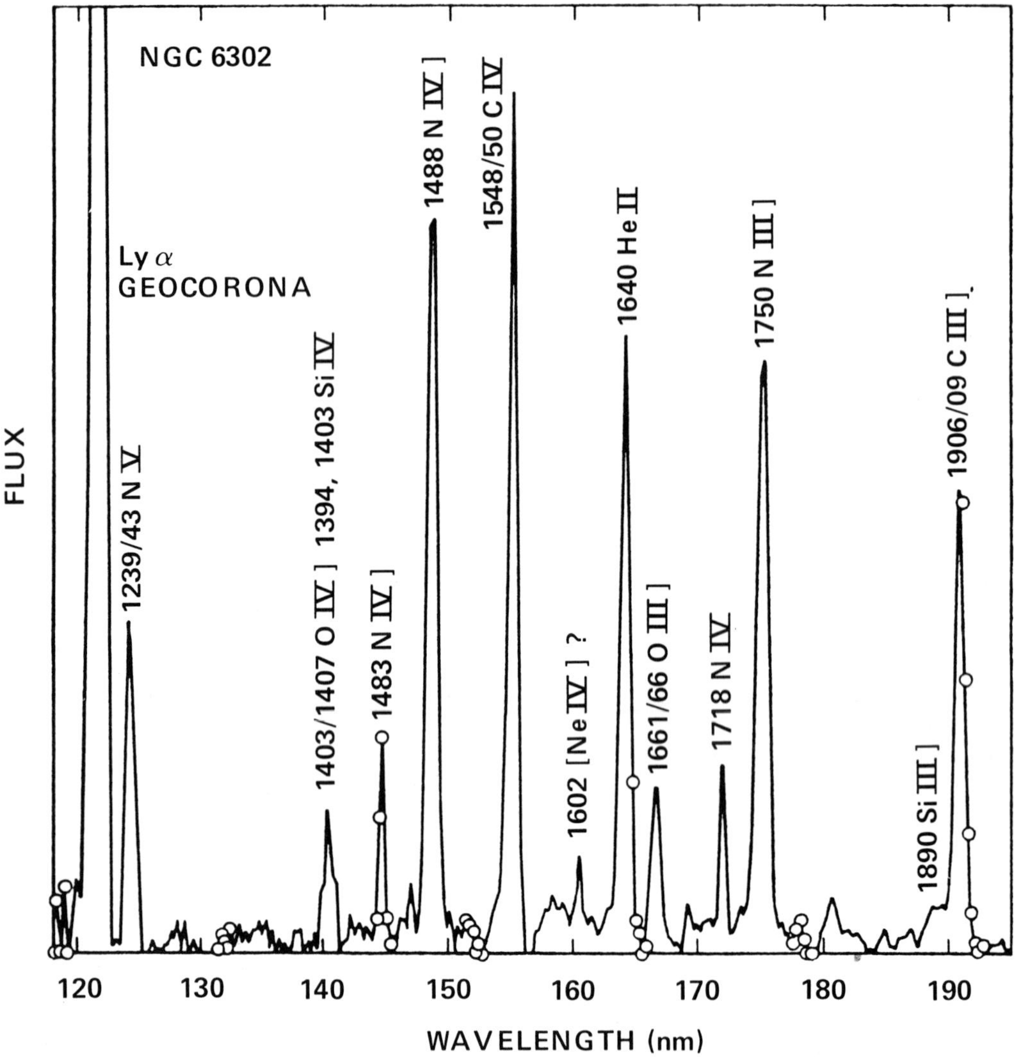
NGC 6302
Ly α
GEOCORONA
1239/43 N V
1403/1407 O IV] 1394, 1403 Si IV
1483 N IV]
1488 N IV]
1548/50 C IV
1602 [Ne IV] ?
1640 He II
1661/66 O III]
1718 N IV
1750 N III]
1890 Si III]
1906/09 C III].
FLUX
120
130
140
150
160
170
180
190
WAVELENGTH (nm)

Figure 2-8

radiating volume is enormous. The huge path length compensates for the feeble emissivity per cm^3. On the other hand, permitted lines of abundant elements observed in the optical region typically involve excitation potentials of 10 to 50 eV. The number of electrons capable of collisionally exciting these levels is negligible. Unless some special stellar or nebular fluorescent mechanism is available, the only means whereby these levels can be excited is photoionization followed by recombination and cascade. Photons of sufficient energy are produced copiously by a star or even a nonthermal source, but the attenuation of the radiation is severe. Collisional excitation rates depend only on density and target areas. Consequently, a tremendous bias factor operates against permitted lines. In a typical planetary nebula, the green nebular lines of [O III] are ten or even twenty times as strong as Hβ. Yet the number of hydrogen ions present exceeds the number of O^{++} ion factors by a factor of thousands!

A different situation holds in the near ultraviolet. Permitted and intercombination lines of ions of abundant elements such as C, N, O and Si dominate this region. They are excited by electron collisions, even though energies typically of the order of 5 to 9 eV are involved (see Fig. 2-8). Many of the transitions are intercombination lines such as N III] λ1749.7, 1752.2 or λ1906.68 1908.73 C III], but some are resonance lines such as λ1548.20, 1550.74 C IV with high A-values. Radiative transfer with its attendant optical depth effects, and absorption by dust within the nebula which complicate resonance scattering, are both factors that have to be taken into account in predictions of line intensities. Forbidden lines such as λ1602, 2422, 2424 [Ne IV] are less prominent in number and intensity in this region.

Forbidden lines of iron in various stages of ionization offer some of the most engaging problems in nebular spectroscopy. These lines are often prominent in the spectra of quasars and they dominate the spectra of objects such as η Carinae and RR Telescopii. Lines of [Fe II] and [Fe III] appear in the spectra of H II regions such as the Orion nebula. Some planetaries, like NGC 6741 (Fig. 2-4), show many forbidden lines of iron in various stages of ionization, but they are often weak, indicating that the concentration of this metal in the gaseous phase is quite low compared to the expected abundance. This apparent depletion has led to the suggestion that much of the iron may be condensed on grains. Lines of other metals of the iron group may appear in the spectra of objects such as NGC 7027, but they are very weak and difficult to detect. Because of their low abundances, no elements beyond the iron groups are detected in the spectra of gaseous nebulae.

The great astrophysical value of forbidden lines is that certain of their intensity ratios yield diagnostics of the nebular plasma, i.e., the electron density, N_ε, and electron temperature, T_ε. We must have this information in order to deduce the chemical composition of the gas (see Chap. 11). The level of excitation and ionization frequently varies from point to point within a nebula, generally dropping with increasing distance from the central star, but often showing capricious variations (see Fig. 2). By measuring suitable line ratios in [N II], [O II], [O III], and [S II], for

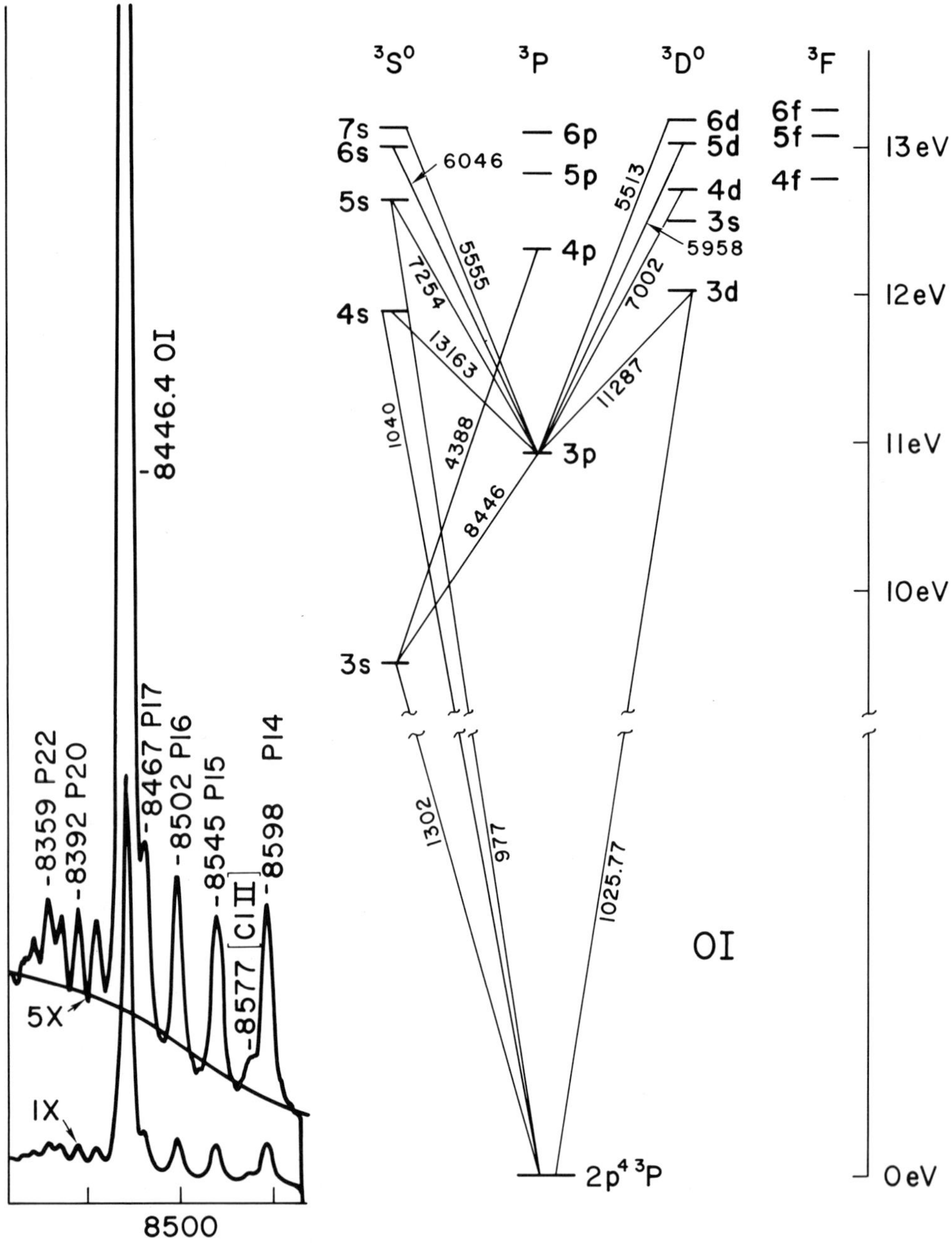
8446.4 OI
-8359 P22
-8392 P20
-8467 P17
-8502 P16
-8545 P15
-8577 [CI II]
-8598 P14
5X
1X
8500
$^3S^0$
^3P
$^3D^0$
^3F
7s
6s
5s
4s
3s
6p
5p
4p
3p
6d
5d
4d
3s
3d
6f
5f
4f
6046
5555
7254
13163
1040
4388
8446
5513
5958
7002
11287
1302
977
1025.77
13 eV
12 eV
11 eV
10 eV
0 eV
$2p^4\ ^3P$
OI

Figure 2-9

example, we can compare $\langle N_\varepsilon \rangle$ and $\langle T_\varepsilon \rangle$ values along different lines of sight through the nebulae.

2C. Permitted Lines in the Optical Region

In addition to their forbidden lines, nitrogen, oxygen and neon are represented by weak ionic lines in the optical regions of the Orion nebula and planetaries such as NGC 6778 or NGC 7009 (Fig. 2-1). Some of these lines appear to be recombination features; others may be excited through fluorescent processes involving radiation from the central star. The most spectacular example of fluorescence is provided by the strong O III and N III lines involved in the celebrated Bowen fluorescent mechanism (see Chap. 3I). This cycle is induced by the strong He III Lyα line produced in high-excitation planetaries. We would expect pure recombination lines to be weak since the abundances of the elements involved lie three or four orders of magnitude below H or He.

Individual possible excitation scenarios have to be examined on a case-to-case basis for virtually all permitted lines in gaseous nebulae. As one example, consider O I λ8446 3p ^3P - 3s $^3S^0$, which attains great strength in dense planetary nebulae, such as IC 4997, and is prominent in the Orion nebula (see Fig. 2-9). It is also observed in quasars. Direct recombination will not work. The spectrum of neutral oxygen, which is based on $2p^3ns$, $2p^3np$, $2p^3nd$ ... configurations, consists of singlets, triplets and quintets. In a recombination scenario, the quintets should appear perhaps stronger than the triplets, as the triplets appear stronger than the singlets in helium. Such is not the case.

The ionization potential of neutral oxygen is almost exactly the same as that of hydrogen. Hence, it is not surprising to find coincidences between some oxygen and Lyman lines. Swings called attention to a coincidence between Lyβ and λ1025.77, $2p^4\ ^3P$ - $2p^33d\ ^3D$ of O I. Thus, one possibility would be for the O I atom to absorb nebular Lyβ, be excited to 3d ^3D, whence it would cascade to 3p ^3P with the emission of λ11287, followed by a jump to 3s ^3S with radiation of a quantum of λ8446. Such a mechanism might work in a high-density quasar, but in a planetary nebula or in a diffuse nebula, we would expect little neutral oxygen in the region where Lyβ quanta might be plentiful. Note that after a few absorptions, Lyβ is almost totally converted to Hα + Lyα.

A third possibility is direct excitation by starlight, which accounts for other permitted lines such as λ7254 which can be emitted after an O I atom has been excited to 5s ^3S by absorption of a quantum of λ977A. A number of these transitions serves to populate 3p ^3P, which is the upper level of λ8446. In Orion, Grandi finds that the flux produced by this process dominates over that produced by direct recombinations by an order of magnitude and represents the observed line ratios. Also, the starlight-excited lines are enhanced towards the Trapezium source. He showed Lyβ fluorescence to be quite inadequate. In other nebulae such as NGC 7027 and 7662, Grandi concluded that the observed intensities of nearly all permitted lines could be explained by a plausible combination of recombination, resonance fluorescence by other nebular lines, or by starlight.

Nebular lines that arise from doubly-excited levels cannot be explained in any satisfactory manner, except perhaps by dielectronic recombination (see Chapter 3E).

Some Suggested References

See also the listings in Chapters 3, 4, and 5.

The primary mechanisms for the excitation of hydrogen and helium lines by photoionization followed by recombination have been discussed by:

Zanstra, H. 1927, Ap. J., 65, 50.
__________ 1931, Dom. Astrophys. Obs. Publ., 4, 209.
__________ 1931, Zeits. f. Astrofis., 2, 1.

A discussion of the helium problem with suitable bibliographical references may be found in:

Osterbrock, D.E. "Astrophysics of Gaseous Nebulae," 1974, San Francisco:
W.H. Freeman Co., p. 25 (hereafter referred to as A.G.N.).

A classical early reference to forbidden lines is:

Bowen, I.S. 1936, Rev. Mod. Phys., 8, 55.

The basic theory is reviewed in:

Condon, E.H., and Shortley, G.H. "Theory of Atomic Spectra," 1935, Cambridge University Press.

The importance of permitted lines of ions of elements heavier than H and He was first pointed out by A.B. Wyse, 1942, Ap. J., 95, 356. A comprehensive discussion of excitation of permitted lines in gaseous nebulae and quasars is given by:

Grandi, S.A. 1975, Ap. J., 196, 465.
__________ 1976, Ap. J., 206, 658.
__________ 1980, Ap. J., 238, 10.

Chapter 2 List of Illustrations

Fig. 2-1. Spectrum of the Bright Ringed Planetary NGC 7009 in the Region λ3340 to λ4782. The spectrum is that of the high-excitation region of the nebula near the minor axis. The following features are of interest:

a. Strong permitted recombination lines of H, He I, and He II. Notice the confluence of the Balmer series and the continuum.
b. The collisionally excited forbidden lines of [O II], [O III], [Ne III], and [Ar IV] are prominent. There also appear some weak forbidden lines, such as λ4227 [Fe V]. Compare Fig. 4.

Chapter 2 List of Illustrations (Continued)

c. Numerous weak permitted lines of C II, N II, O II, N III, and Ne II, which are probably excited by photoionization followed by recombination, although for many of these lines, stellar line fluorescence may play a role. The weak Mg I λ4571 is probably excited by collisions.
d. Lines from the mercury vapor lights of Los Angeles, denoted by LA, fall near λ4358, 4046, and the Balmer limit. (Mt. Wilson Observatory)

Fig. 2-2. The Green Spectral Region of Several Planetaries. Note the pronounced changes in level of excitation from point to point in these nebulae. The effects are most straightforward for NGC 6720, where lower levels of ionization, e.g., [N I], [N II] are most prominent in the outer ring, while ions of higher ionization potential such as [Ar IV] and He II are prominent in the inner regions. In NGC 6543, [N II] is strong in outer knots; [Cℓ III] is more evenly distributed. NGC 2440 is a multi-lobed nebula. The maximum intensity of [N I] falls much further from the star than does that of λ4725 [Ne IV]. Also, in IC 2003, [N II] shows a different concentration from the other elements. The spectrogram of NGC 2392 was secured in bright moonlight. The central brighter knot shows lines of N II, [Cℓ III], [Fe III], He I, and He II, whereas the outer fainter knot shows [N I] as its strongest feature. The strong lines on either side of 5518, 5538 extending across the spectrum with constant intensity are due to mercury vapor lamps in San Jose. These observations were secured at the coudé focus of the Lick 3-m telescope with the Lallemand electronic camera (Aller and Walker, 1970, Ap. J., **161**, 917).

Fig. 2-3. The Near Infrared Spectrum of the Bright Ring in the High Excitation Planetary, NGC 7662. The Paschen series, Pa8 - Pa13, is well displayed. Notice the great strength of the collisionally-excited [S III] lines. These transitions are the analogues of the strong green [O III] lines. Observations were secured with a reticon system with the Shane 3-m telescope at Lick Observatory.

Fig. 2-4. A Portion of the Spectrum of the High-Excitation Nebula NGC 6741 in the Region λ5130 to λ6250 Å. Notice the Pfund 5-n series of He II. Low-excitation lines of [N I] and [N II] appear in the same spectrum with the high-excitation ones of He II, C IV, [K IV], and [Ca V]. Evidently, the nebula is inhomogeneous and has many relatively dense condensations. Numerous lines of iron ions, [Fe III], [Fe VI] and [Fe VII] are present. The strongest lines are off-scale. This tracing was obtained with the image tube scanner on the Shane 3-m telescope at Lick Observatory.

Fig. 2-5. Metastability of the 2s Level in Hydrogen. About one-third of all recombinations in upper levels result in an atom reaching the 2s level whence it can escape by collisions or by two-photon emission.

Chapter 2 List of Illustrations (Continued)

Fig. 2-6. Energy Level Diagram for Helium. Singlet and triplet terms are plotted separately, the energy being given in kaysers (cm^{-1}). Note that both 2^1S and 2^3S are metastable. In an optically thick nebula where the resonance line λ584.4 is strongly scattered, atoms can feed from 2^1P to 2^1S, the A-value for this transition being about 0.001 that of the resonance one. Atoms may escape from 2^1S via a two-photon emission (A = 51.3 sec^{-1}). The 2^3S level is more severely metastable [$A(2^3S - 1^1S) = 1.27 \times 10^{-4}$ sec^{-1}], but if the electron density exceeds about 4000 cm^{-3} as in dense planetaries we can get collisional excitation to 2^1S and 2^1P with a preference of about 6 to 1 in favor of 2^1S (see Osterbrock, "Astrophysics of Gaseous Nebula," p. 25, for detailed discussions). In ions of the He I isoelectronic sequence the situation changes drastically as Z increases. For example, at Z = 5, $A(2^1S_0 - 1^1S_0) = 9.26 \times 10^4$ and $A(2^3S_1 - 1^1S_1) =$ 6.70, but at Z = 10 these probabilities become 1×10^7 and 10,900, respectively!

Fig. 2-7. Typical Forbidden Line Transition Schemes: Prominent transitions observed in the optical and IUE ultraviolet regions. The fine structure separation for each term is not drawn to scale. We give for each term the excitation potential in electron volts, eV, the principal observed lines, the lifetime of the upper level involved for each transition, and the branching ratio expressed as a percentage for the observed lines. The auroral-type transitions for [S II] lie in the infrared, while the nebular and transauroral transitions in [Ne IV] are in the ultraviolet and can be observed with instruments such as the IUE. Transitions analogous to those for [O III] appear in the near ultraviolet for [Ne V] and in the near infrared for the $3p^3$ configuration of [S III].

Fig. 2-8. The Ultraviolet Spectrum of the Irregular Planetary NGC 6302. The observations were obtained with the International Ultraviolet Explorer. This is a nitrogen-rich object (see Chapter 12) in which carbon is relatively depleted, although the C III and C IV lines are still strong. Ionic lines of nitrogen are well-represented as are those of O III, O IV, and Si III. There is no trace of any spectrum of a central star.

Fig. 2-9. Fluorescence and the Excitation of O I λ8446. (Left) In the near infrared of the dense planetary IC 4997, we find that in the region of the confluence of the Paschen lines, the spectrum is dominated by the 8446 O I feature. The two-image tube scanner tracings have relative magnifications of 5 and 1. This line is prominent in many other planetaries. (Right) A partial energy level diagram for O I. The ground term is $2p^4\ ^3P$. The 3d level could be populated by absorptions of Lyman β, which is one mechanism suggested by Grandi for the excitation of λ8446 in quasars. In planetaries such as IC 4997 and the Orion nebula, he suggests that direct excitation by starlight is more likely. Atoms are excited to high s and d configurations whence they cascade to 3p. Telltale permitted lines such as λ7002 and λ7254 are strong in Orion and in IC 4997.

Chapter 3

RADIATIVE EXCITATION, IONIZATION, RECOMBINATION, AND FLUORESCENCE

3A. Basic Relationships. Before we can calculate rate processes for atomic events in gaseous nebulae, we must derive certain basic equations. Our first concern is with radiative processes involving hydrogen.

If a_ν is the atomic line absorption coefficient and f is the absorption oscillator strength, or Ladenburg f, then:

$$\int a_\nu \, d\nu = \frac{\pi \varepsilon^2}{mc} f_{n'n} , \tag{1}$$

where n' is the lower level of the transition and n is the upper level. Then:

$$A_{nn'} = \frac{\omega_{n'}}{\omega_n} \frac{8\pi^2 \varepsilon^2 \nu^2}{mc^3} f_{n'n} , \tag{2}$$

where $A_{nn'}$ is the Einstein coefficient for spontaneous emission. We define the Einstein coefficients in terms of intensity rather than radiation density. Then the following relations hold:

$$\omega_n B_{nn'} = \omega_{n'} B_{n'n} , \quad \frac{\omega_n}{\omega_{n'}} A_{nn'} = B_{n'n} \frac{2h\nu^3}{c^2} , \tag{3}$$

where $\omega_{n'}$ and ω_n are the statistical weights of the lower and upper levels, respectively; ε, m, ν, and c have their usual meanings. Numerically,

$$f_{n'n} = 1.5 \times 10^{-8} \lambda_\mu^2 \frac{\omega_n}{\omega_{n'}} A_{nn'} , \tag{4}$$

where λ_μ is the wavelength expressed in microns. Atomic transition probabilities, expressed as A or f values, may be obtained by appropriate experiments or by theoretical calculations. The latter can become fairly complicated for a complex atom. The situation is simpler in hydrogen where energy levels are degenerate in n and ℓ. In a first reconnaissance of the problem, we need concern ourselves only with transitions between groups of levels of the same n. Following Menzel and Pekeris (1935), the f-value for an n' → n transition is given by:

$$f_{n'n} = \frac{2^6}{3\sqrt{3}\pi} \frac{1}{\omega_{n'}} \frac{1}{\left|\frac{1}{n'^2} - \frac{1}{n^2}\right|^3} \left|\frac{1}{n^3} \frac{1}{n'^3}\right| g_{n'n} . \tag{5}$$

The frequency of any line in the hydrogenic spectrum is given by the Rydberg formula:

$$\nu = RZ^2 \left(\frac{1}{n'^2} - \frac{1}{n^2}\right) , \quad n > n' , \tag{6}$$

where* $R = 2\pi^2\varepsilon^4 m/h^3$. (7)

Here $g_{n'n}$ is a complicated expression involving a product of confluent hypergeometric functions. It is called the Gaunt factor.

An important innovation introduced by Menzel and Pekeris was a generalization of these expressions to include levels in the continuum. In order to apply Equation (5) to the continuum, they put

$$n = i\kappa , \tag{8}$$

so that the Rydberg formula generalized to the continuum becomes:

$$\nu = RZ^2 \left(\frac{1}{n^2} + \frac{1}{\kappa^2}\right) . \tag{9}$$

If an electron is detached from a level n by a quantum of frequency ν such that its kinetic energy is $1/2\ mv^2$, then:

$$\frac{1}{2} mv^2 + \frac{hRZ^2}{n^2} = h\nu . \tag{10}$$

Hence, $1/2\ mv^2 = hRZ^2/\kappa^2$, (11)

and $\{- 2hRZ^2/\kappa^3\} d\kappa = mvdv = hd\nu$. (12)

In applications of these formulae, we have to take absolute values as required since an absorption coefficient cannot be negative. To generalize the f-value to the continuum, differentiate Eq. (1) and require that f be continuous over the series limit. Then,

$$a_\nu = \frac{\pi\varepsilon^2}{mc} \frac{df}{d\nu} . \tag{13}$$

Consider the f-value defined for a unit frequency just to the low frequency side of the series limit. There are Δn lines of mean oscillator strength f per unit frequency interval. Just to the high frequency side, the f-value per unit frequency interval will be $f\Delta\kappa$. Continuity over the series limit exists and therefore demands that $df = f\Delta n$ merges smoothly to

$$df = f\Delta\kappa . \tag{14}$$

*In Chapter 3 we neglect the dependence of the Rydberg constant on the mass of the nucleus. This effect becomes important for radio frequency lines, cf. Chapter 4.

Accordingly:

$$a_\nu = \frac{\pi\varepsilon^2}{mc}\frac{df}{d\kappa}\frac{d\kappa}{d\nu} = \frac{\pi\varepsilon^2}{mc} f \frac{d\kappa}{d\nu} . \qquad (15)$$

Since

$$\left|\frac{d\kappa}{d\nu}\right| = \left|\frac{\kappa^3}{2RZ^2}\right| , \quad a_\nu = \frac{\pi\varepsilon^2}{mc} f \frac{\kappa^3}{2RZ^2} . \qquad (16)$$

Now from Equations (5), (6), (7), and (8) we have:

$$a_n(\nu) = \frac{32}{3\sqrt{3}} \frac{\pi^2\varepsilon^6}{ch^3} \frac{RZ^4}{n^5\nu^3} g_{n\kappa} . \qquad (17)$$

Now consider radiative processes involving line and continuous transitions. Menzel introduced a factor b_n to denote the departure of the population of a level n from that appropriate to thermodynamic equilibrium at the electron temperature, T_ε. Thus, Boltzmann's equation is replaced by an expression of the form:

$$\frac{N(j)}{N_1} = \frac{b_j}{b_1}\frac{\omega_j}{\omega_1} e^{-X_j/kT_\varepsilon} . \qquad (18)$$

In particular, the combined Boltzmann and Saha equation for H now becomes:

$$N_n = b_n N(H^+)N_\varepsilon \frac{h^3}{(2\pi mkT_\varepsilon)^{3/2}} \frac{\omega_n}{2} e^{X_n} , \qquad (19)$$

where $X_n = hRZ^2/n^2kT_\varepsilon$ and $\omega_n = 2n^2$. (20)

In strict thermodynamic equilibrium $b_n \equiv 1$.

The number of radiative excitations from level n' to level n will be per second per unit volume:

$$F_{n'n} = \frac{N_{n'}}{h\nu} \int I_\nu a^*_\nu \, d\nu \, d\omega . \qquad (21)$$

Suppose I_ν does not vary over the line width. We take the average intensity as $J_\nu = (1/4\pi) \int I d\omega$. Here a^*_ν is the absorption coefficient corrected for negative absorptions. Note that we prefer the term negative absorption to stimulated emission since the ejected photon goes in the same direction as the triggering photon. This correction factor, which is $(1 - e^{-h\nu/kT})$ under conditions of LTE, now becomes:

$\left(1 - \frac{\omega_{n'}}{\omega_n}\frac{N_n}{N_{n'}}\right)$. Hence, we have:

$$F_{n'n} = 4\pi\, N_{n'}\left(1 - \frac{\omega_{n'}}{\omega_n}\frac{N_n}{N_{n'}}\right)\frac{J_\nu}{h\nu}\frac{\pi\varepsilon^2}{mc} f_{n'n}, \text{ or} \tag{22}$$

$$F_{n'n} = N_{n'}\, B_{n'n}\, J_\nu\left(1 - \frac{\omega_{n'}}{\omega_n}\frac{N_n}{N_{n'}}\right) , \tag{23}$$

if we prefer to work with the Einstein B's. The corresponding expression for the total energy absorbed may be obtained by multiplying by $h\nu$. The number of spontaneous downward transitions and the emitted energy will be:

$$F_{nn'} = N_n\, A_{nn'} \text{ , and} \tag{24}$$

$$E_{nn'} = F_{nn'}\, h\nu = N_n\, A_{nn'}\, h\nu , \tag{25}$$

respectively.

In gaseous nebulae, the number of atoms in any excited level n is very small compared with the number in the ground level. Hence, for all optical lines we can neglect the negative absorptions. This simplication does not hold for the radio frequency range where n is of the order of 100 and we are concerned with transitions where $\Delta n = n - n' \sim 1$ or 2. Then using Eq. (18) and noting $h\nu_{nn'}/kT$ is small, we find:

$$\left(1 - \frac{\omega_{n'}}{\omega_n}\frac{N_n}{N_{n'}}\right) \sim \left[1 - \frac{b_n}{b_{n'}}\left(1 - \frac{h\nu_{nn'}}{kT}\right)\right] , \tag{26}$$

which may differ markedly from 1. Neglecting for the time being the negative absorptions, the number of photoionizations from the n level to the continuum in the interval ν to $\nu + d\nu$ will be:

$$F_{n\kappa}\, d\nu = 4\pi\, J_\nu\, \frac{a_n(\nu)}{h\nu}\, N_n\, d\nu . \tag{27}$$

Substituting in this equation from Eqs. (17) and (19), we find for number of ionizations from level n to interval ν to $\nu + d\nu$.

$$F_{n\kappa}\, d\nu = N(H^+)\, N_\varepsilon\, \frac{DZ^4}{T_\varepsilon^{3/2}}\, J_\nu\, \frac{b_n\, g_{n\kappa}}{n^3\, \nu^4}\, e^{X_n}\, d\nu , \tag{28}$$

where

$$D = \left\{\frac{2^7}{3\sqrt{3}}\;\frac{\pi^3 \varepsilon^6 R}{ch\,(2\pi mk)^{3/2}}\right\} , \tag{29}$$

is a numerical constant equal to 2.212×10^{41} c.g.s. units.

The corresponding energy absorbed in the interval ν to $\nu + d\nu$ by photoionizations from level n per unit volume per second will be:

$$E_{n\kappa}\, d\nu = F_{n\kappa}\, h\nu d\nu = N(H^+)\, N_\varepsilon \frac{DZ^4}{T_\varepsilon^{3/2}}\, J_\nu \frac{b_n\, g_{n\kappa}}{n^3\, \nu^3}\, e^{X_n}\, h d\nu\,. \quad (30)$$

The recombination rate from the continuum to level n from the velocity range v to v + dv corresponding to ν to $\nu + d\nu$ will be:

$$F_{\kappa n}\, d\nu = N(H+)\, N_\varepsilon\, f(v,\, T_\varepsilon)\, v\sigma_{\kappa n}\, dv\,, \quad (31)$$

where $\sigma_{\kappa n}$ is the cross section for radiative capture and $f(v,\, T_\varepsilon)$ is the Maxwellian velocity distribution, which is valid even under conditions obtaining in a low density gaseous nebula. Now:

$$f(v,\, T_\varepsilon)dv = 4\pi\left(\frac{m}{2\pi kT_\varepsilon}\right)^{3/2} v^2 e^{-mv^2/2kT_\varepsilon} dv\,. \quad (32)$$

Milne showed that a basic relationship independent of temperature, density, type of atom, etc., must exist between the recapture and photoionization coefficients. The derivation of this relationship is closely analogous to that employed to get those between Einstein's A and B coefficients.

We use the principle of detailed balance in thermodynamic equilibrium. Consider a neutral atom in level (0,n) which is photoionized to the continuum in an interval ν to $\nu + d\nu$, leaving the ion in its ground level (1,1). The rate of this process must equal the number of recaptures in this same level (0,n) by an ion in its ground level. The rate of electron recaptures per cm^3 from velocity range v to v + dv in atomic level (0,n) will be the number of ground level ions/cm^3, $N_{1,1}$ multiplied by the target area for recapture, $\sigma_{\kappa n}$, by the total number of electrons N_ε f(v) dv per cm^3, and by v (since each electron sweeps up a volume $\sigma_{\kappa n}$ v per second). The number of photoionizations from level n must equal this recapture rate. Thus:

$$4\pi\, N_{0,n}\, a_{0,n}(\nu)\, B_\nu\left(1 - e^{-h\nu/kT}\right) \frac{d\nu}{h\nu} = N_{1,1}\, N_\varepsilon\, \sigma_{\kappa n}\, f(v, T_\varepsilon)\, v dv\,, \quad (33)$$

where f(v,T) is the Maxwellian velocity distribution and B_ν is the Planck function,

$$B_\nu = \frac{2h\nu^3}{c^2} \frac{1}{e^{h\nu/kT} - 1}\,. \quad (34)$$

Now, $N_{1,1}\, N_\varepsilon/N_{0,n}$ is given by the combined Saha and Boltzmann equation which is of the form:

$$\frac{N_{1,1}N_\varepsilon}{N_{0,n}} = \frac{(2\pi mkT)^{3/2}}{h^3} \frac{\omega_e\,\omega_{1,1}}{\omega_{0,n}} e^{-(I - \chi_n)/kT} , \tag{35}$$

where I is the ionization potential of the atom from its ground level and χ_n denotes the excitation potential of level (0,n). We have further that: $(I - \chi_n) + 1/2\ mv^2 = h\nu$. (36)
Combine these equations. We find that $a_{0,n}(\nu)$ and $\sigma_{\kappa n}$ are related by:

$$\frac{a_{0,n}(\nu)}{\sigma_{\kappa n}(\nu)} = \frac{m^2c^2v^2}{\nu^2h^2} \frac{\omega_{1,1}\,\omega_e}{2\omega_{0,n}} . \tag{37}$$

The statistical weight of the free electron ω_e is 2. Then making use of Eqs. (37), (31), and (32), we find:

$$F_{\kappa n}\,d\nu = N(H^+)\,N_\varepsilon \frac{KZ^4}{T_\varepsilon^{3/2}} \frac{g_{II}}{n^3} e^{X_n} e^{-h\nu/kT_\varepsilon} \frac{d\nu}{\nu} , \tag{38}$$

since for hydrogen $h\nu = hRZ^2/n^2 + 1/2\ mv^2$. Menzel (1937) defined:

$$K \equiv \frac{2hD}{c^2} = 3.257 \times 10^{-6}\ \text{(c.g.s. units)} \tag{39}$$

The total number of recombinations in n is:

$$\int_{\nu_n}^{\infty} F_{\kappa n}d\nu = N(H^+)N_\varepsilon \frac{KZ^4}{T_\varepsilon^{3/2}} \frac{e^{X_n}}{n^3} \int_{\nu_n}^{\infty} g_{II}(\nu)e^{-h\nu/kT_\varepsilon} \frac{d\nu}{\nu}$$

$$= \alpha_n(T_\varepsilon)\ N(H^+)\ N_\varepsilon . \tag{40}$$

Here $\nu_n = RZ^2/n^2$. The integral is sometimes approximated by:

$$\bar{g}_{II} \int_{\nu_n}^{\infty} e^{-h\nu/kT_\varepsilon} d\nu/\nu = \bar{g}_{II} \int_{X_n}^{\infty} \frac{e^{-x}}{x} dx = \bar{g}_{II}\ E_1(X_n) . \tag{41}$$

The total number of recombinations on all levels can be written as:

$$\sum_{n=1}^{n_{max}} \int_{\nu_n}^{\infty} F_{\kappa n}\ d\nu = N(H^+)\ N_\varepsilon\ \alpha_A(T_\varepsilon) , \tag{42}$$

where the total recombination coefficient $\alpha_A(T_\varepsilon)$ is given by:

$$\alpha_A(T_\varepsilon) = \frac{KZ^4}{T_\varepsilon^{3/2}} \sum_{n=1}^{n_{max}} \frac{e^{X_n}}{n^3} \int_{\nu_n}^{\infty} g_{II}(\nu)\ e^{-h\nu/kT}\ d\nu/\nu . \tag{43}$$

The Gaunt factor $g_{II}(\nu)$ has been calculated by Menzel and Pekeris, and by Karzas and Latter. Here, n_{max} is the n value corresponding to the largest, regularly occupied Bohr orbit.

We may also define $\alpha_B(T_\varepsilon)$ as the recombination coefficient corresponding to the total recapture rate on the second and higher levels, viz.:

$$\sum_{n=2}^{n_{max}} \int F_{\kappa n}\, d\nu = N(H^+)\, N_\varepsilon\, \alpha_B(T_\varepsilon) \; ; \tag{44}$$

$$\alpha_B(T_\varepsilon) = \frac{KZ^4}{T_\varepsilon^{3/2}} \sum_{n=2}^{n_{max}} \frac{e^{X_n}}{n^3} \int_{\nu_n}^{\infty} g_{II}(\nu)\, e^{-h\nu/kT} \frac{d\nu}{\nu} \, . \tag{45}$$

Similarly, we should calculate the total number of photoionizations by integrating the expression for $F_{n\kappa}\, d\nu$ over all relevant frequencies, summing over all n. In optically very thick plasmas, this step may be necessary, but in most gaseous nebulae where the populations of the higher levels are low, we need to consider ionizations only from the first level. Using Eq. (28), we find that the total number of ionizations per unit volume per second will be:

$$4\pi\, N_1 \int_{\nu_1}^{\infty} J_\nu\, a_1(\nu) \frac{d\nu}{h\nu} = N(H^+)\, N_\varepsilon \frac{DZ^4}{T_\varepsilon^{3/2}}\, b_1\, e^{X_1} \int_{\nu_1}^{\infty} J_\nu\, g(\nu) \frac{d\nu}{\nu^4} \, . \tag{46}$$

Then, under steady state conditions:

$$4\pi\, N_1 \int_{\nu_1}^{\infty} J_\nu\, a_1(\nu) \frac{d\nu}{h\nu} = N(H^+)\, N_\varepsilon\, \alpha_A(T_\varepsilon) \, . \tag{47}$$

Now, using Eqs. (46), (47), and (42), noting $D/K = c^2/2h$, and dividing out the common factor $KZ^4\, T_\varepsilon^{-3/2}$, we have:

$$b_1 e^{X_1} \frac{c^2}{2h} \int J_\nu\, g(\nu) \frac{d\nu}{\nu^4} = \sum_{n=1}^{n_{max}} \frac{e^{X_n}}{n^3} \int_{\nu_n}^{\infty} g\, e^{-h\nu/kT} \frac{d\nu}{\nu}$$

$$\equiv \sum_{n=1}^{n_{max}} \frac{S_n}{n^3} \equiv G_A(T_\varepsilon) \, , \tag{48}$$

which defines $G_A(T_\varepsilon)$. Here

$$S_n = e^{X_n} \int g\, \exp\left(-\frac{h\nu}{kT}\right) \frac{d\nu}{\nu} \, . \tag{49}$$

For later use we may also define as an analogous quantity $G_B(T_\varepsilon)$ by taking the lower limit of the sum in Eq. (48) as $n = 2$ rather than $n = 1$. In terms of these auxiliary quantities:

$$\alpha_A(T_\varepsilon) = KZ^4\, G_A(T_\varepsilon)\, T_\varepsilon^{-3/2} \;,\quad \alpha_B(T_\varepsilon) = KZ^4\, G_B(T_\varepsilon)\, T_\varepsilon^{-3/2} \;, \qquad (50)$$

Table 3-1. Recombination Functions for Hydrogen

T_ε	G_A	G_B	$\log \alpha_A$	$\log \alpha_B$
5,000	0.074	0.049	- 12.166	- 12.343
10,000	0.128	0.080	- 12.380	- 12.585
20,000	0.218	0.124	- 12.60	- 12.845

N.B.: Units of α are $cm^3\ sec^{-1}$. For example, T dependence of α_B is represented fairly well by the expression $\alpha_B = 4.10 \times 10^{-10}\ T^{-0.8}\ cm^3\ sec^{-1}$.

It is of interest to calculate b_1, the correction factor to the Boltzmann formula for the ground level. From Eq. (48):

$$b_1 = \frac{\left(2h/c^2\right) e^{-X_1} G(T_\varepsilon)}{\int J(\nu)\, g\nu^{-4}\, d\nu} \;. \qquad (51)$$

To assess the order of magnitude of the quantities involved, suppose $J(\nu) = 10^{-14}\ B_\nu(T*)$ where we take $T* = 80{,}000°K$. If we take $T_\varepsilon = 10{,}000°K$, we find $b_1 \sim 3.5 \times 10^7$. The physical meaning of the result is that if we refer all processes to an electron temperature of 10,000°K and the local density, the concentration of atoms in the ground level of the neutral atom exceeds that calculated by the Saha equation by a factor of about 35,000,000. Ionization rates are greatly reduced compared to what they would be at 10,000°K, but recombinations go on at an effective rate fixed only by $\alpha_B(T_\varepsilon) = 4.18 \times 10^{-13}\ cm^3\ s^{-1}$ at $T_\varepsilon = 10{,}000°K$ and the local values of $N(H^+)$ and N_ε.

Note that J_ν is the total radiation field at the point in question, not just the contribution from the star itself; J_ν includes radiation reprocessed within the nebula. Thus, rigorously, one should consider the transfer problem of the radiation.

3B. The Strömgren Sphere. Consider a hot star that ionizes a surrounding volume of gas. Quanta beyond the Lyman limit detach electrons from atoms. As the electrons and protons recombine, the gas emits the characteristic recombination lines. We consider a hydrogen nebula and temporarily neglect the role of other elements, the most important of which is helium. Our ionization equation is (47) where $\alpha_A\ (T_\varepsilon)$ is given by Eq. (43). Now $J(\nu)$ is the specific intensity at the point under consideration in the nebula. The radiation field consists of two components, the attenuated radiation from the star and the "diffuse" Lyman continuum radiation produced by recaptures of the electron on the ground level.

Starlight is simply extinguished in the nebula. At a distance s from the star, its specific intensity is given by:

$$J_s(\nu) = \frac{R^2}{4s^2} F_\nu^*(0)\, e^{-\tau_\nu} , \tag{52}$$

where $\pi F_\nu^*(0)$ is the flux per unit area from the stellar surface and τ_ν is the optical depth measured to the point in question.

$$\tau_\nu = \int_0^s N_1\, a_1(\nu)\, ds . \tag{53}$$

Referring to Eq. (17), we note that $a_1(\nu)$ goes very nearly as ν^{-3},

$$a_1(\nu) = a_1(\nu_1) \left[\frac{g(\nu)}{g(\nu_1)}\right] \nu^{-3} ,$$

where $[g(\nu)/g(\nu_1)]$ varies slowly with frequency. A rigorous calculation of the "diffuse" Lyman continuum radiation field requires that we set up and solve the equations of transfer. The newly created Lyman continuum quantum is fully capable of ionizing an H atom. The number of these created per cm^3 per sec will be $\alpha_1(T_\varepsilon)\, N(H^+)\, N_\varepsilon$. Solution of the transfer equation for the diffuse radiation takes account of the fact that this might occur perhaps at some distance from the volume element where the original ionization occurred. Actually, in many instances, it is probably admissible to assume that the "diffuse" quantum is reabsorbed nearby. This "on the spot" approximation evades the troublesome solution of the transfer equation. We shall use it here, although some workers have criticized it for lack of rigor. If $\alpha_1(T_\varepsilon)$ denotes the recombination coefficient for the ground level, then the ionization equation becomes:

$$4\pi N_1 \int \frac{J_s(\nu)\, a_1(\nu)\, d\nu}{h\nu} + \alpha_1(T_\varepsilon)\, N(H^+)\, N_\varepsilon = \alpha_A(T_\varepsilon)\, N(H^+)\, N_\varepsilon . \tag{54}$$

Now, make use of Eq. (40), the definitions (44) and (45) and note $\alpha_A - \alpha_1 = \alpha_B$, then:

$$4\pi N_1 \int \frac{J_s(\nu)\, a_1(\nu)\, d\nu}{h\nu} = \alpha_B(T_\varepsilon)\, N(H^+)\, N_\varepsilon . \tag{55}$$

For any assumed spatial distribution $N_H(r)$ we can solve the equations numerically. The boundary conditions are that near the star $N_H = N(H^+)$, $N_1 = 0$. At each integration step we employ Eqs. (52), (53), and (55) to calculate N_1, τ_ν, and J_ν until the point is reached where $N_1 \rightarrow N_H$ and the boundary of the ionization zone is established.

Strömgren found that the boundary of the hydrogen ionization zone tended to be very sharp. The physical reason for this sharpness can be fairly easily understood. Near the star the gas is fully ionized because the rate of photoionization is relatively high. The effective rate of recombination, $\alpha_B\, N(H^+)N_\varepsilon$, remains fairly constant throughout the ionized zone as long as T_ε is constant. As we proceed away from the star, the stellar radiation becomes more and more dilute until the ionization rate drops to the point where neutral atoms begin to appear. Then τ_ν increases rapidly and the gas becomes neutral.

Classical sharp-edged Strömgren spheres are seen for example in the southern Milky Way and in the Magellanic clouds. The radius s_0 of a Strömgren sphere can be calculated by a straightforward argument that takes into account the abrupt transition from ionized to neutral zone.

[Total number of Lyc quanta emitted by star] = [Total number of recombinations on second and higher levels]

$$P_{uv} = \int_{\nu_1}^{\infty} \frac{4\pi R^2\ \pi F_\nu^*(0)}{h\nu}\, d\nu = \int_{\nu_1}^{\infty} \frac{L_\nu}{h\nu}\, d\nu = \alpha_B(T_\varepsilon) \int N(H^+)\, N_\varepsilon\, dV\ . \tag{56}$$

If we use the condition for a pure hydrogen nebula that $N_H = N(H^+) = N_\varepsilon$ within the ionized sphere and assume that the density is constant:

$$\alpha_B(T_\varepsilon) \int N(H^+)\, N_\varepsilon\, dV = \frac{4\pi}{3}\, s_0^3\, N_H^2\, \alpha_B(T_\varepsilon)\ . \tag{57}$$

This integral serves to define the so-called excitation parameter U which is a measure of the total number of ionizing quanta produced by the illuminating stars that ionize the region. We define

$$U^3 \cong \int N(H^+)\, N_\varepsilon\, dV \sim s_0^3\, N^2\ , \tag{58}$$

so for $N = 1$, $U = s_0$ (pc cm^{-2}).

In practice, one must include helium, whose first and second ionization potentials are respectively 24.58 and 54.4 eV. Thus, helium is doubly ionized only near the ionizing star, but singly ionized out to a greater distance. The size of the ionization zone compared with that of the hydrogen ionization zone depends on the temperature and ultraviolet flux distribution.

The much lower abundances of other elements have a profound effect on the structures of their ionization zones. In their pioneering studies of low density H II regions in the Milky Way, Struve and Elvey noticed that while the edges of the H-ionization zones were sharp, the [O III] emission which appeared in H II regions near the very hottest stars, faded away gradually with no sharp edge. Qualitatively, this phenomenon is easily understood. Since the oxygen abundance is less than 10^{-3} that of hydrogen, a transition from the O^{++} zone to the O^+ zone takes place gradually. The radiation field is dominated by hydrogen and helium; continuous absorption by oxygen ions plays a rather small role.

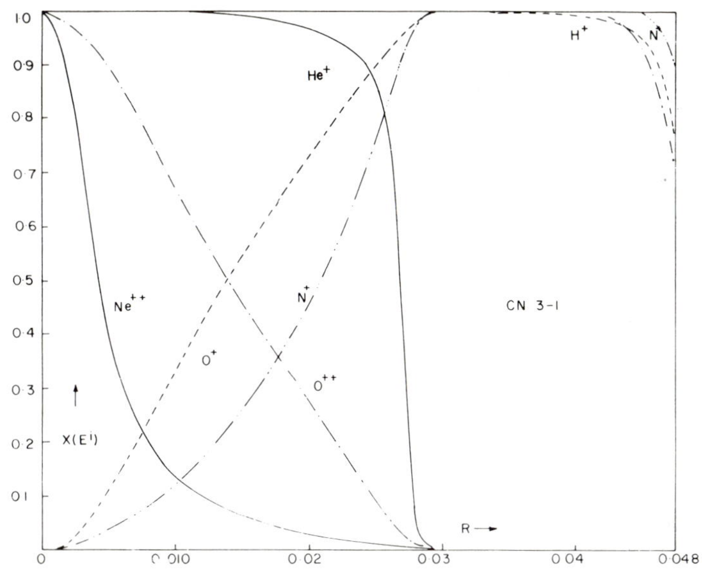

Figure 3-1

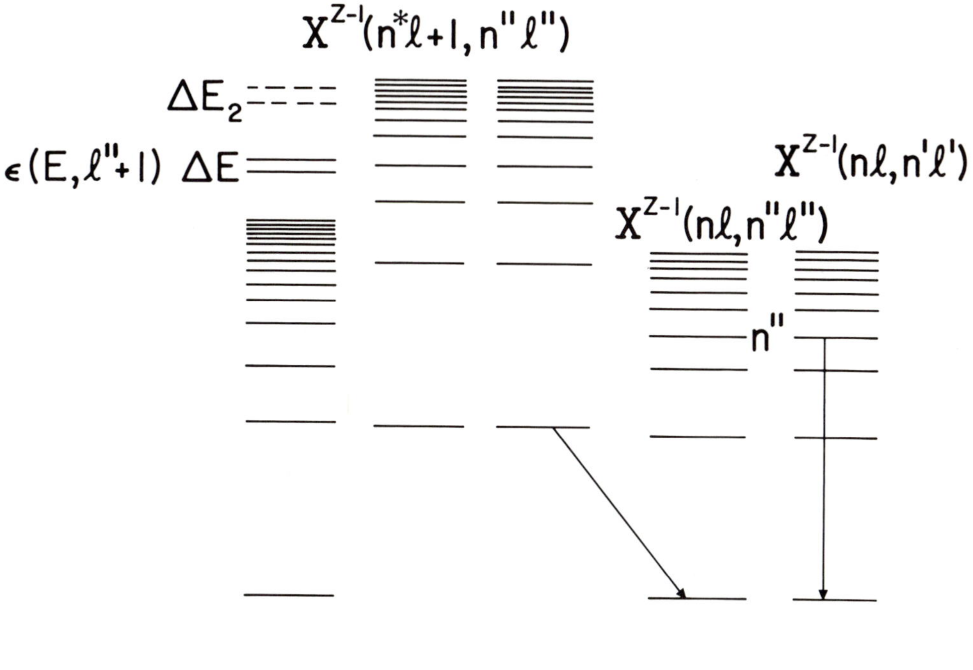

Figure 3-2

Detailed calculations of ionization structures require nebular modeling. Figure 3-1 shows the radial distribution of ions in a low-excitation planetary nebula Cn 3-1. The relatively sharp H^+ and He^+ edges and the slow decline of O^{++} and Ne^{++} with distance from the central star are well illustrated.

3C. Free-Free Emission and Bremsstrahlung. In the framework of the classical theory, when an electric charge is accelerated, at a rate $\ddot{s}$, it must continuously radiate energy in an amount:

$$\frac{dE}{dt} = -\frac{2}{3}\frac{\varepsilon^2(\ddot{s})^2}{c^3} .$$

According to the quantum theory, an electron moving at a velocity v suddenly shifts to a lower velocity v' with the emission of a quantum of energy.

We want to calculate the amount of energy emitted in a transition from a velocity range v to v + dv to a velocity range v' + dv'. First consider the expression for the energy radiated in the range dν when an electron moving with a velocity in a range v to v + dv is recaptured on a level n. From Eq. (38) we find:

$$E_{\kappa n}\, d\nu = F_{\kappa n}\, h d\nu = N(H^+)\, N_\varepsilon \frac{KZ^4}{T_\varepsilon^{3/2}} \frac{g_{II}}{n^3} e^{X_n} e^{-h\nu/kT_\varepsilon} h d\nu . \qquad (59)$$

Let us replace n by iκ' and X_n by $X_{\kappa'}$; then

$$X_{\kappa'} = -\frac{hRZ^2}{\kappa'^2 kT_\varepsilon} , \qquad (60)$$

and Eq. (59) becomes:

$$E_{\kappa\kappa'} d\kappa' d\nu = N(H^+)\, N_\varepsilon \frac{KZ^4}{T_\varepsilon^{3/2}}\, g \exp\left[-\frac{hRZ^2}{\kappa'^2 kT_\varepsilon}\right] \frac{e^{-h\nu/kT_\varepsilon}}{\kappa'^3} h d\nu d\kappa' . \qquad (61)$$

Any particular frequency ν will correspond to an infinite set of (κ, κ') combinations subject only to the condition:

$$\nu_{ff} = RZ^2 \left(\frac{1}{\kappa^2} - \frac{1}{\kappa'^2}\right) . \qquad (62)$$

Recall Eq. (12), and integrate κ' from 0 to ∞:

$$d\nu \int E_{\kappa\kappa'}\, d\kappa' = N(H^+)\, N_\varepsilon \frac{KZ^4}{T_\varepsilon^{3/2}}\, g_{III} \frac{kT_\varepsilon}{2hRZ^2} e^{-h\nu/kT_\varepsilon} h d\nu \int_0^\infty e^{-X_{\kappa'}} dX_{\kappa'} \qquad (63)$$

where we have noted that:

$$dX_{\kappa'} = -\frac{2hRZ^2}{\kappa'^3 kT_\varepsilon} d\kappa' .$$

That is:

$$d\nu \int E_{\kappa\kappa'} d\kappa' = N(H^+) N_\varepsilon \frac{kKZ^2}{2RT_\varepsilon^{1/2}} g_{III} e^{-h\nu/kT_\varepsilon} d\nu , \quad (64)$$

is the emission in the range ν to $\nu + d\nu$. The total emission is then obtained by integrating over all frequencies:

$$\iint E_{\kappa\kappa'}(T_\varepsilon) d\nu \, d\nu_{\kappa'} = N(H^+)N_\varepsilon \frac{g_{III} k^2 KZ^2}{2Rh} T_\varepsilon^{1/2}$$

$$= 1.425 \times 10^{-27} \bar{g}_{III} Z^2 T_\varepsilon^{1/2} N(H^+) N_\varepsilon \text{ ergs cm}^{-3} \text{ s}^{-1} . \quad (65)$$

Notice that the free-free emission or Bremsstrahlung increases towards the infrared and that the total amount of energy radiated grows as T_ε is raised. Thus, in X-ray sources, Bremsstrahlung becomes a very important agent for the dissipation of energy. Free-free emission is the dominant source of thermal emission in the radio frequency region. The problem is complicated because g_{III} is difficult to calculate. In the optical region $g \sim 1$, but in the radio frequency region $g \sim 6$.

In an approximate treatment, He^+ can be included by replacing $N(H^+)$ by $[N(H^+) + N(He^+)]$, but note that if ionized helium is present the appropriate term is $[N(H^+) + N(He^+) + 4N(He^{++})]$.

3D. Ionization Equation for Complex Atoms. From the simple situation of hydrogen we may move to that of more complex atoms in a gaseous nebula in a straightforward manner. We construct the ionization equation from the statement:

$$\begin{array}{c}\text{Number of ionizations} \\ \text{from ground terms}\end{array} = \begin{array}{c}\text{Number of recaptures} \\ \text{on all levels}\end{array}$$

$$N_1(X^i)\int_{\nu_1}^{\infty} \frac{4\pi J_\nu a_1(\nu,X^i)}{h\nu} d\nu = \sum_{j=1}^{\infty} N(X^{i+1})N_\varepsilon \int_{v=0}^{\infty} f(v)v\sigma_{j,X_i}(v)dv \quad (66)$$

Note that generally for complex atoms ionizations may take place from different levels of the ground term. It is customary to treat this ground term as though it were a single level of proper statistical weight. The ionization limit may not be a single level. For example, nitrogen atoms may be photoionized from the $2s^2\ 2p^3\ ^4S_{3/2}$ ground term of N I to the lowest $2s^2\ 2p^2\ ^3\underline{P}$ term at 14.5 eV, to the ^1D term at 16.4 eV or, in much smaller numbers, even to the ^1S term at 19.6 eV. Similarly, Ne^{++} is ionized from the $2s^2\ 2p^4\ ^3\underline{P}$ term to the $2s^2\ 2p^3\ ^4S_{3/2}$ term of Ne IV at 63.5 eV, but it can also be ionized to ^2D and ^2P terms at 68.6 and 71.2 eV, respectively. The threshold photoionization cross sections are 1.8×10^{-18}, 2.5×10^{-18}, and 1.5×10^{-18} cm^2. In hydrogen, the $a_n(\nu)$ function is relatively simple, $a_\nu \sim \nu^{-3}$ because the electron moves in a coulombic field, but the absorption coefficient for equivalent p electrons is more complicated. The reason for the nonhydrogenic behavior of the absorption coefficient is that an outer electron in a partially filled 2p or 3p shell is incompletely screened by its neighbors. Such an electron will move in a field, $Z - \sigma$, where σ, which is a measure of the screening, is less than $(N - 1)$, where N is the number of electrons in the ion. On the other hand, if a 2p electron is raised to a d level, it will move in a nearly hydrogenic field of charge $Z - (N - 1)$. Consequently, in calculating a transition from a bound 2p to a free κd level, one cannot employ the same value of $Z - \sigma$ in the initial and final wave functions. For highly ionized atoms, one may calculate the recapture rates in higher levels as though they were hydrogenic but this approximation may not be adequate for the excited levels of neutral atoms. Tables 3-2 and 3-3 give parameters for interpolation formulae to calculate photoionization coefficients for ground terms of C, N, O, Ne, Si, S, and Ar.

To put Eq. (66) in a more tractable form, we replace $\sigma(v)$ by $a_{j\kappa}(\nu)$ by Milne's Eq. (37) and use Eq. (32). Let $\omega_{i,1}$ be the statistical weight of the ground level of the ion X_i and $\omega_{i+1,1}$ that of the ground level of X_{i+1}. Then:

$$\frac{N(X^{i+1})\,N_\varepsilon}{N(X^i)} = \frac{\left(2\pi m k T_\varepsilon\right)^{3/2}}{h^3}\,\frac{\omega_\varepsilon\,\omega_{i+1,1}}{\omega_{i,1}} \times \frac{\omega_{i,1}\displaystyle\int_{\nu_1}^{\infty}\frac{J_\nu\,a_1(\nu, X^i)}{h\nu}\,d\nu}{\displaystyle\sum_{j=1}^{\infty}\frac{m}{h}\frac{2}{c^2}\,\omega_j\int_0^{\infty} a_j(\nu, X^i)\,e^{-mv^2/2kT_\varepsilon}\,\nu^2 v\,dv} . \tag{67}$$

Application of this rather complex-looking equation requires a knowledge of the radiation field and atomic continuous absorption coefficients. We can write it in the form:

$$\frac{N(X^{i+1})\,N_\varepsilon}{N(X^i)} = \frac{\Gamma(I_\nu)}{\alpha(T_\varepsilon)} , \tag{68}$$

where $\Gamma(I_\nu)$ expresses the involvement of the radiation field and $\alpha(T_\varepsilon)$ is the recombination term which depends on only the local electron temperature and particulars of the absorption coefficients. Table 3-4 gives recombination coefficients for radiative captures. Suppose that J_ν can be approximated as $J_\nu = W_\nu B_\nu(T_1)$. That is, we assume that the dilute radiation field can be described roughly as that of a black body characterized by some temperature T_1 and dilution factor W_ν.

Since the ionization frequency ν_1 lies in the ultraviolet for all atoms and ions of astrophysical interest, we can usually reduce this expression to a simpler form, making use of the Wien approximation to Planck's law, viz:

$$J_\nu \sim W_\nu \frac{2h\nu^3}{c^2} \exp\left[-\frac{h\nu}{kT_1}\right] , \tag{69}$$

valid for $h\nu/kT_1 \gg 1$. Then:

$$e^{h\nu/kT_1} - 1 \rightarrow \exp\left|\frac{h\nu}{kT_1}\right| \sim \exp\left[\frac{\chi_i}{kT_1} + \frac{mv^2}{2kT_1}\right] . \tag{70}$$

The equation may now be written in a form analogous to that developed by Strömgren (1948):

$$\frac{N(X^{i+1})\, N_\varepsilon}{N(X^i)} = \left[\frac{2\pi mk}{h^2}\right]^{3/2} \frac{2\omega_{i+1,1}}{\omega_{i,1}} T_1^{3/2} \exp\left\{\frac{-\chi_i}{kT}\right\} \left[\frac{T_\varepsilon}{T_1}\right]^{1/2} D_r , \tag{71}$$

where χ_i is the ionization potential of the ion X^i, and:

$$D_r = \frac{\omega_1 \int_0^\infty W_\nu\, \alpha_1(\nu, X^i)\, \nu^2\, e^{-\xi}\, d\xi}{\sum_{j=1}^{\infty} \omega_j \int_0^\infty \alpha_j(\nu, X^i)\, \nu^2\, e^{-x}\, dx} , \tag{72}$$

where:

$$x = \frac{mv^2}{2kT_\varepsilon} \quad \text{and} \quad \xi = \frac{mv^2}{2kT_1} . \tag{73}$$

This form of the equation has often been employed to evaluate the ionization status. To calculate the actual ionization equilibrium, we need to consider two additional effects:

a) dielectronic recombination

b) charge exchange.

Although these processes are of no importance for hydrogen, they can become significant for heavier atoms.

3E. Dielectronic Recombination. This process, which is of great astrophysical interest, is the inverse of autoionization. In that phenomenon, two electrons in the outer shell of an atom are excited simultaneously, such that the total energy exceeds the first ionization potential. The electron excited to the higher level finds itself in a state above the first ionization potential. It may undergo a transition to an unbound state, i.e, escape.

Dielectronic recombination may be illustrated by the following example: Consider an ion of charge Z in a state $3s^2\, 3p^3$ SL which collides with an electron whose angular momentum is $(\ell'' + 1)$ and whose energy is E. Suppose now that this total energy E corresponds to the ion of charge Z − 1 in which a 3s electron is excited to 3p and an outer electron is held in a configuration $n''\ell''$. Then we may have the reaction:

Table 3-2. Photionization parameters for ground terms of ions of C, N, O, and Ne.

$a_\lambda = \sigma_{th} \{\alpha[\lambda/\lambda_0(L'S')]^s + (1-\alpha)[\lambda/\lambda_0(L'S')^{s+1}\} \; 10^{-18}\ cm^2.$

$\lambda_0(L'S')$ = threshold wavelength corresponding to transition LS → L'S'.

Transition (LS→L'S')	s	a	σ_{th}	λ_0(A)
$C(^3P)-C^+(^2P)$	2.0	3.317	12.19	1100.0
$N(^4S)-N^+(^3P)$	2.0	4.287	11.42	852
$N^+(^3P)-N^{++}(^2P)$	3.0	2.860	6.65	418.7
$N^{++}(^2P)-N^{3+}(^1S)$	3.0	1.626	2.06	261.4
$O(^3P)-O^+(^4S)$	1.0	2.661	2.94	910
$O(^3P)-O^+(^2D)$	1.5	4.378	2.26	732
$O(^3P)-O^+(^2P)$	1.5	4.311	2.26	665
$O^+(^4S)-O^{++}(^3P)$	2.5	3.837	7.32	352.7
$O^{++}(^3P)-O^{3+}(^2P)$	3.0	2.014	3.65	225.6
$O^{3+}(^2P)-O^{4+}(^1S)$	3.0	0.831	1.27	160.2
$Ne(^1S)-Ne^+(^2P)$	1.0	3.769	5.35	575
$Ne^+(^2P)-Ne^{++}(^3P)$	1.5	2.717	4.16	301.8
$Ne^+(^2P)-Ne^{++}(^1D)$	1.5	2.148	2.71	280.0
$Ne^+(^2P)-Ne^{++}(^1S)$	1.5	2.216	0.52	258.3
$Ne^{++}(^3P)-Ne^{3+}(^4S)$	2.0	2.277	1.80	194.5
$Ne^{++}(^3P)-Ne^{3+}(^2D)$	2.5	2.346	2.50	180.2
$Ne^{++}(^3P)-Ne^{3+}(^2P)$	2.5	2.225	1.48	173.5
$Ne^{3+}(^4S)-Ne^{4+}(^3P)$	3.0	1.963	3.11	127.6
$Ne^{4+}(^3P)-Ne^{5+}(^2P)$	3.0	1.471	1.40	98.0
$Ne^{5+}(^2P)-Ne^{6+}(^1S)$	3.0	1.145	0.49	78.5

R.J. Henry, 1970, Ap. J., 161, 1153.

Table 3-3. Photionization parameters for ground terms of ions of Si, S, and Ar.

Transition ($SL \rightarrow S'L^1$)	s	α	β	σ_{th}	λ_0(A)
$Si(^3P)-Si^+(^2P)$	5	4.420	+8.934	39.16	1521
$Si^+(^2P)-Si^{++}(^1P)$	1.5	2.305	-2.107	1.41	758
$S(^3P)-S^+(^4S)$	3.0	21.595	3.062	12.62	1197
$S(^3P)-S^+(^2D)$	2.5	0.135	5.635	19.08	1016
$S(^3P)-S^+(^2P)$	3.0	1.159	4.743	12.70	925
$S^+(^4S)-S^{++}(^3P)$	1.5	1.695	-2.236	8.20	530
$S^{++}(^3P)-S^{3+}(^2P)$	2.0	18.427	0.592	0.38	353.7
$S^{3+}(^2P)-S^{4+}(^1S)$	2.0	6.837	4.459	0.29	262
$Ar(^1S)-Ar^+(^2P)$	3	38.738	+2.849	24.26	787
$Ar^+(^2P)-Ar^{++}(^3P)$	0.5	0.082	-0.665	28.63	448.8
$Ar^+(^2P)-Ar^{++}(^1D)$	1	0.623	-1.528	15.28	422.2
$Ar^{++}(^3P)-Ar^{3+}(^4S)$	2.5	5.929	-4.231	2.23	303.1
$Ar^{++}(^3P)-Ar^{3+}(^2D)$	2	9.375	-4.204	0.86	284.9
$Ar^{++}(^3P)-Ar^{3+}(^2P)$	2.5	7.825	-4.094	0.92	274.1
$Ar^{3+}(^4S)-Ar^{4+}(^3P)$	2	10.797	+5.687	0.59	207.3
$Ar^{4+}(^3P)-Ar^{5+}(^2P)$	2.5	7.452	+4.232	0.63	165.3
$Ar^{5+}(^3P)-Ar^{6+}(^1S)$	2	2.438	+3.237	0.34	135.8

Here a more elaborate interpolation formula is needed:

$$\alpha_\lambda = \sigma_{th}\left[\alpha W^s + (\beta-2\alpha) W^{S+1} + (1+\alpha-\beta) W^{S+2}\right] \times 10^{-18} \text{ cm}^2$$

where

$$W = \lambda/\lambda_0(S'L')$$

λ_0 in A is given in the last column.

R.D. Chapman and R.J.W. Henry, 1971, Ap. J., 168, 169, and 1972, Ap. J., 173, 243.

Table 3-4. Total radiative recombination coefficients, α_{rad}, for the direct process.

The first column gives recombined species, α_G (Gould 1978), α_{AP} (Aldrovandi and Pequignot 1973), η = parameter in the latter's formula $\alpha_{rad} = A_{rad}\ t^{-n}$ where $t = (T_\varepsilon/10{,}000)$.

$A^{+(Z-1)}$ (recombined species)	α_G (10^{-13} cm^3 s^{-1})	α_{AP} (10^{-13} cm^3 s^{-1})	η 0.624
C..................	4.66	4.7	0.624
C^+..................	24.5	23	0.645
C^{+2}..................	50.5	32	0.770
C^{+3}..................	84.5	75	0.817
N..................	3.92	4.1	0.608
N^+..................	22.8	22	0.639
N^{+2}..................	54.4	50	0.676
N^{+3}..................	95.5	65	0.743
O..................	3.31	3.1	0.678
O^+..................	20.5	20	0.648
O^{+2}..................	54.3	51	0.666
O^{+3}..................	103	96	0.670
Ne..................	2.83	2.2	0.759
Ne^+..................	17.1	15	0.693
Ne^{+2}..................	44.4	44	0.675
Ne^{+3}..................	98.1	91	0.668
Si..................	6.48	5.9	0.601
Si^+..................	14.4	10	0.786
Si^{+2}..................	36.5	37	0.693
Si^{+3}..................	73.2	55	0.821
S..................	4.65	4.1	0.630
S^+..................	24.2	18	0.686
S^{+2}..................	35.3	27	0.745
S^{+3}..................	70.3	57	0.755
Ar..................	3.30	3.77	
Ar^+..................	24.5	19.5	
Ar^{+2}..................	43.1	32.3	
Ar^{+3}..................	74.0	40.1	

Data for argon are from Aldrovandi, S.M.V., and Péguignot, D., 1974 Revista Brasileiro de Fisica, 4, 491.

Table 3-5. Coefficients for dielectronic recombinations. Nussbaumer and Storey (1983)

	$\alpha_{rec} \times 10^{12}$ (cm^3 s^{-1}) T_ε = 10,000	20,000		$\alpha_{rec} \times 10^{12}$ (cm^3 s^{-1}) T_ε = 10,000	20,000
$C^+ \to C^0$	0.185	0.143	$O^+ \to O^0$	0.076	0.077
$C^{++} \to C^+$	6.04	4.08	$O^{++} \to O^+$	1.66	1.07
$C^{3+} \to C^{++}$	13.1	7.18	$O^{3+} \to O^{++}$	11.4	10.07
			$O^{4+} \to O^{3+}$	34.5	28.8
$N^+ \to N^0$	0.522	0.270	$O^{5+} \to O^{4+}$	6.02	9.65
$N^{++} \to N^+$	2.04	2.10			
$N^{3+} \to N^{++}$	21.6	17.6			
$N^{4+} \to N^{3+}$	15.4	12.1			

$$X^Z(3s^2\ 3p^3\ SL) + \varepsilon(E_i;\ \ell'' + 1) \to X^{Z-1}(3s\ 3p^4\ S'L';\ n''\ell'')\ . \quad (74)$$

Most of the time the captured electron will bounce back to the continuum and the status quo is restored. Sometimes, though, before the electron escapes to the continuum, the excited inner electron may drop back to its initial state. Compare Fig. 3-2. Thus:

$$X^{(Z-1)}\ (3s\ 3p^4\ S'L';\ n''\ell'') \to X^{Z-1}\ (3s^2\ 3p^3\ SL;\ n''\ell'') + h\nu_1\ . \quad (75)$$

The captured electron cannot now escape since the energy of level ($n''\ell''$) in this configuration falls below the first ionization potential. The ion simply decays to a lower level, $n'\ell'$.

$$X^{Z-1}(3s^2\ 3p^3\ SL,\ n''\ell'') \to X^{Z-1}(3s^2\ 3p^3\ SL;\ n'\ell') + h\nu_2\ . \quad (76)$$

There are three stages in the dielectronic recombination process. We consider the simplest situation.

I. Resonance Capture or Auger Decay

$$X^{+Z}(n\ell) + \varepsilon(E,\ell''+1) \rightleftarrows X^{Z-1}\ (n^*\ell + 1;\ n''\ell'')\ . \quad (77)$$

The incoming electron must have exactly the right amount of energy to correspond to a doubly excited bound level, in order for a capture to take place. The inverse effect, ionization of an atom in a doubly excited level, is autoionization (sometimes called the Auger effect).

II. Stabilization Process

$$X^{+Z-1}(n^*\ell+1;\ n''\ell'') \to X^{Z-1}(n\ell,\ n''\ell'') + h\nu_1\ . \quad (78)$$

In order for capture to take place, the transition ($n^*\ell+1$; $n''\ell''$) $\to$ ($n\ell$; $n''\ell''$) must occur before the electron escapes. The inner electron

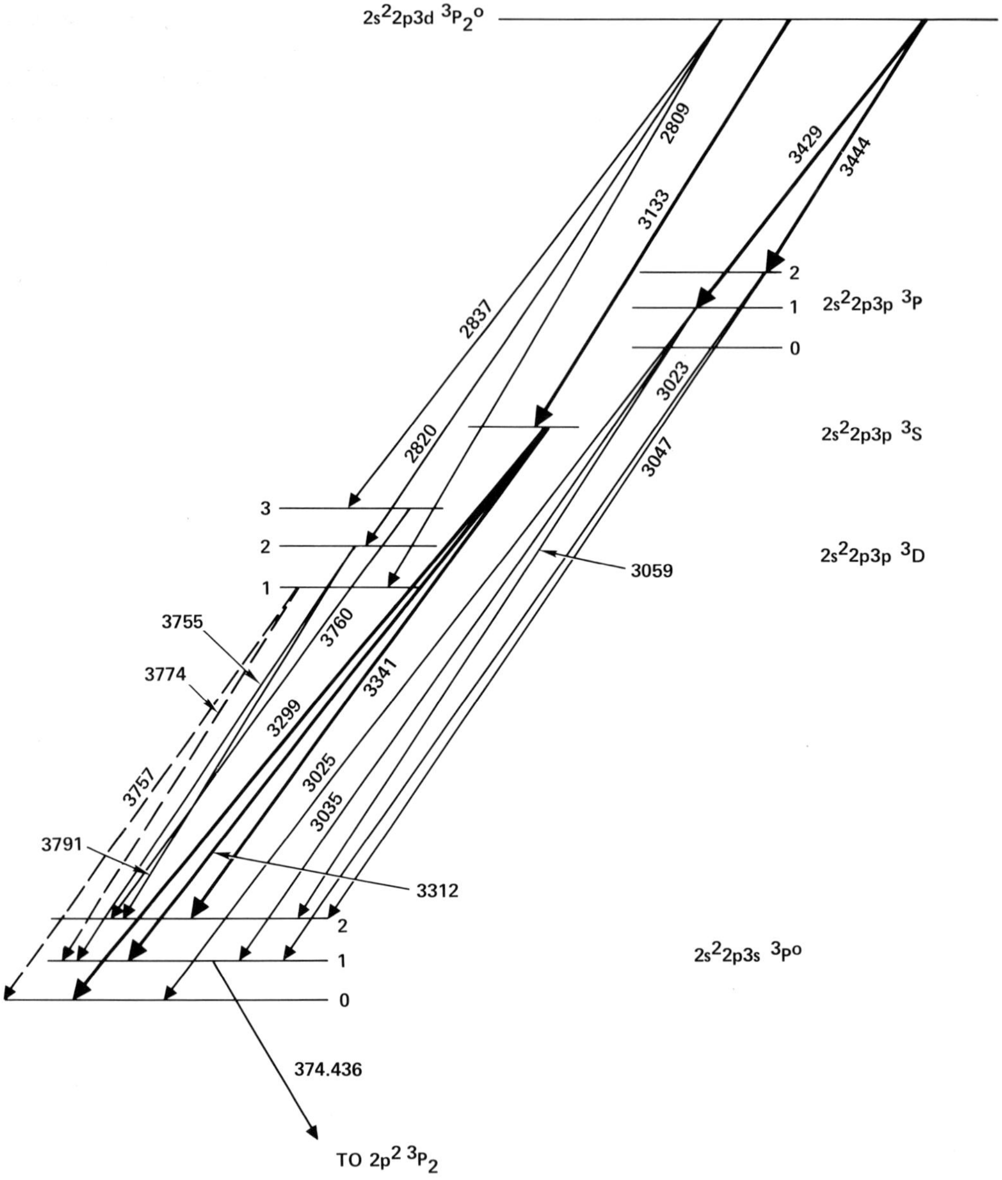
2s²2p3d ³P₂ᵒ
2809
3133
3429
3444
2837
2820
2
1
0
2s²2p3p ³P
3023
2s²2p3p ³S
3047
3
2
1
2s²2p3p ³D
3059
3755
3774
3760
3341
3299
3757
3025
3035
3791
3312
2
1
0
2s²2p3s ³Pᵒ
374.436
TO 2p² ³P₂

Figure 3-3

actually is not required to return to the specific $n\ell$ level whence it arose; the only requirement for stabilization is that the energy of the electron in the excited level falls below the first ionization potential. A helium isoelectronic sequence ion, $1s^2$, may capture an electron in perhaps a 7s level while one of the inner electrons jumps to a 2p level. The final term may be (1s 2p ^1P) 7s ^2P. When the 2p electron jumps back, there occurs a transition 1s 2p $\rightarrow$ $1s^2$, i.e., a resonance-type jump. The emitted line ν_1 will fall close to the resonance transition of the helium-like ion. Such "satellite" lines have been observed in the solar corona.

III. Cascade

In the last step, the ion decays from the $n''\ell''$ level with the emission of one or several quanta, usually to a level in the ground term, but it may reach a metastable level, viz:

$$X^{(Z-1)}(n\ell;\, n''\ell'') \rightarrow X^{Z-1}(n\ell,\, n'\ell') + h\nu_2 \;. \qquad (79)$$

Let us describe some basic principles underlying the calculation of the rate of the process. We use the following schematic notation:

$$X^+ = X^{(+Z)}(n\ell);\; X^{**} = X^{(Z-1)}(n^*\ell{+}1;\, n''\ell''),\; X^* = X^{(Z-1)}(n\ell,\, n'\ell')$$

In thermodynmamic equilibrium, the principle of detailed balance applies and the number of recombinations will equal the number of Auger decays.

$$N_\varepsilon N_o(X^+)\,\alpha_{od} = N_o(X^{**})\,A_A \;, \qquad (80)$$

where A_A denotes the Einstein A coefficient for an Auger decay, and α_{od} is the dielectronic recombination coefficient under conditions of thermodynamic equilibrium. Note that then ions can be excited to X^{**} from X^* as well as reach this state by recombination. The relative number of ions $N(X^+)$, doubly excited atoms $N(X^{**})$, and electrons N_ε is given by the Saha equation:

$$N_o(X^{**}) = \frac{\omega^{**}}{2\omega^+}\left(\frac{h^2}{2\pi m k T_\varepsilon}\right)^{3/2} \exp\left[-\frac{E(X^{**})}{kT_\varepsilon}\right] N_\varepsilon\, N_o(X^+) \;. \qquad (81)$$

Under nebular conditions, we note that ions escape from X^{**} both by Auger decay and stabilization transitions. Let A_S denote the transition probability for a stabilization transition. Essentially, as is analogous to the situation in ordinary recombination, $\alpha_d = \alpha_{od}$, since it depends on T_ε and energy level parameters. If we put

$$N(X^{**}) = b(X^{**})N_o(X^{**}), \qquad (82)$$

then:

$$b(X^{**}) = \frac{A_A}{A_A + A_S} \;. \qquad (83)$$

The recombination rate $N(X^+)N_\varepsilon\alpha_d$ will equal the rate of recombined ions trickling down in stabilizing transitions, i.e., $N(X^{**})\ A_s$. Hence:

$$N_\varepsilon N(X^+)\ \alpha_d = N_o(X^{**})\ b(X^{**})\ A_s\ , \qquad (84)$$

so

$$\alpha_d = \frac{N_o(X^{**})}{N_\varepsilon N(X^+)}\ b(X^{**})\ A_s \qquad (85)$$

That is:

$$\alpha_d = \frac{\omega^{**}}{2\omega^+}\left(\frac{h^2}{2\pi m\ kT_\varepsilon}\right)^{3/2} \exp\left[-\frac{E(X^{**})}{kT_\varepsilon}\right] \frac{A_A\ A_S}{A_A + A_S} \qquad (86)$$

For any given initial state i we have to sum over various possible recombination scenarios to get

$$\alpha_d(i,tot) = \Sigma\alpha_d(i\ n^*\ \ell^*,\ n''\ell'')\ , \qquad (87)$$

and the recombination rate is then obtained by summing over all possible initial states, i. Usually, one state will dominate. Then the recombination rate is:

$$N(X^Z(i))\ N_\varepsilon\ \alpha_d(i,\ tot)\ .$$

Historically, the astrophysical importance of dielectronic recombination was first recognized in the solar corona. The high temperature of this attenuated outer envelope of the sun was indicated by the presence of broad emission lines of [Fe X] to [Fe XIV] and other elements which were identified by Edlén. The line profiles and other data indicated electron temperatures near 2,000,000°K. Ionization equilibrium is maintained by electron collisions, so we have that:

$$N\left(Fe^i\right)\ N_\varepsilon\ q_{Fe}\left(i \rightarrow i+1,\ T_\varepsilon\right) = N\left(Fe^{i+1}\right)N_\varepsilon\ \alpha_{Fe}\left(i+1 \rightarrow i,\ T_\varepsilon\right)\ . \qquad (88)$$

Here q is the collisional rate coefficient for ionization and i denotes the stage of ionization of iron. Repeated applications of this equation yields, in particular, the ratio of N(Fe X) to N(Fe XIV), viz:

$$\frac{N(Fe^{+9})}{N(Fe^{+13})} = \frac{\prod_9^{13} \alpha_{Fe}(i+1 \rightarrow i,\ T_\varepsilon)}{\prod_9^{13} q_{Fe}(i \rightarrow i+1,\ T_\varepsilon)} \qquad (89)$$

Even when allowance is made for uncertainties in collisional ionization rate factors and recombination coefficients, one finds that if only ordinary radiative recombinations are taken into account, the electron temperature required to fit the observed ratio of ions is much lower than 2,000,000°K. If dielectronic recombination is considered, the ionization equilibrium at any fixed temperature is shifted in favor of the lower ionization stage. Thus, a higher temperature is required to fit the observed ion ratios and, in fact, the discordance is removed. In calculations of ionization equilibria for very hot high-density plasmas (such as occur in the solar corona), one also has to take into account processes such as the collisional excitation of an inner electron to a level lying above the first continuum, followed by autoionization.

Dielectronic recombinations proceed rapidly at high temperatures because, at the energy domain where high n-value states become closely crowded together, many free electrons are poised for capture. See Figure 3-2. On the other hand, at a gas kinetic temperature of 10,000°K, which is characteristic of many gaseous nebulae, only a limited number of autoionization levels are available. P.V. Storey (1981) has calculated dielectronic recombination coefficients in ions of carbon and nitrogen. Among the examples he has considered is the recombination of C^{+3} via the 2p $n\ell$ levels of C^{2+}. Here ($n\ell$) refers to the captured electron. Within 1.25 eV above the ionization limit for C^{+3}, 2s(^{2}S), there are Auger levels belonging to configurations 2p 4p, 2p 4d, and 2p 4f. Recombination can occur when an electron is captured in one of these levels and undergoes radiative decay. For example, the 2p 4d $^1F°$ level can decay at once to $2p^2$ $^1D°$ which decays to 2s 2p with emission of a C III λ2297 photon.

Thus, at temperatures well above 10,000°K, one must consider the very high levels where the captures will be most numerous, but at temperatures near 10,000°K, we are concerned only with terms just above the ionization limit and for these we have the simplifying situation that $b(^{**}) = 1$. One calculates the transition probabilities for all stabilizing transitions and subsequent cascade processes. It is possible to evaluate an effective recombination coefficient $\alpha_{eff}(\gamma SLJ)$ for any particular level (γSLJ) as well as α_d for the entire number of dielectronic recombinations. One takes into account all cascades that lead to state γSLJ, both from bound and autoionization levels. The rate of populating a level (γSL) is N_ε $N(X^*)$ $\alpha_{eff}(\gamma SL)$ and the emissivity in any line, λ, originating from (γSL) will be obtained by multiplying this quantity by R(λ) where R(λ) is the branching ratio for this transition. Intensities of some permitted lines, such as C II λ1335, in the ultraviolet region λ900 to λ3000 are influenced by dielectronic recombination which must be taken into account as well as collisional excitation effects.

Nussbaumer and Storey (1983) have calculated total dielectronic recombination coefficients for ions of C, N, and O for the range $1000 < T < 60{,}000$°K. They find dielectronic recombination to be more important than direct radiative recombination for most of these ions over most of this range. See Table 3-5.

In summary, specific dielectronic recombination rates have to be calculated for individual ions and energy levels. A popular general formula proposed by Burgess (1964) is not applicable at T = 10,000°K. At low temperatures it tends to badly underestimate recapture rates.

3F. Collisional Ionization. In regions which are heated by shock waves rather than by photoionization, as in supernova remnants, ionization by collision may become important. As previously noted, such a situation prevails in the solar corona. Under these special circumstances, the ambient electron temperature will be higher than in a region whose temperature is governed by photoionization. Collisional ionizations and excitations exert a profound cooling effect on the gas; mechanical energy in the form of shock wave dissipation is presumably supplied at a rate sufficient to heat the gas. Under such situations, time-dependent effects become important since the gas cools as energy is radiated. See Chapter 9.

3G. Charge Exchange. Ionization of a gaseous nebula depends not only on photoionization, but also on the phenomenon of charge exchange. For singly charged species, the most important interactions are those that involve charge exchange with neutral H or occasionally helium

$$X^+ + H^o \rightleftarrows X^o + H^+ \ .$$

The best known of these reactions is that between ground level O^+ and H^o, viz.:

$$O^+(^4S_{3/2}) + H^o(^2S_{1/2}) \rightleftarrows O^o(^3P) + H^+ \ , \tag{91}$$

the astrophysical importance of which was first pointed out by J.W. Chamberlain (1956). The interaction here is almost resonant!

The reaction rate coefficients

$$\begin{aligned} &O^+ + H^o \rightarrow O^o + H^+, \quad (1.04 \times 10^{-9}\ \text{cm}^3\ \text{s}^{-1}) \text{ and} \\ &O^o + H^+ \rightarrow O^+ + H^o, \quad (9.1 \times 10^{-10}\ \text{cm}^3\ \text{s}^{-1} \end{aligned} \tag{92}$$

are in the ratios of statistical weights, i.e., $\omega(O^o)\ \omega(H^+)/\omega(O^+)\ \omega(H^0) = \omega(^3P)/\omega(^4S)\omega(^2S) = 9/8$. The rate coefficient for the corresponding nitrogen charge reaction is lower because this reaction is endothermic with an energy defect of 0.94 eV (the ionization potential of nitrogen is greater than that of hydrogen).

$$\begin{aligned} &N^o + H^+ \rightarrow N^+ + H^o\ (1 \times 10^{-12}\ \text{cm}^3\ \text{s}^{-1}) \ , \\ &N^+ + H^o \rightarrow N^o + H^+\ (7 \times 10^{-13})\ \text{cm}^3\ \text{s}^{-1} \ . \end{aligned} \tag{93}$$

The rate of the corresponding reaction of $S^+ + H^o$ would appear to be negligible. Reaction rates have to be calculated for individual processes. The charge exchange process resembles the formation of a quasimolecular ion. If E_i is the energy of the initial system and E_f that of the final system, charge transfer at thermal energies is improbable unless $E_i - E_f$ falls to kT or less at some point of closest approach.

Interactions between multicharged ions and neutral H or He are of great importance. Examples of such reactions and their rates ($cm^3\ s^{-1}$) at T = 10,000°K are:

$$\begin{array}{llll} N^{+3} + H^o \rightarrow N^{+2} + H^+ & (2.93 \times 10^{-9}) \\ C^{+3} + H^o \rightarrow C^{+2} + H^+ & (3.58 \times 10^{-9}) \\ O^{++} + H^o \rightarrow O^+ + H^+ & (0.77 \times 10^{-9}) \\ N^{++} + H^o \rightarrow N^+ + H^+ & (0.86 \times 10^{-9}) \\ Ne^{+3} + H^o \rightarrow Ne^{+2} + H^+ & (5.68 \times 10^{-9}) \end{array} \quad (94)$$

As the neutral hydrogen atom approaches the ion, only polarization effects occur, but after charge swapping has taken place, the proton and ion strongly repel each other. Charge exchange is often dominated by transitions to excited product ions, and radiation can be emitted, as in the $O^{++} + H^o$ reaction noted above. Sometimes even molecular ions can be formed, e.g., $C^+ + H \rightarrow CH^+ + h\nu$, although most molecular ions may be destroyed subsequently by dissociative recombinations.

Charge exchange can be incorporated into the ionization equation in a straightforward manner. Consider, e.g., the ionization equilibrium of a neutral and singly ionized element. If β_{CE} denotes the charge exchange coefficient for the reaction that ionizes this element $H^+ + X^o \rightarrow H^o + X^+$ while β^1_{CE} is the rate coefficient for the inverse reaction, then:

$$4\pi\ N(X^o) \int_{\nu_1}^{\infty} \frac{a_{1\nu}\ J_\nu\ d\nu}{h\nu} + \beta_{CE}\ N(H^+)\ N(X^o) = \beta^1_{CE}\ N(X^+)\ N(H^o) + \alpha_T\ N(X^+)\ N_\varepsilon\ , \quad (95)$$

where the first term on the left gives the ionization rate by photoionization and the last term on the right indicates the number of ordinary plus dielectronic recombinations.

In calculations of ionization equilibria and structural models for gaseous nebulae, it is important to consider charge exchange. In nebulae of moderate-to-high excitation, ions may be found in zones where there is an appreciable concentration of H^o, so the production of O^+ ions by charge exchange can become significant (see Chapter 7).

3H. Stellar Far Ultraviolet Radiation Fields. In the discussion of the primary mechanism in Chapter 2, we presented Zanstra's argument that if a nebula is optically thick, all Lyman quanta will be broken down into Balmer quanta and Lyα plus 2-photon emission quanta, thus:

$$\int_{\nu_1}^{\infty} \frac{L^*(\nu)}{h\nu}\ d\nu = Q_{uv} = Q_{BA} = Q_{Ly\alpha} + Q(2q) \quad (96)$$

where Q_{BA} is the sum of all line and continuum Balmer quanta, viz.:

$$Q_{BA} = \sum \frac{I(Hn)}{h\nu_n} + \int_{\nu_2}^{\infty} \frac{I(Bac)}{h\nu} d\nu = \int \alpha_B(T_\varepsilon) N(H^+) N_\varepsilon dV$$

$$= KZ^4 \int G_B(T_\varepsilon) \frac{N(H^+)N_\varepsilon}{T_\varepsilon^{3/2}} dV \ . \tag{97}$$

Fortunately, the ratio of the number of quanta in Hβ or any line of the Balmer series to the total number of Balmer quanta Q(Ba) varies slowly with T_ε. Anticipating the results of Chapter 4, we have that the total number of quanta in Hβ is given by:

$$Q(H\beta) = \int \frac{E(H\beta)}{h\nu_{H\beta}} d\nu = \int \alpha_{H\beta}(T_\varepsilon) N(H^+) N_\varepsilon dV = \int \frac{N(H^+)N_\varepsilon}{h\nu_{H\beta}} E^0_{4,2} \times 10^{-25} dV \ . \tag{98}$$

Now, $\alpha_{H\beta}/\alpha_B$ is the fraction of recombinations to the second and higher levels that results in the production of Hβ, and $E^0_{4,2}$ is a factor of the order of unity. See Eq. 4-15, 16. Then:

$$Q_{BA} = (\alpha_B/\alpha_{H\beta})\ Q(H\beta) \ , \tag{99}$$

where $\alpha_B/\alpha_{H\beta}$ = 8.35, 8.55, and 8.88 for T_ε = 5000°K, 10,000°K and 20,000°K, respectively. The number of stellar photons in a frequency interval Δν at Hβ is $L^*(\nu_{H\beta})\ \Delta\nu/h\nu_{H\beta}$. Note that we can write:

$$\left[\frac{F^*(H\beta)\ \Delta\nu}{F^n(H\beta)}\right]_{OBS'D} = \left[\frac{L^*(\nu_{H\beta})\ \Delta\nu}{Q(H\beta)\ h\nu_{H\beta}}\right] = \left[\frac{\alpha_B(T_\varepsilon)}{\alpha_{H\beta}(T_\varepsilon)} \frac{L(\nu_{H\beta})\ \Delta\nu}{h\nu_{H\beta} \int_{\nu_1}^{\infty} \frac{L_\nu}{h\nu} d\nu}\right] \ . \tag{100}$$

The left hand side of the equation is an observed quantity measurable for any nebula where we can compare the total monochromatic Hβ nebular flux $F^n(H\beta)$ with the stellar flux at Hβ in an interval Δν. It is independent of atmospheric and interstellar extinction, although it could be affected in principle by dust absorption within the nebula. The practical effect is small. The right hand side of the equation is obtained by theory. The stellar flux distribution may be computed by model atmosphere theory. The ratio is a unique function of T_{eff}, surface gravity, and assumed stellar chemical composition. Realistic model calculations for planetary nebulae nuclei and other very hot stars are difficult since it is necessary to consider the curvature of the layers and deviations from local thermodynamic equilibrium (LTE).

Zanstra and his contemporaries, particularly Ambarzumian, recognized that one could employ λ4686 of He II to determine $T(^*)$ for any nebula that is optically thick in the He II Lyman continuum, i.e., for λ < 228 Å. Here we must compare the total number of quanta that can give rise to λ4686 (Paschen α of He II) with the total number of recombinations on the second level.

$$\left[\frac{F^{*}(\nu_{4686})\,\Delta\nu}{F^{n}\,(\nu_{4686})}\right] = \frac{\alpha_B^{\mathrm{He\,II}}}{\alpha(4686)}\;\frac{L^{*}(\nu_{4686})\,\Delta\nu}{h\nu_{4686}\int_{4\nu_1}^{\infty}\frac{L_\nu^{*}}{h\nu}\,d\nu}\;. \tag{101}$$

At $T_\varepsilon \sim 20{,}000$°K, α_B(He II)/α(4686) ~ 50, according to data by Brocklehurst (1971). See Chapter 4. Strictly speaking, in a real nebula one must consider the interactions of the H and He radiation fields. Photoionization of He and He^+ removes quanta otherwise available for ionization of H, but ground-state recombinations in He^+ and He^{++} ions, the resonance $1^1S - 2^1S$ He I line and Lyα He II all provide quanta capable of ionizing H. The two effects have a tendency to cancel; in most instances, we can forget about interlocking.

When temperature determinations are carried out by these two methods, it is found that although Zanstra-methods He II temperatures, T_Z(He II) are frequently consistent with Zanstra-method H I temperatures T_Z(H I) (especially at T > 100,000 K), they are often higher. The seemingly most straightforward explanation of this discordance is that the nebula is optically thin in H, but thick in He^+. Hydrogenic Lyman continuum quanta are incompletely absorbed, $Q_{BA} < Q_{UV}$ and a spuriously low temperature is derived.

Is this explanation always the correct one? Should we accept T_Z(He II) as the true temperature of the central star and use it to compute bolometric corrections, total luminosities, etc.? Some compelling counterevidence may be cited. It has been pointed out often that a number of planetary nebulae show O-type absorption line spectra very similar to those of main-sequence O-type stars. Even when non-LTE atmospheric theory is used to interpret the data, the derived T_{eff} is often much lower than T_Z(He II) (Méndez et al. 1981). Pottasch (1983) found that the T_Z(H I) - T_Z(He II) discordance may well arise from the use of an inappropriate model atmosphere. In some stars, extended hot coronae or winds may play appreciable roles or close binaries may be involved. Some planetary nuclei are binaries; yet others are suspected binaries. The visible region flux is contributed by a star of relatively low temperature while the nebula is ionized by a hotter companion. Clearly, Zanstra's procedure cannot be applied in such situations. Until these problems are resolved, it would appear better to avoid use of T_Z(He II).

An analogous treatment can be given for He I, using λ5876 (Zanstra 1959), but we must consider the interaction with the H I radiation field. Methods based on energy balance were proposed by Zanstra (1931) and by Stoy (1933). See Chapter 6. Recently, Preite-Martinez and Pottasch (1983) have determined the temperatures of many planetary nebular nuclei by a generalized Stoy method.

In principle, the best method for getting the far ultraviolet radiation fields of exciting stars is to construct theoretical models that reproduce the observed line intensities (see Chapter 7). Thus, Harrington et al. find T(*) ~ 95,000 K for the nucleus of NGC 7662, a value in good accord with other determinations.

3I. Line Excitation; the Bowen Fluorescent Phenomenon in O III and N III

Except for H, He I, and He II, permitted lines involving high excitation potentials are normally weak, as we would expect on the basis of abundance considerations. See Chapter 11. A spectacular exception to this rule are certain strong permitted lines of O III in the near ultraviolet, between λ2800 and λ3800. Less intense, but still much stronger than predicted by recombination theory and subsequent cascade, are N III lines 4097, 4103, and λ4634, 4641. A curious feature is that other prominent O III and N III lines seen in laboratory sources are missing or extremely weak in nebulae.

Bowen noticed three significant clues: a) these strong O III and N III lines appear only in high-excitation nebulae with strong λ4686; b) all the O III lines originated from a single upper level, $2s^2\ 2p\ 3d\ ^3P_2$, either directly or by cascade; c) the N III lines likewise all originated from the 3d ^2D term by cascades. The appearance of these lines constitutes one of the most remarkable coincidences in nature.

An O III atom in the $2s^2\ 2p^2\ ^3P_2$ level of the ground term requires a photon of wavelength λ303.799 Å to reach the $2s^2\ 2p\ 3d\ ^3P_2$ level. This transition almost coincides with the resonant Lyα λ303.780 of He II. Small kinematical and Doppler shifts could easily bring the two frequencies into coincidence. Once in the $^3P_2$ level it has a 98.13% probability of returning directly to the ground term, but a 1.87% probability of cascading to a level of the $2s^2\ 2p\ 3p$ configuration and thence to the $2s^2\ 2p\ 3s$ configuration (assuming only one electron is "active"). The transition scheme is as indicated in Fig. 3-3. To compound this remarkable coincidence the downward transition $2s^2\ 2p\ 3s\ ^3P_1^0 - 2s^2\ 2p^2\ ^3P_2$ λ374.436 O III produces a fluorescence in N III! One N III transition, 2p $^2P_{3/2}$ - 3d $^2D_{5/2}$, λ374.434, nearly coincides with the O III line. The other, 2p $^2P_{3/2}$ - 3d $^2D_{3/2}$ falls very near at λ374.442. Hence, the N III 3d ^2D term is selectively populated by the absorption of O III photons whose frequencies are slightly changed by Doppler shifts. The N III atom may cascade from $2s^2$ 3d ^2D to $2s^2$ 3p ^2P to produce the prominent λ4634, 4641, 4642 lines. Then transitions from $2s^2$ 3p ^2P to $2s^2$ 3s ^2S give λ4097 and λ4103 which flank Hδ From $2s^2$ 3s ^2S the atom escapes to the $2s^2$ 2p ^2P level.

The He II Lyα radiation which attains a high intensity in high-excitation planetaries provides a copious source of energy to O III ions upon which only the $2s^2\ 2p^2\ ^3P_2$ to $2s^2\ 2p\ 3d\ ^3P_2$ transition can draw. Although in any given event the probability is less than 2% that quanta of the Bowen florescent mechanism BFM will be produced, repeated scattering results in the gradual degradation of the $2s^2\ 2p^2\ ^3P - 2s^2\ 2p\ 3d\ ^3P_2$ quanta. Not all He II Lyα quanta will be transformed to BFM quanta. We can define an efficiency factor:

$$R = \frac{\lambda(A)}{4686}\,\frac{I(\lambda)}{I(4686)}\,\frac{P}{P(\lambda)} \times \frac{\alpha_{eff}(P\alpha,\ \mathrm{He\ II})}{\alpha_{eff}(Ly\alpha,\ \mathrm{He\ II})}\ , \tag{102}$$

where $\alpha_{eff}(P\alpha$, He II) and $\alpha_{eff}(Ly\alpha$, He II) are the effective recombination coefficients for recaptures that result in the production of λ4686 and Lyα, respectively. P is P(cascade) = 0.0187; P(λ) is the probability for the emission of a particular line. See Table 3-6 which is due to Saraph and Seaton (1980). At T = 12,000°K in NGC 7662, they find R ~ 0.32 ± 0.10 for λ3133. The ratio of the α_{eff} is 0.328 and 0.277 for T_ε = 10,000°K and 20,000°K, respectively.

Given accurate A-values for the transitions involved, the various lines of the BFM should have well-defined intensity ratios if the formulation is correct. Weymann and Williams (1969) found that the predicted intensity ratio λ3341/λ3444 is twice the observed ratio and suggested that the $2s^2$ 2p 3p ^3S level is depopulated not only by the expected transitions to $2s^2$ 2p 3s $^3P^o$ but also by the transition $2s^2$ 2p 3p ^3S to 2s $2p^3$ $^3P^o$ involving <u>two</u> active electrons. This effect apparently turns out to be important. Another possible similar transition is $2s^2$ 2p 3d → $2p^4$ but Saraph and Seaton find it to be of

Table 3-6.

Probabilities P(λ) for emission of photons in lines, following excitation O III 2p 3d $^3P_2$. The corresponding intensities, I, are normalized to I = 1.000 for λ3132.86. Wavelengths are λ(air) for these primary and secondary cascades.

	^3L	J	$^3L'$	J	λ	P(λ)	I
Primary Cascades 2p 3d ^3L_J 2p 3p $^3L'_{J'}$	$^3P^o$	2	^3P	2	3444.10	3.74,-3	0.277
				1	3428.67	1.25,-3	0.093
			^3S	1	3132.86	1.23,-2	1.000
			^3D	3	2837.17	1.16,-3	0.104
				2	2819.57	2.08,-4	0.019
				1	2808.77	1.38,-5	0.0013
Secondary Cascades 2p 3p ^3L_J 2p 3s $^3L'_{J'}$	^3P	2	$^3P^o$	2	3047.13	2.14,-3	0.179
				1	3023.45	7.12,-4	0.060
		1		2	3059.30	3.95,-4	0.033
				1	3035.43	2.37,-4	0.020
				0	3024.57	3.17,-4	0.027
	^3S	1	$^3P^o$	2	3340.74	1.79,-3	0.136
				1	3312.30	1.07,-3	0.082
				0	3299.36	3.57,-4	0.028
	^3D	3	$^3P^o$	2	3759.87	5.39,-4	0.037
		2		2	3791.26	2.41,-5	0.0016
				1	3754.67	7.22,-5	0.0049
		1		2	3810.96	1.80,-7	0.00012
				1	3774.00	2.67,-6	0.00018
				0	3757.21	3.57,-6	0.00024

From H.E. Saraph and M.J. Seaton, 1980, <u>M.N.R.A.S.</u>, <u>193</u>, 617.

little consequence. In ordinary and dielectronic recombination processes involving complex atoms, similar effects involving two active electrons might be significant.

A further refinement in the theory was noted by Dalgarno and Sternberg (1982) who showed that charge exchange can affect the $2s^2\ 2p\ 3p\ ^3S$ and $2s^2\ 2p\ 3p\ ^3D$ levels. The effect on the intensities of λ3299, 3312 and 3341 which arise from $^3S_1$ is negligible, but $2s^2\ 2p\ 3p\ ^3D_1$ (from which λ3774 and λ3757 arise) is populated almost entirely by charge exchange

$$O^{3+} + H^o \rightarrow O^{2+}(2p\ 3p\ ^3D_1) + H^+ \ .$$

The BFM mechanism makes a negligible contribution since λ2809, which feeds $2p\ 3p\ ^3D_1$, has a very low A-value. The λ3744 and 3757 transitions are indicated by dotted lines in Fig. 3-3.

Selected Bibliography

Our discussion of absorption and emission processes involving H is taken from D.H. Menzel, 1937, Ap. J., 85, 330. This and other papers of the Harvard series on physical processes in gaseous nebulae are reprinted in:

Physical Processes in Ionized Plasmas, ed. D.H. Menzel (New York: Dover Publications, 1962, hereafter referred to as PPIP).

Hydrogenic line and continuous absorption coefficients have been discussed by:

Menzel, D.H., and Pekeris, C.L. 1935, M.N.R.A.S., 96, 77.
Karzas, W.V., and Latter, R. 1961, Ap. J. Suppl., 6, 167.
Burgess, A. 1964, M.N.R.A.S., 69, 1.

The basic theoretical references to the ionization of hydrogen and the formation of H II regions with sharp boundaries are:

Strömgren, B. 1939, Ap. J., 89, 526; 1948, Ap. J., 108, 242.

For a discussion of continuous absorption coefficients for complex atoms see:

Burke, P.G. Atomic Processes and Applications, ed. P.G. Burke and B.L. Moiseiwitsch (Amsterdam: North-Holland Publications, 1976).

Calculations for individual cross sections have been given by many workers. A few examples of this work are:

Hidalgo, M.B. 1968, Ap. J., 153, 981. (C, N, O, Ne).
Henry, R.J.W. 1970, Ap. J., 161, 1153. (C, N, O, Ne).
Chapman, R.D., and Henry, R.J.W. 1971, Ap. J., 168, 169. (S)

Selected Bibliography (Continued)

Chapman, R.D., and Henry, R.J.W. 1972, Ap. J., 173, 243. (Al, Si, Ar)
Reilman, R.F., and Manson, S.T. 1979, Ap. J. Suppl., 40, 815; 1981, 46, 115.
Sakhibullih, N., and Willis, A.J. 1976, Astron. Astrophys. Suppl., 31, 11. (C IV)
Pradhan, A.K. 1980, M.N.R.A.S., 190, 5P. (Ne II, Ne III, Ne IV)

Radiative recombination of complex atoms has been discussed by:

Gould, R.J. 1978, Ap. J., 219, 250.

Approximate formulae and coefficients have been given by:

Tarter, C.B. 1971, Ap. J., 168, 313; 1972, 172, 251.
Aldrovandi, S.M.V., and Pequignot, D. 1973, Astron. Astrophys., 25, 137.

The cross sections given by these authors for dielectronic recombination appear to be systematically too small.

Dielectronic Recombination. A general review with references to previous work is given by:

Seaton, M.J., and Storey, P.G. Atomic Processes and Applications, ed. P.G. Burke and B.L. Moiseiwitsch (Amsterdam: North-Holland Publications, 1976, p. 134).

Most of the discussions have pertained to high temperatures as in the solar corona and high densities as in quasars where the effects become more complex. See, e.g.:

Burgess, A. 1964, Ap. J., 139, 776. (Corona)
Burgess, A., and Summers, H.P. 1969, Ap. J., 157, 1008.
Davidson, K. 1975, Ap. J., 195, 285. (Quasars)

Dielectronic recombination rates appropriate to nebular temperatures and densities are discussed, for example, by:

Jacobs, V.L., Davis, J., Rogerson, J.E., and Blaha, M. 1979, Ap. J., 230, 627.
Storey, P.J. 1981, M.N.R.A.S., 195, 27P. See also I.A.U. Symposium No. 103.
Nussbaumer, H., and Storey, P.J. 1983, Astron. Astrophys.

Charge Exchange. A brief summarizing account of the problems of charge exchange, with references to earlier work and timely warnings pertaining to inherent uncertainties in theoretical calculations, is given by:

Selected Bibliography (Continued)

Dalgarno, A. Planetary Nebulae (ed. Y. Terzian, International Astronomical Union Symposium No. 76, Dordrecht, Reidel, 1978, p. 139).

The astrophysical importance of oxygen-hydrogen charge exchange reactions was noticed by:

Chamberlain, J.W. 1956, Ap. J., 124, 390.

Its effect on nebular ionization structure was examined by:

Williams, R.E. 1973, M.N.R.A.S., 164, 111.

Some representative calculations of charge exchange coefficients are:

Dalgarno, A., Butler, S.E., and Heil, T.G. 1980, Ap. J., 241, 442.
Dalgarno, A., and Butler, S.E. 1980, Ap. J., 241, 838.

The Zanstra method has been applied to planetary nebulae symbiotic stars, Be stars, and diffuse galactic nebulae. See:

Zanstra, H. 1931, Zeits. f. Astrofis., 2, 1, 329, 1239.
__________ 1930, Publ. Dom. Ap. Obs. Victoria, 4, 209.

A systematic formulation is given by:

Harman, R.J., and Seaton, M.J. 1966, M.N.R.A.S., 132, 15.

A summary of the earlier work with descriptions of investigations by: Ambarzumian (1932), Stoy (1933), Wurm (1951), and others are given by:

Aller, L.H., and Liller, W. 1968, Stars and Stellar Systems, 7, 546, Nebulae and Interstellar Matter, ed. by Barbara Middlehurst and L.H. Aller, Chicago, Univ. Chicago Press.

Modernized versions of the Stoy method are given by:

Kaler, J.B. 1976, Ap. J., 210, 843.
Preite-Martinez, A., and Pottasch, A.S. 1983, Astron. Astrophys., 126, 31.

The pioneering ultraviolet study (with the ANS satellite is):

Pottasch, R., Wesselius, P.R., Wu, C.C., Fieten, H., and van Duinen, R.J. 1979, Astron. Astrophys., 62, 95,

see also:

Pottasch, R. 1981, Astron. Astrophys., 94, L13.
1983, Planetary Nebulae, Dordrecht, Reidel Publ. Co.
Méndez, R. et al. 1981, Astron. Astrophys., 101, 323.

Selected Bibliography (Continued)

Bowen proposed his fluorescent mechanism in: 1935, Ap. J., **81**, 1. The earliest theoretical treatments by D.H. Menzel and L.H. Aller, 1941, Ap. J., **94**, 436, by T. Hatanaka, 1947, J. Astr. Geophys. Japan, **21**, 1, and W. Unno, 1955, Publ. Astr. Soc. Japan, **7**, 81, were hampered by poor atomic and observational data. The first accurate models were given by R.J. Weymann and R.E. Williams, 1969, Ap. J., **157**, 1201, and by J.P. Harrington, 1972, Ap. J., **176**, 127. Improved atomic data by H. Saraph and M.J. Seaton, 1980, M.N.R.A.S., **193**, 617, replace earlier calculations by H. Nussbaumer, 1969, Astrophys. Lett., **4**, 183. A. Dalgarno and A. Sternberg, 1982, Ap. J., discuss the role of charge exchange. The BFM may play a role in many astrophysical sources. T. Kallman and R. McCray, 1980, Ap. J., **242**, 615, discuss its possible importance in X-ray sources where much of the soft X-ray luminosity absorbed by the nebula appears as Lyα (He II) which controls the ionization and temperature structure of the plasma until degraded. The BFM is expected to have a major effect on the fate of these He II photons.

List of Illustrations

Fig. 3-1. Ionization Structure of a Low-Excitation Planetary Nebula, CN 3-1. We plot the fraction $X(E^i)$ of the numbers of ions in the ith ionization stage of element E as a function of the distance from the central star. Notice that O^{++} [IP(O^+) = 35.11 eV] and Ne^{++} [IP(Ne^+) = 41.07 eV] are restricted to the zone where helium (IP = 24.58 eV) is ionized. In the zone where helium is neutral ($r > 0.029$ pc), O^+ and N^+ exist, but their concentrations drop to zero where H becomes neutral. In this nebular model there is no overlap between the O^{++} and H° zones. (Astrophys. Space Sci., **67**, 351, 1980, Fig. 2) The distance of the nebula is assumed to be 3300 pcs.

Fig. 3-2. Dielectronic Recombination. A free electron of energy E with $\ell = \ell'' + 1$ encounters an ion of net charge Z with an electron in configuration ($n\ell$). The incoming electron is captured in a configuration $n''\ell''$ while an excitation of the ion from $n\ell$ to an excited configuration $n^*\ell + 1$ occurs. Before the electron escapes to the continuum in an autoionization process, the inner electron decays from $n^*\ell + 1$ to $n\ell$. The recently captured electron cannot now escape and decays to some lower level (n'l'). Here ΔE denotes a small range of energies near the ionization limit of X^Z. The Auger levels of the doubly excited ion X^{Z-1} are fairly well spread out at this energy. At a higher temperature (e.g., 25,000°K) many electrons will occupy an energy interval such as ΔE_2, accessible to numerous Auger levels of X^{Z-1}. Hence, the recombination rate will be greatly enhanced.

Fig. 3-3. Observed Transitions of Bowen Fluorescent Mechanism. Ions are excited to the $2s^2\,2p\,3d\;{}^3P^o_2$ level whence they may cascade to levels in the $2s^2$ 2p 3p and eventually $2s^2$ 2p 3s configurations. Strong transitions in optical regions are indicated by heavy lines.

Captions for Illustrations (Continued)

Weaker or UV transitions are indicated by medium lines. The 2p 3p $^3D_1$ level is populated mostly by charge exchange. The postions and separations of the levels are not drawn to scale.

Problems

1) Show that if the volume emission in Hβ is given by:

$$E(H\beta) = N(H^+)\, N_\varepsilon\, E^o_{4,2}(T_\varepsilon)\, 10^{-25}, \text{ then}$$

$$\frac{Q(BA)}{Q(H\beta)} = 1.332 \times 10^8 \frac{G_B\,(T_\varepsilon)}{T_\varepsilon^{3/2}\, E^o_{4,2}} .$$

2) In some planetary nebulae, the spectrum of the central star is visible only in the ultraviolet. Compare the nebular λ1640 Hα He II emission with the nearby stellar continuum. Derive an expression similar to Eq. (101) for determining T_Z(He II).

3) Examine the Strömgren sphere problem for a nebula that contains both hydrogen and helium. Consider three situations:

a) The central star temperature $T(^*)$ is about 30,000° so that the helium Strömgren sphere is much smaller than that of hydrogen (see Fig. 3-1).

b) $T(^*)$ = 48,000° so that helium and hydrogen Strömgren spheres are about the same size. There is no He^{++}.

c) $T(^*)$ = 80,000° so the radius of the He^{++} sphere is much smaller than that of H^+ or He^+ which are assumed equal in size.

The abundance ratio is assumed to be He/H = 0.10. Note that recombination on the ground level of He° produces quanta that can ionize either H or He. Assume that the absorption coefficient from the ground $1s^2\ ^1S_o$ level of He is $a_\nu = 7.6 \times 10^{-18}\ (\nu/1.81\ \nu_1)^{-2}\ cm^2$, while that of ground-level H is $a_\nu = 6.3 \times 10^{-18}\ (\nu/\nu_1)^{-3}\ cm^2$. At T = 10,000°, the ground-level recombination coefficient for helium is $1.59 \times 10^{-13}\ cm^3\ s^{-1}$, while the recombination coefficient for the second and higher levels (Case B) is $2.73 \times 10^{-13}\ cm^3\ s^{-1}$. Note that resonance line transitions and some 2-photon decays from 2^1S and 2^3S levels can produce radiation capable of ionizing hydrogen. Osterbrock (AGN p. 25) estimates that captures on the second and higher levels have probabilities between 56 and 96 percent (depending on N_ε) of producing quanta capable of ionizing H. Hence, most of the photons produced by recombinations of helium ions and electrons to produce neutral helium, lead to the ionization of hydrogen. Many quanta produced by recombination of electrons and He^{++} ions can ionize both helium and hydrogen. In situation c) we can regard the ionization of He^+ as all proceeding in a domain where H is all ionized.

Consider an illuminating star of 20 solar radii that shines as a black body at T = 40,000°. Suppose the density of the surrounding cloud is $N_H = 1000\ cm^{-3}$; n(he)/n(H) = 0.10. What will be the size of the Strömgren sphere and what will be the error made in assuming that the nebula is made of pure hydrogen?

Chapter 4

LINE AND CONTINUOUS SPECTRA OF HYDROGEN AND HELIUM

4A. Statistical Equilibrium, the Elementary Theory of the Balmer Decrement.

In H II regions, planetary nebulae and perhaps radio galaxies, Seyfert galaxies and quasars, the H and He lines are produced by a process of photoionization followed by recombination and cascade. Expressions for calculating the rates of the various atomic processes, recapture and cascade (in this instance) are available. We should be able to calculate the population in each level and find the relative intensities of observable lines in the H and He spectrum. In particular, we should be able to compute the relative intensities of the Balmer lines, the so-called Balmer Decrement.

First, we impose the steady state or statistical equilibrium condition which requires that the population of each discrete level be constant with time. We further suppose that there is no direct line excitation from the central star. We need only require that the ionization is produced by a photoionization process, not by collisions. Hydrogen atoms are photoionized from the ground level by radiation beyond the Lyman limit and the electrons are recaptured on all levels, ($n\ell$). We shall follow the historical path and, in the first approximation, neglect the dependence of cascade processes on ℓ.

The solution of the problem involves finding the constants, b_n, which measure the departure of the population of each level, n, from that appropriate to local thermodynamic equilibrium (LTE) at the electron temperature, T_ε. In an optically thin nebula where there is not only no line radiation from the central star, but also where all Lyman line radiation produced in the nebula promptly escapes, we can write:

Number of atoms cascading from higher levels, n" to level n	+	Number of direct recaptures on level n	=	Number of cascades downward from level n to lower levels

$$\sum_{n''=n+1}^{\infty} F_{n''n} + \int_{\nu_n}^{\infty} F_{\kappa n}\, d\nu = \sum_{n'=1}^{n-1} F_{nn'} \qquad (1)$$

where n" denotes a level above n, while n' denotes a level below n.

The second term on the left hand side is given by Eq. 3-40. To calculate the terms under the summation signs, we employ Eq. 3-2, 3-5, 3-6 and 3-19 to obtain, for example:

$$F_{nn'} = N_n A_{nn'} = N(H^+) N_\varepsilon \frac{KZ^4}{T_\varepsilon^{3/2}} e^{X_n} b_n \frac{g}{n'} \frac{2}{n} \frac{1}{\left(n^2 - n'^2\right)} . \quad (2)$$

In their solution of the problem, Menzel and Baker took advantage of the fact that as n increases without bound, b_n approaches 1. Hence, if one starts with a sufficiently high level of n, sets $b_n = 1$, then solves for b_{n-1} and proceeds by iteration, a solution can be found.

A more satisfactory procedure is Seaton's generalization of an approach originally due to Plaskett. Let $C_{n''n}$ denote the probability that a capture on level n" is followed by a cascade to n by all possible paths. Also, following Seaton, let us define:

$$A_n = \sum_{n'=1}^{n-1} A_{nn'} \quad \text{and} \quad P_{nn'} = \frac{A_{nn'}}{A_n} , \quad (3)$$

It follows that:

$$C_{n''n} = \sum_{m=n}^{n''-1} P_{n''m} C_{mn} , \quad (4)$$

and of course $C_{nn} = 1$.

Let $\alpha_{n''}(T_\varepsilon)$ denote the recombination coefficient for the n" level; that is,

$$\int_{\nu_{n''}}^{\infty} F_{\kappa n''} \, d\nu = N(H^+) N_\varepsilon \alpha_{n''}(T_\varepsilon).$$

Then define:

$$F_n = N(H^+) N_\varepsilon \sum_{n''>n} \alpha_{n''}(T_\varepsilon) C_{n''n} . \quad (5)$$

The equation of statistical equilibrium can now be written as:

$$F_n = N_n A_n . \quad (6)$$

The great advantage of the $C_{n''n}$ factors is that they involve only atomic constants (A-values) and can be calculated once and for all. The resultant set of equations of statistical equilibrium can be solved, starting most conveniently with the higher levels and proceeding to the lower ones.

This optically thin nebula which Menzel and Baker called Case A would have a very low surface brightness and would be difficult to observe. A situation more likely to be encountered is that of an optically thick nebula, wherein Lyman quanta will be created by degradation of ultraviolet quanta. Thus, there can occur direct line excitation from the ground level at a rate F_{1n}. The equation of statistical equilibrium must now be written:

$$\sum_{n''=n+1}^{\infty} F_{n''n} + \int_{\nu_n}^{\infty} F_{\kappa n}\, d\nu + F_{1n} = \sum_{n'=1}^{n-1} F_{nn'} \; . \tag{7}$$

Rigorously, one would have to solve the equation of transfer and allow for the breakdown of ultraviolet into Lyman line quanta. A simplification can be justified. In Menzel and Baker's Case B, it is assumed $F_{1n} = F_{n1}$, hence, Eq. (1) must be replaced by: (8)

$$\sum_{n''=n+1}^{\infty} F_{n''n} + \int_{\nu_n}^{\infty} F_{\kappa n}\, d\nu = \sum_{n'=2}^{n-1} F_{nn'} \; . \tag{9}$$

Yet a third situation is illustrated in what is called "Case C," an optically thin nebula illuminated by a star that shines as a black body in the lines as well as the continuum. We now have:

$$\sum_{n''=n+1}^{\infty} F_{n''n} + \int_{\nu_n}^{\infty} F_{\kappa n}\, d\nu + F^{*}_{1n} = \sum_{n'=1}^{\infty} F_{nn'} \; , \tag{10}$$

where:

$$F^{*}_{1n} = 4\pi N_1 \frac{J_\nu}{h\nu} \frac{\pi\varepsilon}{mc} f_{1n} = N(H^+)N_\varepsilon \frac{KZ^4}{T_\varepsilon^{3/2}} \left[W \frac{b_1 e^{h\nu/kT_\varepsilon}}{e^{h\nu/kT_1} - 1}\right] \frac{2g_{1n}\, e^{X_n}}{n(n^2 - 1)} \; , \tag{11}$$

if we assume J_ν to be dilute black body radiation, viz.:

$$J_\nu = \frac{2h\nu^3}{c^2} W \frac{1}{e^{h\nu/kT_1} - 1} \tag{12}$$

Case C would correspond also to a nebula of low surface brightness. It and Case A are of interest in giving insight to recombination problems for less abundant elements, which often might constitute optically thin cases. For example, if the illuminating star has strong resonance absorption lines of C II, N II and O II, then the nebular permitted lines of these ions might tend to follow Case A, whereas if the star radiates as a black body, they might approach Case C. In the latter instance, direct starlight excitation could greatly increase the intensities of permitted lines involving lower n-values.

It is often useful to derive an effective recombination coefficient for the number of quanta emitted in a particular transition, n − n', e.g., Hβ.
Thus:

$$F_{nn'} = P_{nn'} F_n = N(H^+)\, N_\varepsilon\, P_{nn'} \sum_{n''>n} \alpha_{n''} C_{n''n} = N(H^+)\, N_\varepsilon\, \alpha_{nn'} \tag{13}$$

The total energy emitted per unit volume in the nn' transition would then be:

$$E_{nn'} = N(H^+)N_\varepsilon \, \alpha_{nn'} \, h\nu_{nn'} \; . \tag{14}$$

In particular, we have for Hβ:

$$E(H\beta) = \alpha(H\beta)\, N(H^+)\, N_\varepsilon \, h\nu_{H\beta} \; ,$$

where $\alpha(H\beta)$ is the effective recombination coefficient for Hβ. Following William Clarke we can define $E^o_{4,2} = \alpha(H\beta)\, h\nu_{H\beta} \times 10^{25}$, then:

$$E(H\beta) = N(H^+)\, N_\varepsilon \, E^o_{4,2} \times 10^{-25} \; (\text{erg cm}^{-3} \text{ sec}^{-1}) \; . \tag{15}$$

Table 4-1

The Hβ Emission Coefficient $E^o_{4,2} = \alpha(H\beta)\, h\nu_{H\beta} \times 10^{25}$

(as based on the improved Balmer decrement theory)

Temperature	5000	10,000	15,000	20,000	
Case A	1.525	0.8210	0.557	0.418	Clarke (1965)
Case B	2.222	1.241	0.863	0.660	Clarke (1965)
N_ε = 100	2.202	1.235		0.658	Brocklehurst (1971)
10^4	2.224	1.240		0.659	Brocklehurst (1971)
10^5	2.249	1.246		0.660	Brocklehurst (1971)
10^6	2.285	1.256		0.661	Brocklehurst (1971)

$h\nu(H\beta) = 4.086 \times 10^{-12}$ ergs

Case B values may be approximated by an interpolation formula:

$$E^o_{4,2}(t) = 1.387 \, t^{-0.983} \, 10^{-0.0424/t} \; ; \; t = \frac{T_\varepsilon}{10,000} \; . \tag{16}$$

The dependence of $\alpha(H\beta)$ or $E^o_{4,2}$ on N_ε is shown explicitly by Brocklehurst's calculations (1971).

4B. Improved Balmer Decrement Theory.

The original theory of the Balmer decrement did not take into account the dependence of b on both n and ℓ. A neglect of the dependence of b on ℓ amounts to assuming that:

$$N(n,\ell) = \frac{2\ell + 1}{n^2} N(n), \text{ where } N(n) = \sum_{\ell=0}^{n-1} N(n,\ell) \; . \tag{17}$$

Table 4-2

Hydrogen-Balmer Line Intensities Relative to $I_{4,2}$(H I0 = 100.

	$T_e = 5 \cdot 10^3$°K $N_e(cm^{-3})$					T_e 10^4°K $N_e(cm^{-3})$					$T_e = 2.10^4$°K $N_e(cm^{-3})$			
n	10^2	10^4	10^5	10^6	n	10^2	10^4	10^5	10^6		10^2	10^4	10^5	10^6
3	303.2	300.3	296.8	291.9	3	285.9	284.7	283.1	280.7	3	274.4	274.0	273.2	272.4
4	100.0	100.0	100.0	100.0	4	100.0	100.0	100.0	100.0	4	100.0	100.0	100.0	100.0
5	45.85	45.98	46.14	46.47	5	46.85	46.90	46.98	47.14	5	47.55	47.56	47.58	47.64
6	25.15	25.25	25.39	25.76	6	25.91	25.95	26.01	26.22	6	26.43	26.44	26.46	26.56
7	15.41	15.48	15.60	16.05	7	15.91	15.94	16.00	16.28	7	16.26	16.26	16.28	16.44
8	10.18	10.23	10.36	10.92	8	10.51	10.53	10.60	10.96	8	10.73	10.73	10.76	10.99
9	7.097	7.141	7.281	7.939	9	7.306	7.335	7.417	7.832	9	7.460	7.462	7.505	7.764
10	5.157	5.200	5.377	6.111	10	5.304	5.326	5.436	5.884	10	5.401	5.405	5.472	5.737
11	3.871	3.916	4.127	4.896	11	3.975	3.997	4.131	4.574	11	4.038	4.046	4.130	4.367
12	2.983	3.036	3.285	4.066	12	3.055	3.085	3.243	3.674	12	3.099	3.113	3.212	3.414
13	2.349	2.409	2.689	3.442	13	2.401	2.436	2.608	3.011	13	2.430	2.450	2.554	2.722
14	1.885	1.958	2.267	2.960	14	1.921	1.967	2.149	2.517	14	1.942	1.969	2.075	2.214
15	1.536	1.622	1.947	2.561	15	1.562	1.616	1.800	2.127	15	1.576	1.610	1.708	1.827
16	1.268	1.371	1.708	2.230	16	1.288	1.352	1.536	1.817	16	1.297	1.338	1.427	1.528
17	1.060	1.175	1.515	1.948	17	1.074	1.145	1.325	1.563	17	1.081	1.125	1.205	1.291
18	0.8958	1.025	1.359	1.713	18	0.9057	0.9845	1.158	1.354	18	0.9097	0.9583	1.029	1.102
19	0.7641	0.9047	1.224	1.513	19	0.7709	0.8545	1.019	1.181	19	0.7732	0.8229	0.8866	0.9487
20	0.6575	0.8110	1.108	1.347	20	0.6620	0.7510	0.9037	1.038	20	0.6629	0.7140	0.7710	0.8246
21	0.4980	0.7324	1.003	1.210	21	0.5727	0.6645	0.8045	0.9196	21	0.5727	0.6230	0.6745	0.7226
22	0.4980	0.6698	0.9079	1.100	22	0.4992	0.5942	0.7180	0.8228	22	0.4984	0.5482	0.5942	0.6364
23	0.4379	0.6152	0.8244	1.012	23	0.4379	0.5340	0.6437	0.7428	23	0.4366	0.4846	0.5260	0.5658
24	0.3879	0.5698	0.7508	0.9394	24	0.3868	0.4840	0.5796	0.6760	24	0.3849	0.4315	0.4681	0.5065
25	0.3454	0.5285	0.6855	0.8772	25	0.3429	0.4402	0.5235	0.6185	25	0.3411	0.3857	0.4183	0.4561
26	0.3095	0.4925	0.6287	0.8211	26	0.3069	0.4028	0.4750	0.5682	26	0.3041	0.3470	0.3755	0.4126
27	0.2786	0.4588	0.5792	0.7685	27	0.2752	0.3691	0.4327	0.5230	27	0.2722	0.3131	0.3383	0.3747
28	0.2524	0.4283	0.5368	0.7185	28	0.2484	0.3396	0.3962	0.4821	28	0.2450	0.2839	0.3063	0.3413
29	0.2295	0.3994	0.5001	0.6710	29	0.2249	0.3128	0.3643	0.4449	29	0.2213	0.2581	0.2784	0.3116
30	0.2103	0.3730	0.4683	0.6268	30	0.2050	0.2889	0.3365	0.4112	30	0.2009	0.2356	0.2541	0.2853

Brocklehurst (1971)

Table 4-3.

Hydrogen, Paschen Series Intensities. $I(H\beta) = 100$

		T_ε = 5000°K		T_ε = 10,000°K		T_ε = 20,000°K	
λ	n	$N_\varepsilon=10^2$	$N_\varepsilon=10^4$	$N_\varepsilon=10^2$	$N_\varepsilon=10^4$	$N_\varepsilon=10^2$	N_ε 10^4
12818.1	5	18.4	18.2	16.3	16.3	14.500	14.6
10938.1	6	10.0	9.87	9.04	9.04	8.11	8.11
10049.4	7	6.1	6.01	5.53	5.54	5.04	5.04
9546.0	8	3.95	3.93	3.65	3.65	3.35	3.35
9229.0	9	2.72	2.71	2.54	2.54	2.35	2.35
9014.9	10	1.95	1.95	1.84	1.84	1.72	1.72
8862.8	11	1.46	1.46	1.38	1.38	1.28	1.28
8750.5	12	1.12	1.13	1.06	1.07	0.99	0.985
8665.0	13	0.883	0.89	0.833	0.833	0.775	0.775
8598.4	14	0.71	0.72	0.665	0.671	0.621	0.621
8545.4	15	0.595	0.592	0.540	0.549	0.504	0.507
8502.5	16	0.474	0.494	0.446	0.456	0.415	0.420
8467.26	17	0.395	0.421	0.372	0.384	0.346	0.352
8438.0	18	0.334	0.363	0.313	0.327	0.291	0.298
8413.3	19	0.285	0.317	0.267	0.282	0.247	0.256
8392.4	20	0.245	0.281	0.229	0.246	0.212	0.221
8374.5	21	0.185	0.252	0.198	0.210	0.183	0.193
8359.6	22	0.184	0.228	0.172	0.192	0.159	0.169
8345.6	23	0.162	0.206	0.151	0.172	0.140	0.150

Brocklehurst (1971).

At low-to-moderate densities such a situation cannot be expected to hold, at least for the lower levels and ℓ-values. Consider the energy level diagram for H (Fig. 2-5). Atoms which find themselves in p-levels are quickly drained to the ground level, since the A-values for 1s - np transitions are much higher than those for other $n\ell$ - $n'\ell'$ combinations. The s and d excited electrons descend less quickly into 2p. Jumps involving high values of ℓ, e.g., 7i to 6h, have relatively low A-values, since ν is small, even though f may be large. Hence, if we plot log $b(n\ell)$ against ℓ for some n-value, such as $n > 7$, we find that b(np) is much smaller than other b-values. Therefore, the theory must be developed to take the ℓ values into account. Detailed calculations have been made by Clarke (1965) and by Brocklehurst (1971). Equation (1) now becomes:

$$\sum_{n''=n+1}^{\infty} \sum_{\ell''=\ell\pm1} N(n''\ell'', T_\varepsilon)\, A(n''\ell'', n\ell) + N(H^+)\, N_\varepsilon\, \alpha(n\ell; T_\varepsilon) =$$

$$N(n\ell; T_\varepsilon)\, A(n,\ell)\ , \qquad (18)$$

where:

$$A(n,\ell) = \sum_{n'=n_0}^{n-1} \sum_{\ell=\ell\pm1} A(n\ell; n'\ell')\ , \qquad \begin{matrix} n_0 = 1 \text{ Case A} \\ n_0 = 2 \text{ Case A} \end{matrix} \qquad (19)$$

and

$$P(n''\ell'';\ell n) = \frac{A(n''\ell''; n\ell)}{A(n''\ell'')}\ . \qquad (20)$$

Defining:

$$C(n''\ell''; n\ell) = \sum_{n^j=n}^{n''-1} \sum_{\ell^j=\ell\pm1} P(n''\ell''; n^j\ell^j)\, C(n^j\ell^j, n\ell)\ , \qquad (21)$$

the equation of statistical equilibrium becomes:

$$N(n,\ell) = \frac{N(H^+)N_\varepsilon}{A(n\ell)} \sum_{n'=n}^{\infty} \sum_{\ell'=0}^{n''-1} \alpha(n''\ell'')\, C(n''\ell''; n\ell)\ . \qquad (22)$$

The solution is carried out as before.

Calculation of the $A(n\ell, n'\ell')$'s and $\alpha(n\ell)$'s pose formidable numerical problems (cf. Burgess 1964). Quantum mechanical expressions for the A-values involve terms that nearly cancel one another. It is possible to develop recursion formulae to overcome these difficulties, however. The Balmer decrement is calculated from the expression:

Table 4-4.

He II Paschen Series Intensities $I_{n,3}$; I(λ4686) = 100

		T_ε = 10,000			T_ε = 20,000		
λ	n	N_ε = 0	10^4	10^6	0	10^4	10^6
4686.1	4	100	100	100	100	100	100
3204.1	5	39.8	40.3	42.5	43.8	45.2	46.0
2734.1	6	20.1	20.3	21.1	23.2	23.9	24.3
3512.0	7	12.0	12.2	12.7	14.0	14.4	14.8
2386.2	7	7.77	8.0	8.3	9.18	9.5	9.7
2307.0	9	5.38	5.6	5.8	6.4	6.6	6.8
2253.0	10	3.92	4.1	4.3	4.7	4.8	4.9
He II Balmer Series Intensities I(λ4686) = 100							
1640.2	3	625	660	681	714	745	760
1215.2	4	189	201	213	234	246	253
1085.0	5	84.1	90.4	98.1	106	113	116
1025.3	6	45.6	49.1	51.9	58.3	61.8	63.9
992.4	7	27.8	30.0	31.9	35.8	37.9	39.2
972.1	8	18.5	19.8	21.1	23.7	25.0	25.9
958.7	9	12.9	13.8	14.7	16.5	17.4	18.1
949.4	10	9.3	10.0	10.7	12.0	12.7	13.2

Seaton, M.J., 1978, M.N.R.A.S., 185, 5P.

Table 4-5.

Predicted Intensities in the He II Pickering Series, $I_{4,3} = I(\lambda 4686) = 100$

	n	BL	T_ε = 10,000		T_ε = 20,000	
λ			10^4	10^6	10^4	10^6
6560.1	6	H 12	13.56	13.69	13.61	13.66
5411.5	7		7.76	7.96	8.09	8.19
4859.3	8	Hβ	4.91	5.07	5.22	5.31
4541.6	9		3.33	3.55	3.58	3.64
4338.7	10	Hγ	2.37	2.47	2.57	2.63
4199.8	11		1.76	1.84	1.91	1.96
4100.0	12	Hδ	1.34	1.41	1.46	1.50
4025.6	13	He I	1.05	1.11	1.15	1.18
3968.4	14	H_ε†	0.84	0.90	0.91	0.95
3923.5	15		0.68	0.74	0.74	0.78
3887.4	16	H, He	0.56	0.62	0.61	0.64
3858.1	17		0.47	0.53	0.51	0.54
3833.8	18	H9	0.39	0.46	0.43	0.465
3813.5	19		0.33	0.403	0.365	0.40
3796.3	20	H10	0.29	0.36	0.313	0.35
3781.6	21	H11	0.248	0.32	0.271	0.31
3768.93	22	H12	0.217	0.29	0.236	0.28
3748.5	23		0.190	0.266	0.207	0.251
3732.8	24	H	0.169	0.244	0.183	0.227
3720.3	25		0.150	0.224	0.162	0.206
3710.4	26	H	0.135	0.206	0.145	0.188
3706.2	27		0.122	0.191	0.130	0.172
3702.4	28	H, He	0.110	0.177	0.117	0.158
3697.2	29		0.100	0.164	0.106	0.146
3695.5	30	H	0.092	0.152	0.096	0.134
$\alpha_{4,3} \times 10^{14}$ cm^3 s^{-1}			34.92	33.95	16.90	16.57

Brocklehurst, M. (1971).

†usually a blend H_ε + [Ne III].

$$\frac{I(Ba_n)}{I(H\beta)} = \frac{4861}{\lambda(Ba_n)} e^{X_n - X_4} \frac{\sum\limits_{\ell'=0}^{1} \sum\limits_{\ell=\ell'\pm1} (2\ell+1) b(n,\ell,T_\varepsilon,N_\varepsilon) A(n\ell;2\ell')}{\sum\limits_{\ell'=0}^{1} \sum\limits_{\ell=\ell'\pm1} (2\ell+1) b(4,\ell,T_\varepsilon,N_\varepsilon) A(4\ell,2\ell')} . \tag{23}$$

Here $I(Ba_n) = I(n \rightarrow 2)$ denotes the intensity of a Balmer line. Note that for the higher levels and at high densities b may also depend on N_ε. See Table 4-2. Analogous expressions may be derived to compare Paschen and Brackett line intensities with Hβ. See Table 4-3. The neglect of the ℓ dependence on the calculation of $b(n,\ell)$ would appear to be more severe for Case A than for Case B where the 1s - np transitions are balanced by np - 1s transitions.

Reference to Eq. (2) shows that for the hydrogenic isoelectronic sequence the number of n - n' transitions (or the A-value) goes as the fourth power of Z! We can use the results of the hydrogenic calculations for ionized helium if we replace T_ε by (T_ε/Z^2). Thus, the hydrogenic Balmer decrement computed for T_ε = 5000 applies to He^+ at T_ε = 20,000. Since the frequency of a (n,n') transition goes as Z^2, alternate members of the He II Balmer series very nearly coincide with Lyman lines of hydrogen. See Table 4-4. The intensity ratio λ1640 He II Hα/λ4686 He II Pa α is a useful datum for calibrating ultraviolet intensities. For objects of small angular size the ratio is valuable for getting the extinction (see Chapter 8); for other objects I(1640) serves as a "standard" intensity as Hβ serves in the visual region. Tables 4-5 and 4-6 give intensities of He II Pickering and Pfund series referred to I(4686) = 100.

Up to this point the theory has been developed for purely radiative processes. For the higher levels, one must take into account the possibility that level populations may be modified by collisions with protons and electrons that shift atoms primarily from (n,ℓ) to $(n,\ell\pm1)$ and from (n,ℓ) to $(n\pm1;\ \ell\pm1)$. As density increases, collisions with protons that shift atoms from one ℓ-level to another without changing n are the first to become important. Later, electron collisions that change both n and ℓ become significant. Changes involving $\Delta n = \pm 1$ $\Delta\ell = \pm 1$ are the most frequent. The equations now become involved as the Balmer decrement depends on density.

The effect of collisions is to redistribute atoms among the levels more nearly in proportion to their statistical weights. Thus, as n increases, $N(n,\ell)$ approaches $(2\ell + 1)N(n)/n^2$, i.e., that appropriate to the approximation wherein the ℓ-dependence was totally neglected. So for the large n-values, the approximation inherent in the elementary theory tends to be less inaccurate. A(n,n') falls off rapidly as n increases, but the collisional effects increase steeply with n. In fact, for high levels, collisional processes of the type $H(n,\ell) + H^+ \rightarrow H(n,\ell\pm1) + H^+$ operate much faster than radiative processes. For each density we can always specify a critical value of n, n_c, at which collisions become important. Alternately, for a given n, we can follow Pengelly and Seaton (1964) to define a critical electron density N_ε^{crit} for a level (n,ℓ) at which the radiative rate equals the collisional transition rate, viz.:

Table 4-6

He II Pfund Series Intensities, I_{n-5}; $I_{4.3}$ = I(λ4686) = 100

			T_ε = 10,000		T_ε = 20,000	
λ	n	blend	$N_\varepsilon = 10^4$	10^6	10^4	10^6
8236.6	9		2.22	2.30	2.21	2.22
7592.8	10		1.57	1.60	1.59	1.61
7177.8	11		1.15	1.18	1.18	1.19
6890.8	12		0.88	0.90	0.90	0.91
6683.2	13	He I	0.68	0.70	0.70	0.71
6570.0	14	H α	0.54	0.56	0.56	0.57
6406.3	15		0.44	0.46	0.49	0.46
6310.8	16	[S IV]	0.36	0.38	0.37[2]	0.38
6233.8	17		0.30	0.32	0.31	0.32
6170.3	18		0.250	0.273	0.260	0.27
6118.2	19		0.206	0.236	0.221	0.232
6074.3	20		0.182	0.208	0.189	0.200
6037.7	21		0.157	0.184	0.163	0.175
6004.8	22		0.137	0.165	0.142	0.159
5977.0	23		0.120	0.143	0.124	0.138
5953.0	24		0.106	0.134	0.110	0.124
5932.2	25		0.094	0.121	0.097	0.111
3913.5	26		0.084	0.116	0.086	0.101
5896.9	27		0.076	0.106	0.077	0.092
5882.4	28		0.068	0.094	0.069	0.084
5867.8	29		0.062	0.087	0.063	0.077
5858.3	30		0.056	0.080	0.057	0.071

Brocklehurst, M. 1971

Table 4-7.

Predicted Intensities of Triplet Lines in Helium (Case A) referred to $I(2^3P - 4^3D) = I(\lambda 4771)$ as 1.00

$I(2^3P-n^3D)$		$N_\varepsilon = 10^4$		$2^3P - n^3S$		$N_\varepsilon = 10^4$	
λ	n	10,000	20,000	λ	n	10,000	20,000
5875.6	3	2.764	2.58	7065.2	3	0.328	0.477
4026.2	5	0.474	0.487	4713.2	4	0.092	0.139
3819.7	6	0.264	0.274	4120.8	5	0.038	0.058
3705.1	7	0.163	0.170				
3634.3	8	0.108	0.113				
3587.3	9	0.075	0.078	$2^3S - n^3P$			
3554.4	10	0.054	0.057	1083.0	2	4.416	5.011
3530.5	11	0.041	0.043	3889.8	3	2.263	2.789
3512.5	12	0.031	0.033	3187.7	4	0.917	1.157
3498.6	13	0.025	0.026	2945.1	5	0.440	0.560
3487.7	14	0.020	0.021	2829.1	6	0.245	0.312
3479.0	15	0.016	0.017	2763.8	7	0.149	0.190
3471.8	16	0.013	0.014	2723.2	8	0.098	0.125
$\alpha(4^3D-2^3P) \times 10^{-14}$				2696.1	9	0.068	0.086
$cm^{-3}\ s^{-1}$		1.367	0.6633	2677.1	10	0.049	0.062

Predicted Singlet Line Intensities for Helium referred to $I(2^1P - 4^1D) = I(4922)$ as 1.00

		$N_\varepsilon = 10^4\ cm^{-3}$ (Case A)		$N_\varepsilon = 10^4$ (Case B)	
2^1P-n^1D	n	$T_\varepsilon = 10{,}000$	20,000	10,000	20,000
6678.2	3	2.905	2.713	2.89	2.70
4921.9	4	1.000	1.000	1.00	1.00
4387.9	5	0.466	0.480	0.468	0.481
4143.7	6	0.258	0.269	0.259	0.270
4009.3	7	0.159	0.167	0.159	0.167
3926.5	8	0.105	0.111	0.105	0.111
n^1S-2^1P		$N_\varepsilon = 10^4$	Case A	$N_\varepsilon = 10^4$	Case B
7281.3	3	0.279	0.426	0.508	0.696
5047.7	4	0.103	0.157	0.146	0.206
4437.5	5	0.048	0.074	0.061	0.088
4169.0	6	0.027	0.041	0.032	0.046
$n^1P - 2^1S$					
5015.7	3	0.049	0.058	2.151	2.55
3964.7	4	0.029	0.035	0.855	1.03
3613.6	5	0.0166	0.020	0.418	0.506
3447.6	6	0.0106	0.013	0.236	0.285
$\alpha_{eff}(4^1D-2^1P) \times 10^{-14}$		0.4014	0.1919	0.4117	0.1976 $cm^3\ s^{-1}$

(Brocklehurst, 1972, M.N.R.A.S., 157, 211)

Predicted I(4922)/I(4471)	$T_\varepsilon = 5000$	10,000	20,000
Case A	0.2692	0.2660	0.2628
Case B	0.2755	0.2736	0.2706

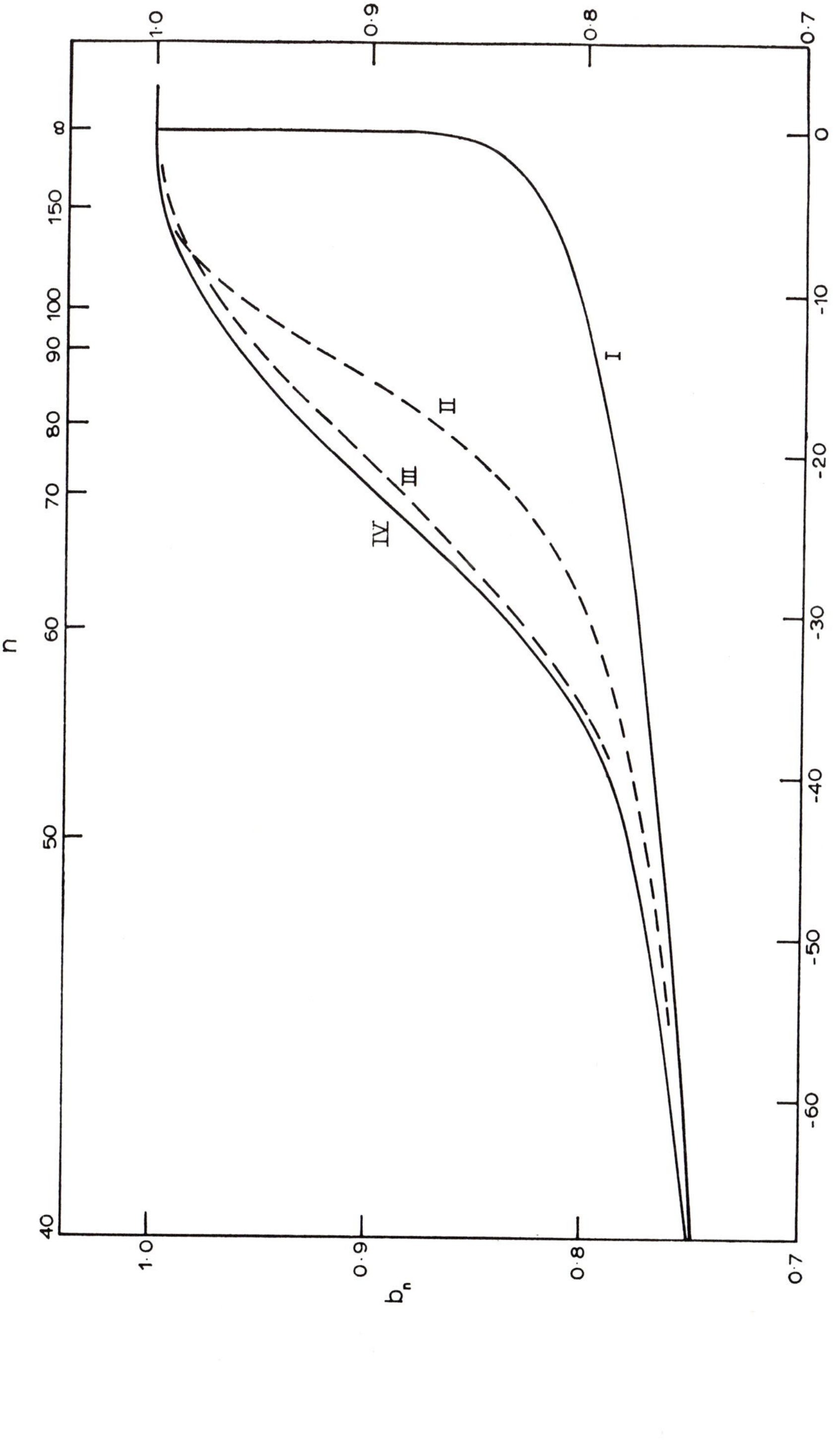
n
40
50
60
70
80
90
100
150
∞
1·0
0·9
0·8
0·7
b_n
0
-10
-20
-30
-40
-50
-60
$\Delta\sigma$, cm^{-1}
I
II
III
IV

Figure 4-1

$$N_{\varepsilon}^{crit}(n,\ell) \sum_{\ell'} q(n\ell,n'\ell') = \sum_{n'\ell'} A_{n\ell,n'\ell'} \tag{24}$$

The corresponding $N^{crit}(n) = A_n q_n^{-1}$ is found by averaging $A_n^{-1}\, q(n\ell,n'\ell')$ over ℓ. For N_ε exceeding this critical value, the $N(n,\ell)$ populations will satisfy Eq. (17).

At T = 10,000°K, S. Drake finds N_{ε}^{crit} values for low n as follows:

level	2p	3	4	5	6	7
$N_{\varepsilon}^{crit}(n)$	4.6×10^{12}	2.5×10^{9}	5.7×10^{8}	1.4×10^{8}	3.7×10^{7}	1.1×10^{7}

At such large densities, optical depths can become large, especially in the Lyman lines and we must solve the equation of transfer (or at least calculate escape probabilities). Then the net radiative rates of decay of any level are reduced and the critical electron densities are decreased.

We have to solve a complicated set of equations:

$$N(H^{+})\, N_{\varepsilon}\, \alpha_{n\ell}(T_{\varepsilon}) + \sum_{n''>n}^{\infty} \sum_{\ell=\ell''\pm 1} N(n''\ell'')\, A(n''\ell'',n\ell) +$$

$$\sum_{\ell'=\ell\pm 1} N(n\ell')N(H^{+})q_{p}(n\ell,n\ell\pm 1) + \sum_{m=n-1}^{m=n+1} \sum_{\ell'=\ell\pm 1} N(m\ell')N_{\varepsilon}q_{\varepsilon}(m\ell',n\ell)$$

$$= N_{n\ell}\Big\{ \sum_{n'=n_{o}}^{n-1} \sum_{\ell'=\ell\pm 1} A(n\ell;n'\ell') +$$

$$\sum_{\ell'=\ell\pm 1} N(H^{+})q_{p}(n\ell,n\ell\pm 1) + \sum_{m=n-1}^{m=n+1} \sum_{\ell'=\ell\pm 1} q_{\varepsilon}(n\ell,m\ell')N_{\varepsilon}\Big\} \tag{25}$$

Here q_p and q_ε are the rate coefficients for collisional redistributions by proton and electron collisions, respectively. Electron reshuffling collisions become important only at relatively high n values and/or densities. Actually, by the time they become significant, collisional reshuffling has largely produced a distribution dependent on statistical weights as described by Eq. (17). Ultimately, we have to add terms to account for collisional ionization from level n to the continuous and three-body recaptures.

As n increases to its limiting value, b_n approaches 1. Figure 1, due to Seaton (1964), shows the importance of the collisional terms. More refined calculations giving results as a function of N_ε and T_ε , have been obtained by Brocklehurst (1970) and by Brocklehurst and Seaton (1972).

In comparing theory and observation, there are two additional effects we must remember (besides the effect of interstellar extinction which always tends to steepen the decrement):

1) In an optically thick nebula, the intensity of Lyman α may build up to the point were the 2p level may develop a nonnegligible population. Also, when atoms are trapped in the 2s level they may be

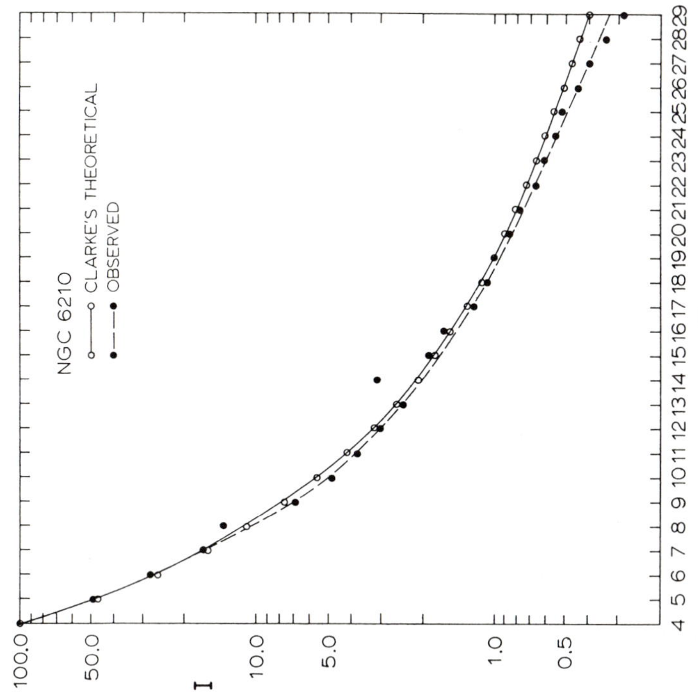
NGC 6210
CLARKE'S THEORETICAL
OBSERVED
I
n

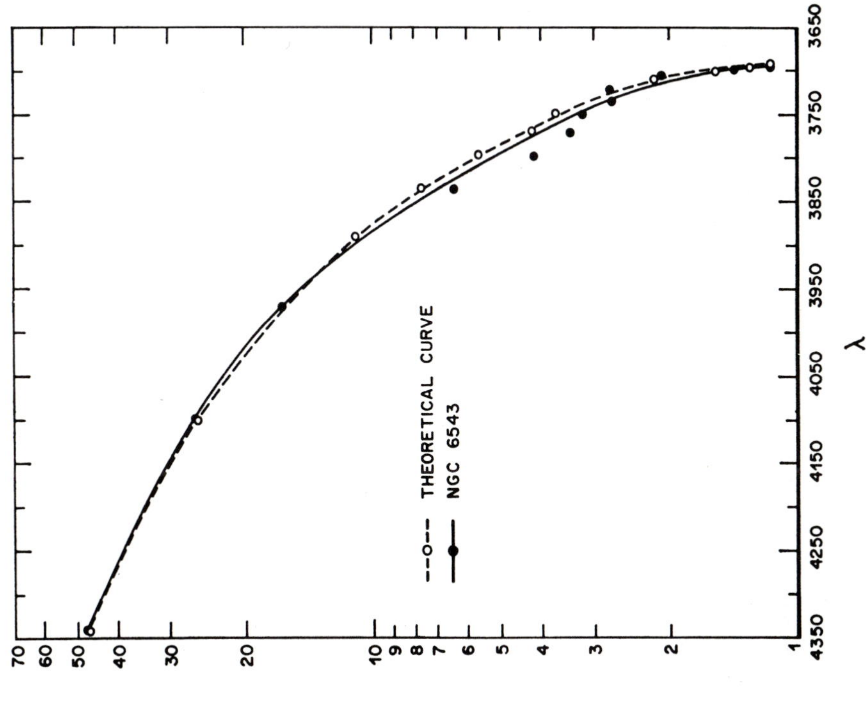
THEORETICAL CURVE
NGC 6543
λ

Figure 4-2

able to absorb Balmer quanta. Hence, in an optically thick nebula, a Balmer quantum might be absorbed before it escapes from the nebula. This self-reversal effect would tend to make the Balmer decrement for $n \geq 4$ less steep. Under extreme conditions the $n = 2$ level could act like a pseudo-ground level. Then Hα could be scattered (compare Case B) and increase strongly in intensity with respect to Hβ;

2) On the other hand, if collisional excitations become important, the Balmer decrement will be steepened; i.e., the intensity ratio I(Hα)/I(Hβ) will be increased, since the number of collisional excitations falls off rapidly as n increases!

Theories of recombination followed by cascade can be developed for other ions besides H I and He II. The most important application is to He I. Brocklehurst (1972) used accurate atomic data and took into account collisional redistribution of energy and angular momentum among higher levels at nebular densities in order to predict intensities of both singlet and triplet lines. See Table 4-7. Note that the metastability of 2^1S and 2^3S causes some interesting complications. Robbins (1968, 1970) has investigated optical depth and self-absorption effects for a number of He I lines, assuming planetary nebulae to be spherical structures with constant velocity gradients. He then solved for line intensity ratios as a function of electron temperature, optical depth in λ3889, and expansion rates. The maximum effects are found for V(R) = 0. Since λ3889 is blended with H7 λ3889, it is not a practical line to use to assess optical depth effects. These can be seen more easily in λ7065 or λ3187. See Table 4-8.

Table 4-8. Dependence of intensities of λ3187 and λ7065 as a function of optical depth in He I λ3889 for a static nebula (Robbins 1968).

		I(λ)/I(λ4471)				
	T_ϵ	$\tau_0(\lambda3889) = 0$	10	20	50	75
λ3187	10,000	0.93	0.62	0.44	0.22	0.15
	20,000	1.22	0.82	0.58	0.29	0.20
λ7065	10,000	0.48	1.19	1.53	1.89	1.98
	20,000	0.66	1.60	2.04	2.02	2.63

4C. The Observed Balmer Decrement

Early photographic spectrophotometry of Hα, Hβ, Hγ, Hδ showed that observed Balmer decrements in diffuse and planetary nebulae followed Menzel and Baker's Case B. Photoelectric measurements by Miller (1970) for NGC 7027 and NGC 7662 confirmed that Case B was also valid for higher series members. Figure 4-2 shows observed Balmer decrements for NGC 6210 and IC 3568, planetaries that do not appear to be much affected by interstellar reddening. Since the Balmer decrement is relatively insensitive to temperature, a measurement of it often helps to assess the interstellar extinction, which always steepens the decrement in a predictable way (see Chapter 8). Collisional excitation and optical depth effects will affect the Balmer decrement in different ways. These phenomena will appear in dense and/or optically thick nebulae. Their detection and assessment require accurate observations and determinations of extinction. If far ultraviolet observations are available, extinction estimates may often be made relatively accurately for nebulae of small angular size.

4D. The Line and Continuous Spectrum of Hydrogen in the Radio-Frequency Region

Recombination lines of hydrogen and helium and a few other elements, particularly carbon, involving large values of the principal quantum number n, have been observed in H II regions, planetary nebulae and the centers of galaxies. If a transition occurs between a level with a quantum number n and one with a quantum number n + Δn, then:

$$\nu = R_c\, Z^2 \left(1 - \frac{5.486 \times 10^{-4}}{m_a}\right) \left(\frac{1}{n^2} - \frac{1}{[n+\Delta n]^2}\right) \simeq$$

$$R_c\left(1 - 5.486 \times 10^{-4}/m_a\right) Z^2\left(\frac{2\Delta n}{n^3}\left[1 - \ldots\right]\right) \sim 2R_{corr}\, Z^2\, \Delta n/n^3 . \tag{26}$$

Here m_a is the mass of the emitting atom in atomic mass units and R_c is the Rydberg constant for an atom of "infinite" mass. A transition $n + \Delta n \to n$ is called an nα transition if $\Delta n = 1$, while $\Delta n = 2$ and 3 would be the nβ and nγ lines. The hydrogen recombination line, H 158 γ line, corresponds to the 161 → 158 transition in hydrogen. Neutral helium behaves hydrogenically at high n-values, the corresponding neutral helium transitions would be denoted as He°158γ or He I 158γ. Relatively high frequency transitions (n,n') ~ 50 fall in the mm range, while transitions (n,n') ~ 250 would appear in the meter range. Of course, different series of lines will overlap. For example, transitions such as 110α, 138β, 158γ, and 186ε fall near 6 cm and can be observed with the same combination of antenna and receiver and thus with identical angular resolution.

Before we examine the problem of spectral line formation in the radio frequency domain, we must consider the underlying continuum. In the radio frequency region:

$$B_\nu = \frac{2\nu^2\, kT}{c^2} \quad \text{(Rayleigh-Jeans approximation)} \tag{27}$$

Thermal continuum emission arises from free-free processes (see Eq. [3-64]). The coefficient of absorption per unit mass may be obtained from this expression with the aid of Kirchhoff's law:

$$j_\nu = 4\pi\, k_\nu^{ff}\, B_\nu(T) \tag{28}$$

Taking into account the negative absorptions, and comparing Eq. (3-64) we obtain for the absorption coefficient per unit volume:

$$\rho k_\nu'^{ff} = \frac{8\pi\, \varepsilon^6\, Z^2}{3\sqrt{6\pi}\; c(mk)^{3/2}} \frac{N(H^+)\, N_\varepsilon}{T_\varepsilon^{3/2}\, \nu^2}\, g_{III} = 0.0177 \frac{N(H^+)\, N_\varepsilon\, g_{III}}{T_\varepsilon^{3/2}\, \nu^2} \tag{29}$$

The g_{III} factor has been calculated by Oster (1961). It depends on T_ε, N_ε, and ν in a rather complex manner, which depends on the discriminant:

$$d = \frac{8.1 \times 10^{-6}\, \nu}{T_\varepsilon^{1/2}\, N_\varepsilon^{1/3}} \,. \tag{30}$$

In a typical gaseous nebula, where $T_\varepsilon \sim 10{,}000°K$, $N_\varepsilon \sim 1000$, d will exceed unity for a frequency of 1 GHz. Then:

$$g_{III} = \frac{\sqrt{3}}{\pi} \ln \left[\frac{(2k)^{3/2}}{4.22\pi\, m^{1/2}\, \varepsilon^2}\right] \frac{T_\varepsilon^{3/2}}{Z\nu} = \frac{\sqrt{3}}{\pi} \ln \left[4.97 \times 10^7 \frac{T_\varepsilon^{3/2}}{Z\nu}\right]$$

$$\equiv \frac{\sqrt{3}}{\pi}\, g^o_{III} \,. \quad [\nu \text{ in Hz}] \tag{31}$$

At higher densities, higher temperatures and lower frequencies such that $d \ll 1$, then:

$$g_{III} = \frac{\sqrt{3}}{\pi} \ln \frac{420\, T_\varepsilon}{Z\, N_i^{1/3}} \,. \tag{32}$$

In most situations, we will have $d > 1$ and Eq. (31) will be the proper formula to apply.

If we express frequencies in GHz (i.e., 10^9 c/s), the distance along the path in parsecs (1 parsec = 3.09×10^{18} cms):

$$\tau_c = \int (k\rho)\, dl = 0.0301\, T_\varepsilon^{-3/2} \left(\frac{\nu}{GHz}\right)^{-2} g^o_{III} \int N_\varepsilon^2 \left(\frac{dl}{parsec}\right)$$

$$\cong 0.08235\, a(\nu,T)\, T^{-1.35}\, \nu^{-2.1} \int N_\varepsilon^2 \left(\frac{dl}{parsec}\right) . \tag{33}$$

where $a(\nu,T)$ is a slowly varying function of ν and T_ε that does not differ much from unity. A more elaborate expression for τ_c has been given by Mezger and Henderson (1967). The integral in Eq. (33) is called the emission measure (N_ε in electron/cm^3).

Consider an ionized plasma of uniform temperature, T_ε, and optical thickness, τ_c, at some frequency, ν. Since the source function is $\underline{J}_\nu(T) = B_\nu(T)$, we have for the radiation produced by the cloud.

$$I_\nu = \int_0^{\tau_c} B_\nu(T_\varepsilon) \exp(-\tau_\nu)\, d\tau_\nu = B_\nu(T) \left[1 - \exp(-\tau_c)\right] . \tag{34}$$

Consider a cloud of finite optical thickness in the continuum, for example, the Orion nebula or a relatively dense planetary, such as NGC 6572 or NGC 7027. Since τ_c varies essentially as ν^{-2}, such clouds will be optically thick at low frequencies. Then the emergent intensity I_ν will be simply $B_\nu(T_\varepsilon)$ and the brightness temperature, T_b, will equal T_ε. At higher frequencies, the optical depth decreases until $I_\nu = \tau_\nu B_\nu(T_\varepsilon)$. If the geometry of the nebula is known or assumed, we can predict I_ν for various assumptions about the N_ε and T_ε variation along the line of sight, convolve the thus-derived surface brightness function $I_\nu(\alpha,\delta)$ with the antenna pattern and compare with observations. In principle, although not necessarily in practice, the continuum is relatively simple to treat since the electrons follow a Maxwellian distribution at T_ε and emission and absorption processes can be calculated for local thermodynamic equilibrium (LTE). See Figure 4-3.

The situation is more complicated for the lines because of deviations from thermodynamic equilibrium (non-LTE effects). The population of each level is given by Eq. (3-19) where the b_n's must be evaluated before the problem can be solved. Initially, let us adopt the oversimplified assumption that $b_n = 1$, in order to lay some of the groundwork for the physics of the problem and to emphasize the large differences that will appear when actual b_n values are employed. Imagine a flat isothermal slab of uniform density and level of ionization, much larger than the projected antenna pattern. Let T_c be the brightness temperature of the background continuum and $\underline{J_c}$ its source function which includes the 2.7°K radiation. For LTE, the source function for the nebula will be the Planckian function $B_\nu(T)$ which is proportional to the temperature. Then we can write:

$$\frac{\int \underline{J}_\nu(t_L)\, e^{-t_L} dt_L}{\int \underline{J}_\nu(\tau_c)\, e^{-\tau_c} d\tau_c} = \frac{(T_L + T_c)}{T_c} = 1 + T_L/T_c \ . \tag{35}$$

Here $t_L = \tau_L + \tau_c$ where τ_L is the line optical depth computed strictly from the line absorption coefficient. It will vary across the line in accordance with the line broadening mechanism.

If k_ν is the mass absorption coefficient as ordinarily defined and k'_ν denotes the same corrected for negative absorptions (stimulated emissions), we may define the total corrected volume absorption coefficient integrated over the line profile by means of the following expression:

$$\begin{aligned} N_{n'}\, 4\pi\, I \int a'_\nu\, d\nu &= h\nu\left[B_{n'n}\, N_{n'} - B_{nn'}\, N_n\right] I \\ &= h\nu\, B_{n'n}\, N_{n'} \left\{1 - \frac{\omega_{n'}}{\omega_n}\frac{N_n}{N_{n'}}\right\} I \ , \end{aligned} \tag{36}$$

where it is assumed that the intensity I does not vary steeply over the line. Under conditions of LTE, with $h\nu/kT_\varepsilon \ll 1$, the term in $\{\ \}$ reduces to $h\nu/kT$. We find:

$$\int a'_\nu\, d\nu = \frac{h^2\, \nu^2}{4\pi k T_\varepsilon} B_{n'n} = \frac{c^2 h}{8\pi k T \nu}\frac{\omega_n}{\omega_{n'}} A_{nn'} \ . \tag{37}$$

We have made use of Eqs. 3-1,3. The relation between the absorption Einstein B and the Ladenburg f is:

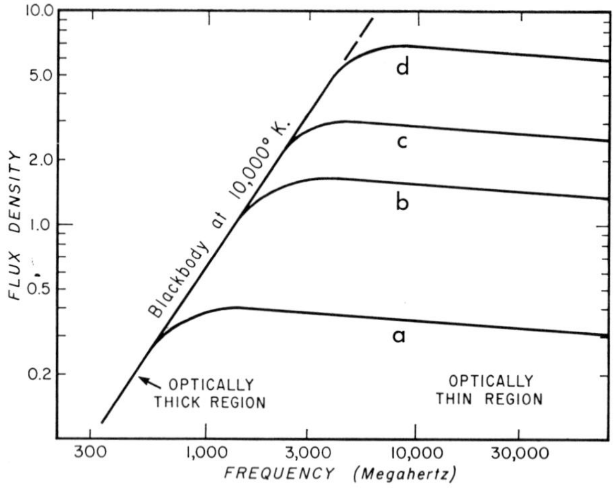

Figure 4-3

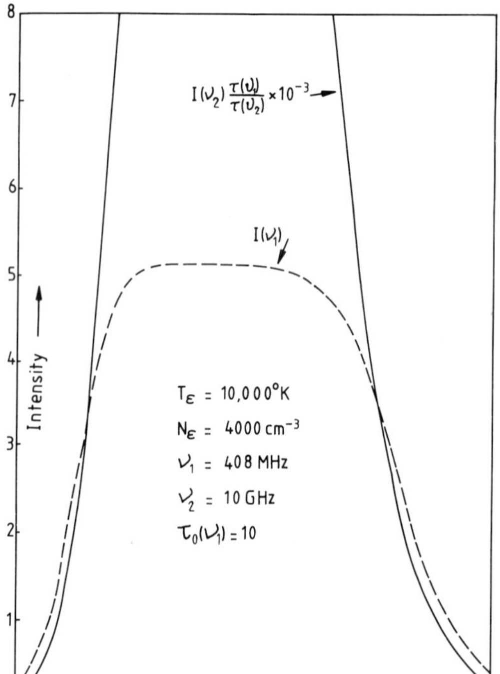

Figure 4-4

$$\frac{h\nu}{4\pi} B_{n,n+\Delta n} = \frac{\pi\varepsilon^2}{mc} f_{n,n+\Delta n} \ . \tag{38}$$

For high quantum numbers, one may use an expression for $f_{n,n+\Delta n}$, due to Menzel (1969) and involving hypergeometric functions to obtain transition probabilities for

$$n + \Delta n \rightarrow n, \ \Delta n = 1,2,3 \ldots \alpha, \ \beta, \ \gamma \ . \tag{39}$$

At any frequency ν, the absorption coefficient will be:

$$a_\nu' = \phi_\nu \int a' \, d\nu \ , \tag{40}$$

where ϕ_ν equals the normalized line absorption coefficient.

For the situation of pure thermal broadening, the absorption coefficient would be:

$$a_\nu' \sim \exp\left(- \left[\frac{\Delta\nu}{\Delta\nu_0}\right]^2\right), \text{ where } \Delta\nu_0 = \frac{\nu_0}{c} \sqrt{\frac{2kT_\varepsilon}{M}} \ . \tag{41}$$

Actually, T_ε is replaced by $T_0 > T_\varepsilon$ in order to allow for the effects of broadening by mass motions, commonly called turbulence. If $T_\varepsilon \sim 10{,}000°K$, T_0 would be about 20,000°K, typically.

Transitions involving high n values are widened by collisions with electrons and ions, that is, by the interatomic Stark effect. If $\Delta\omega$ is the distance from the line center in circular frequency units, i.e, $\Delta\omega = 2\pi \, \Delta\nu$, we may define a discriminator time, Δt_d, by the relation $\Delta\omega \, \Delta t_d \sim 1$, so that if $\Delta\omega$ is small as in r.f. lines, Δt will be large. During this time the configurations of the surrounding charges will change significantly. Consequently, we must regard the line broadening events as discrete collisions and apply the impact broadening theory which widens a line according to a dispersion relation:

$$a_\nu \sim \left[(\nu - \nu_0)^2 + \left(\frac{\gamma}{4\pi}\right)^2\right]^{-1} \ . \tag{42}$$

In practice, the two sources of broadening act simultaneously. The Doppler and collisional broadening functions have to be convolved.

It is useful to define a quantity called the full width at half-maximum FWHM for these absorption processes. For pure Doppler broadening,

$$(\Delta\lambda)_{FWHM} = \frac{2\lambda_o}{c} \sqrt{2k \ln 2} \left[\frac{T_P}{M}\right]^{1/2} \ . \tag{43}$$

where T_p is the gas kinetic temperature of the atom or ion, p. Usually, $T_p = T_\varepsilon$. For a line arising from a high-level transition $n + \Delta n \rightarrow n$, $T_\varepsilon = 10{,}000°K$ and $T_0 = 20{,}000°K$, Brocklehurst and Leman (1971) find the ratio of the (FWHM)'s for collisional and Doppler broadening to be:

$$\left[\frac{\Delta\nu_{coll}}{\Delta\nu_0}\right]_{FWHM} = 0.14 \left(\frac{n}{100}\right)^{7.4} \left(\frac{N_\varepsilon}{10^4}\right) . \tag{44}$$

Note the steep rise in the importance of collisional broadening with n.

Turning to the non-LTE situation, we find that the existence of a b_n that increases slowly with n introduces complications: first, because the populations of successive levels will deviate in a significant fashion from that expected in a simple theory; second, we must consider the transfer problem. An increase in population with n permits a masering action to occur. Let LTE values be denoted by superscript (o).

The dependence on b_n on n is shown, for example, by the calculations of Brocklehurst (1970). Its value is sensitive to electron density. For the radio frequency domain, a quantity of considerable importance will be d ln b_n/dn. To set up the equation of transfer, we follow Goldberg (1966).

Note that the line absorption coefficient requires two corrections:

a) N_n^o is replaced by $N_n^o\, b_n$ (since LTE no longer exists);
b) The correction factor for the negative absorptions:

$$\left[1 - \exp\left(-\frac{h\nu}{kT}\right)\right] \text{ in LTE becomes } \left[1 - \frac{b_n}{b_{n'}} \exp\left(-\frac{h\nu}{kT_\varepsilon}\right)\right] \text{ in non-LTE.}$$

Let κ_c' denote the coefficient of continuous absorption per unit volume for the continuum as corrected for negative absorptions (the LTE expression applies here). Within a spectral line:

$$\kappa'_{total} = \kappa'_c + \kappa'_L , \tag{45}$$

where L refers to the line proper. For a transition from n' to n:

$$\kappa'_{total} = \kappa'_c + b_{n'}\, \kappa_L'^{o} \left[1 - \frac{b_n}{b_{n'}} \exp\left(-\frac{h\nu}{kT_\varepsilon}\right)\right] \left[1 - \exp\left(-\frac{h\nu}{kT_\varepsilon}\right)\right]^{-1} , \tag{46}$$

where $\kappa_L'^{o}$ is the absorption coefficient defined for LTE. We have:

$$\kappa_L'^{o} = N^o_{n'} \left(1 - e^{-h\nu/kT_\varepsilon}\right) \int a_{n'n}(\nu)\, d\nu;\; N^o_n = N_n/b_n . \tag{47}$$

Let:

$$\beta_{n'n} = \frac{\left[1 - (b_n/b_{n'}) \exp(-\, h\nu/kT_\varepsilon)\right]}{\left[1 - \exp(-\, h\nu/kT)\right]} . \tag{48}$$

For small $h\nu/kT_\varepsilon$ the negative absorption factor becomes:

$$\left[1 - \frac{b_n}{b_{n'}} \left(1 - \frac{h\nu}{kT_\varepsilon}\right)\right] = \frac{b_{n'} - b_n}{b_{n'}} + \frac{b_n}{b_{n'}} \frac{h\nu}{kT_\varepsilon} \sim \frac{b_{n'} - b_n}{b_{n'}} + \frac{h\nu}{kT_\varepsilon}$$

$$= \frac{h\nu}{kT_\varepsilon} \left[1 - \frac{kT_\varepsilon}{h\nu} \left(\frac{b_n - b_{n'}}{b_{n'}}\right)\right] \simeq \frac{h\nu}{kT_\varepsilon} \left[1 - \frac{kT_\varepsilon}{h\nu} \frac{d \ln b_n}{dn} \Delta n\right] . \tag{49}$$

The equation of radiative transfer is:

$$\frac{dI_\nu}{k\rho ds} = I_\nu - S_\nu , \tag{50}$$

where S_ν is the source function and $d\tau = k\rho ds = \kappa ds$ where k is the mass absorption coefficient. Let $d\omega$ denote the element of solid angle. Then:

$$\frac{dI_\nu}{ds} d\omega = - \left\{\kappa_c' + b_{n'} \kappa_L'^{o} \beta_{nn'}\right\} I_\nu d\omega + N_n A_{nn'} h\nu \frac{d\omega}{4\pi} + \kappa_c' B_\nu(T_\varepsilon) d\omega . \tag{51}$$

Now, using Eqs. (47) and (3-1,3,18) we find:

$$N_n A_{nn'} h\nu \frac{d\omega}{4\pi} = \frac{b_n}{b_{n'}} \frac{\omega_n}{\omega_{n'}} \exp\left(-\frac{h\nu}{kT}\right) N_{n'} A_{nn'} h\nu_{nn'} \frac{d\omega}{4\pi} =$$

$$b_n \frac{2h\nu^3}{c^2} \frac{1}{e^{h\nu/kT}-1} \left[\frac{N^o_{n'} b_{n'}}{b_{n'}}\right] \left[1 - \exp\left(-\frac{h\nu}{kT}\right)\right] \int a_{n'n} d\nu = b_n B_\nu(T_\varepsilon) \kappa_L'^{o} \tag{52}$$

Hence:

$$\frac{dI_\nu}{ds} = - \left\{\kappa_c' + b_{n'} k_L'^{o} \beta_{nn'}\right\} I_\nu + (\kappa_c' + b_n \kappa_L'^{o}) B_\nu(T_\varepsilon) , \tag{53}$$

which can be put in the form:

$$\frac{d\bar{I}_\nu}{d\bar{\tau}_\nu} = \bar{I}_\nu - \bar{S}_\nu , \tag{54}$$

where we use the bar to indicate that an integration has been performed over the line profile.

$$d\bar{\tau}_\nu = - \left[\kappa_c' + \kappa_L'^{o} b_{n'} \beta_{nn'}\right] ds , \quad \bar{S}_\nu = \eta_\nu B_\nu ,$$

$$\bar{\eta}_\nu = \frac{\kappa_c' + b_n \kappa_L'^{o}}{\kappa_c' + b_{n'} \kappa_L'^{o} \beta_{nn'}} . \tag{55}$$

Now, introducing the line profile function, ϕ_ν, we have:

$$\kappa_\nu^L = b_{n'} \beta_{n'n} \kappa_L'^{o} \phi_\nu . \tag{56}$$

Under conditions actually occurring in gaseous nebulae, the variation of b_n and $\beta_{n'n}$ is such that κ_ν^L is negative! Hence, a masering action will occur.

$$\text{Let } \tau = \tau_c + \tau_L(\nu), \text{ where } d\tau_L(\nu) = \kappa_\nu^L \, ds \ , \tag{57}$$

$$\eta_\nu(L) = [\kappa_c' + b_n \kappa_L'^o \phi_\nu]/[\kappa_c' + b_{n'} \kappa_L'^o \beta_{nn'} \phi_\nu] \ . \tag{58}$$

Then I_ν^L is obtained by computing the total contribution from line plus continuum and subtracting off the continuum.

$$I_\nu^L = \int_0^\infty \eta_\nu (L) B_\nu (s) e^{-\tau} d\tau - \int_0^\infty B_\nu(s) \, e^{-\tau_c} d\tau_c \tag{59}$$

Following Brocklehurst and Seaton (1972) one may introduce further the condition that the source function in the line is much smaller than the source function in the continuum.

What makes the analysis of radio recombination lines tricky is that most gaseous nebulae are heterogeneous structures. The denser part of the nebula makes the most important contribution in both lines and continuum at high frequencies, while the outer less dense regions are the most important contributors at low frequencies. Thus, at a fixed frequency, recombination lines and continua may arise from different zones as may recombination lines of different series $n_1\alpha$, $n_3\gamma$, etc. Of course, recombination lines at different frequencies may be produced in different zones. The observational problem is further exacerbated by the role of the antenna pattern which means that the spatial resolution varies from one frequency to another.

One of the parameters sought is the temperature, T_ε of the H II regions, which is found from the ratio $\int T_L \, d\nu / T_c$ within the framework of a number of assumptions. It is often supposed that the lines are formed and radiative transfer occurs under LTE conditions with no maser action in a plane parallel homogeneous slab. Also, all optical depths are assumed to be small and collisional broadening effects are neglected. Seaton has described a situation for a line n66 in Orion, where T_ε = 10,000°K, N_ε = 5000 and τ = 0.0037. The b_{66} factor is 0.922 but maser action introduces a multiplicative factor of 1.078. The two factors cancel each other out to 0.6%! Thus, Seaton concludes that observations may imitate predictions of LTE theory even when b_n factors, masering action and line broadening effects are important. With certain allowable combinations of ν, N_ε and T_ε, the optical depth in the line can actually become negative. Maser action can then amplify the background continuum, the amplification being greater near the core of the line than in the wings. Thus, the profile will be distorted.

Since the contribution to both line and continuum from each volume element of a heterogeneous nebula depends on T_ε, N_ε and integration path through the nebula, the best procedure is to construct a model. Comparison of model prediction with observation allows modification until a fit is obtained. Often, observational data of sufficient angular resolution or frequency coverage are lacking. There yet remains the problem of uniqueness. Lockman and Brown (1976) have summarized difficulties in getting T_ε from recombination line observations. They conclude that you need a specific knowledge of the density and nebular geometry before the gas kinetic temperature can be obtained from a limited set of recombination line observations.

A favorite object for radio frequency observations has been the Orion nebula. Consider first the continuum observations. Most H II regions are optically thin at all practical frequencies; hence, all we can obtain is a measurement of $\tau_\nu \bar{B}_\nu$ convolved with the antenna pattern. In Orion, however, the nebula is optically thick at 408 MHz (73 cm), the wavelength used in the Mills cross which give a resolution of 2'. Mills and Shaver (1968) combined their observations at 408 MHz with data of similar resolution obtained at 10.5 GHz (2 cm), where the nebula is optically thin. See Figure 4-4. Let subscript a denote quantities pertaining to 408 MHz and b denote those pertaining to 10.5 GHz. Then at each point on the surface of the nebula:

$$I_a = \int B_a(T_\varepsilon)\, e^{-\tau_a}\, d\tau_a, \qquad I_b = \tau_b\, B_b(T_\varepsilon)\,. \tag{60}$$

Since τ(408 MHz)/τ(10.5 GHz) is a known function of T_ε, it is possible to solve for both emission measure, $\int N_\varepsilon^2\, ds$, and T_ε from an isophotal map by a process of iteration, provided we make some assumptions about the structure of the nebula. By this procedure Shaver found T_ε = 8550 ± 500°K (1969), which appears to be in accord with later results by Chaisson and Dopita (1977) and by Pauls and Wilson (1977), who obtained 9000 ± 500°K and 8200 ± 300°K, respectively. They derived T_ε from line to continuum ratios. These results are in harmony with T_ε derived from the [O III] lines (see Chapter 10). Lockman and Brown (1975), however, concluded that an isothermal model would not fit Orion; the densest part of the nebula seems to be coolest. Their model comprised a dense core: i) N_ε = 32,000/cm^3, radius ~ 0.043 pc, T_ε = 7500°K; a shell ii) r ≅ 0.56 pc, N_ε = 3200/cm^3, T_ε = 10,000°K; and envelope iii) N_ε ~ 200/cm^3, r = 2.5 pc, and T_ε = 12,500°K. Optical data do not appear to support the Lockman-Brown model. One cannot represent the total continuum flux in the Orion nebula with a model of constant density. Brocklehurst and Seaton (1972) were able to obtain a satisfactory fit from 0.09 to 11 GHz with a spherically symmetrical model in which N_ε decreased from 10^4 cm^{-3} near the center to about 100 at 1 parsec. Further, by allowing for masering action and for the broad wings found on lines of large n, they were able to fit the observed line profiles, and line-to-continuum ratios as well. The model is not completely satisfactory in that the predicted size and the corresponding volume are therefore eight times too small. Evidently, the material is clumpy; perhaps the bulk of it is concentrated in blobs occupying about 10% of the volume.

Recombination lines produced when singly ionized helium atoms capture electrons are also observed. Line formation processes for He and H are similar, hence, from a comparison of the line fluxes, the He/H ratio can be found provided the radiating volumes are the same. The latter condition is not necessarily fulfilled, especially in dusty, clumpy regions. By concentrating on high-frequency r.f. observations of ionized gas clouds associated with very hot stars, it appears possible to obtain a reliable measurement of n(He)/n(H). The observed ratio derived from radio data appears to be about 0.095 and does not seem to depend on distance from the galactic center. Some optical observers have found their data to indicate a slight dependence of He/H on distance, however.

Recombination lines of heavier elements have also been observed. The transitions differ in frequency because the Rydberg constant changes with atomic weight (see Eq. 26). Also, the observed lines tend to be formed in regions often contiguous with, but not identical to, H II regions, either in planetaries or the interstellar medium. Consider carbon as an illustration in point. The ionization potential of carbon, 11.26 eV, is less than that of H, 13.6 eV. Hence, radiation longward of λ912 escapes from H II regions into a neighboring H I zone. Carbon recombination radiation was discovered first in the H I region behind the Orion nebula and adjacent to the molecular cloud. It has been found subsequently in other objects, where it gives a velocity that agrees with molecular lines rather than with recombination hydrogen lines. Furthermore, the line widths are appropriate to these cooler zones; they are very much narrower than lines from H II regions. Some of these cool neutral zones must have densities as high as $n_H \sim 10^5$ cm^{-3}, if we assume normal C/H abundance ratios. The level of ionization and gas kinetic temperatures is low, the dynamics and chemistry can be complex.

Sometimes C II radiation is found in dense regions containing embedded B stars. These stars are invisible optically but often are surrounded by compact H II regions (observable at radio frequencies) surrounded by a "cocoon" of dust. Stellar radiation heats the dust so that the object is observable as an infrared source. In such a cloud the C^+ region is very narrow, it is cut off abruptly by the CO zone on the outside and by the H II zone on the interior. Throughout most of the C II zone, T ~ 40°K and hydrogen exists as H_2.

4E. The Optical Continuum

In previous sections we have referred to radiative processes involving hydrogen and helium that produce quanta in the continuum. These include: 1) recombination of ions and electrons; 2) free-free emission or bremsstrahlung; and 3) the two-photon emission. Because all other elements are relatively rare, we may confine our attention to hydrogen and helium.

The reality of a continuous nebular spectrum is displayed most strikingly by the confluence of hydrogen lines to form the Balmer continuum (see Fig. 2-1).

At the series limit ~ λ3650, one may measure the emission in a strip Δλ corresponding to an interval Δν in the Balmer continuum. One must estimate the position of the underlying continuum which arises mostly from recombinations on the third and higher levels of H, with some contribution from free-free and two-photon emissions. The thus-obtained I(Bac)Δλ may be compared with I(Hβ). Referring to Eq. 3-38, we see that the general expression for the emission of radiation in an interval ν to ν + dν resulting from recombination of free electrons to the level n in an hydrogenic ion is:

$$E_{\kappa n}\, d\nu = N(H^+)\, N_\varepsilon\, hKZ^4\, T_\varepsilon^{-3/2}\, e^{X_n}\, n^{-3}\, g_{II}\, \exp\left(-\frac{h\nu}{kT_\varepsilon}\right) d\nu\ . \quad (61)$$

For the Balmer continuum, in particular, we have that:

$$E_{\kappa 2}\, d\nu = 2.7\times 10^{-33}\, N\left(H^+\right) N_\varepsilon\, T_\varepsilon^{-3/2}\, g_{II}\, e^{X_2}\, \exp\left(-\frac{h\nu}{kT_\varepsilon}\right) d\nu\ . \quad (62)$$

At the Balmer limit, $h\nu/kT_\varepsilon = X_2$, $g_{II} = 0.88$ and we get:

$$E_{BAC}\,\Delta\nu = 2.37 \times 10^{-33}\, N(H^+)\, N_\varepsilon\, T_\varepsilon^{-3/2}\, \Delta\nu\ , \qquad (63)$$

whence:

$$\frac{E_{BAC}\,\Delta\nu}{E(H\beta)} = 2.37 \times 10^{-8} \frac{\Delta\nu}{E^{o}_{4,2}(T_\varepsilon)\, T_\varepsilon^{3/2}} = 5.81 \times 10^{-22} \frac{\Delta\nu}{\alpha(H\beta)\, T_\varepsilon^{3/2}}\ . \qquad (64)$$

This ratio depends on T_ε and slowly on N_ε at high densities so it could give at least a consistency check on T_ε determinations by other methods.

To calculate the total bound-free and free-free contribution to the continuum, we must include all recaptures that can produce emission at a frequency, ν. We sum Eq. (61) over all levels for which $\nu_n = hRZ^2/n^2 < \nu$ and add the free-free emission which is given by Eq. III-64. We thus obtain:

$$E(bf + ff) = N(H^+)\, N_\varepsilon \frac{hKZ^4}{T_\varepsilon^{3/2}} e^{-h\nu/kT_\varepsilon} \{\sum_{n_m}^{\infty} \frac{g_{II}}{n^3} e^{X_n} + \frac{g_{III}\, kT_\varepsilon}{2h\, RZ^2}\}\ , \qquad (65)$$

which is valid for hydrogen and ionized helium.

Following Brown and Mathews (1970) we define an emission coefficient γ(ergs cm^3 Hz^{-1} s^{-1}) by the expression:

$$E\, d\nu = N(\text{ion})\, N_\varepsilon\, \gamma\, d\nu\ . \qquad (66)$$

Now, γ is composed of several terms and if we define γ_T by the expression:

$$E_c\, d\nu = N(H^+)\, N_\varepsilon\, \gamma_T\, d\nu\ , \qquad (67)$$

where E_c is the continuum emission per cm^3 in all directions:

$$\gamma_T = \gamma(\text{H I}) + \gamma(2q) + \gamma(\text{He I})\, N(He^+)/N(H^+)$$

$$+\ \gamma(\text{He II})\, N(He^{++})/N(H^+)\ . \qquad (68)$$

Note that γ(H I) is calculated directly from Eq. (65) by setting Z = 1, while γ(He II) is obtained by setting Z = 2. In the optical region, the Brackett, Pfund, and higher series all contribute.

In the two-photon emission process a hydrogen atom escapes from the 2s level with the emission of two photons of frequency:

$$\nu' = y\nu(2s - 1s) \text{ and } \nu'' = (1 - y)\, \nu(2s - 1s)\ , \qquad (69)$$

where $\nu(2s - 1s) \sim \nu(Ly\alpha)$. Let $P_{2q}(\nu)$ be the probability of emission of a photon between ν and $\nu + d\nu$. Consider the limiting situation of low density and Menzel and Baker Case B. Then every atom that reaches the 2s level will emit the two-photon continuum. The total number of recaptures resulting in atoms reaching the n = 2 level will be $\alpha_B(T_\varepsilon)N(N^+)N_\varepsilon$ where $\alpha_B(T_\varepsilon)$ is given by Eq. III-45. The effective recombination coefficients for the 2s level may be denoted as $X_o\ \alpha_B$, where X_o is of the order of 0.3. Thus, the intensity in the low-density limit will be given by:

$$\gamma_\nu^o(2q)\ d\nu = \alpha_B\ X_o\ 2\ h\nu\ P_{2q}(\nu)d\nu\ . \tag{70}$$

The factor 2 arises because two photons are created in each event. Note that although $P_{2q}(\nu)$ is symmetrical about $\nu_{2q}^c = 1/2\nu(\mathrm{Ly}\alpha)$, corresponding to $\lambda 2421$, $\gamma_\nu^o(2q)$ shows a skewed distribution with a maximum at $\nu = 1.85 \times 10^{15}$, corresponding to $\lambda 1620$.

The density of the plasma has a profound effect on the 2-photon emission. At moderate densities, atoms may escape to the 2p level by collisions with protons and electrons. If the number of H atoms escaping from the 2s level by collisions with electrons and protons are $N(2s)\ N_\varepsilon\ q_\varepsilon(2s - 2p)$ and $N(2s)N(H^+)q_p(2s - 2p)$, respectively, where q_p and q_e denote the rate coefficients for 2s - 2p collisional interchanges; then only the fraction:

$$\beta_{2q} = \frac{A_{2q}}{A_{2q} + N_\varepsilon q_e + N(H^+)q_p} = \frac{1}{1 + N\ r(t)} \tag{71}$$

of atoms reaching the 2s level will escape by the two-photon process. Proton collisions are about an order of magnitude more effective than electron collisions. Thus, $q_p(2s\text{-}2p)$ and $q_\varepsilon(2s\text{-}2p) = 4.74 \times 10^{-4}$ and $0.57 \times 10^{-4}\ \mathrm{cm^3\ s^{-1}}$, respectively, at 10,000°K. They decrease slowly with temperature, but it is sufficiently accurate to use the values at 10,000°K for most purposes. Hence:

$$\gamma(2q)\ d\nu = \beta_{2q}\ \gamma_o\ d\nu\ . \tag{72}$$

Note that the Balmer discontinuity, $F(4650^+)/F(4650^-)$ will depend on the contribution of the two-photon emission which tends to be suppressed in nebulae of higher density, $N_\varepsilon > 10^5$.

In optically thick nebulae, more complicated processes can occur. As a consequence of repeated scaterings of Lyα quanta, a sizable population can be built up in the 2p level and collisional transfers from 2p to 2s can occur. In fact, the production of two-photon emissions could become an important destructive agent for the Lyα quanta if the number of scatterings is very large. Thus, in the limiting situation, X would not only exceed X_o but could actually approach 1; i.e., every recombination would ultimately produce two continua quanta. Actually, such an extreme situation rarely occurs. What is more likely to happen is that Lyα quanta will be destroyed by dust grain absorption or by escape from the nebula. The problem of the two-photon emission under conditions of large optical depth has been treated by Cox and Mathews (1969); Drake and Ulrich (1981) have discussed the problem under conditions of high density.

Table 4-9

Coefficients for Continuous Absorption

in units of 10^{-40} ergs cm^3 s^{-1} Hz^{-1} T_ε = 5000°K

λ(A)	$\nu \times 10^{-14}$	γ_H	γ_{2q}	$\gamma_{He\ I}$	$\gamma_{He\ II}$
1000.0	29.9790	0.000	0.000		0.000
1200.0	24.9830	0.000	0.000		0.017
1300.0	23.0610	0.000	9.122		0.106
1400.0	21.4140	0.000	13.391		0.509
1500.0	19.9860	0.001	14.974		1.988
1600.0	18.7370	0.003	15.411	0.228	6.539
1800.0	16.6550	0.023	14.951	0.925	47.489
2051.0	14.6170	0.159	13.711	3.510	329.556
2053.0	14.6030	0.161	13.700	3.549	0.340
2200.0	13.6270	0.408	12.921	6.283	0.863
2400.1	12.4910	1.199	11.892	12.104	2.548
2599.4	11.5330	2.970	10.970	19.442	6.343
2600.8	11.5270	2.987	10.964	5.048	6.379
2725.4	11.0000	4.917	10.442	12.553	10.531
2855.2	10.5000	7.884	9.912	19.675	16.937
2997.9	10.0000	12.633	9.382	26.796	27.229
3121.4	9.6044	18.334	8.956	32.430	39.633
3122.0	9.6026	18.365	8.954	27.667	39.700
3331.0	9.0000	32.354	8.301	53.754	70.282
3421.4	8.7623	40.435	8.044	64.044	88.022
3422.0	8.7607	40.496	8.042	7.223	88.155
3527.0	8.5000	51.698	7.757	9.832	112.818
3642.0	8.2315	66.459	7.460	12.519	145.422
3648.0	8.2180	0.327	7.445	12.654	5.320
3679.2	8.1483	0.349	7.368	13.052	5.685
3679.9	8.1467	0.350	7.367	0.406	5.694
4000.0	7.4948	0.649	6.646	0.759	10.576
4282.8	7.0000	1.038	6.091	1.363	16.912
4499.9	6.6622	1.429	5.710	1.775	23.292
4996.5	6.0000	2.670	4.967	4.278	43.577
5096.0	5.8829	2.982	4.838	4.721	48.673
5450.8	5.5000	4.277	4.415	6.169	69.850
5695.8	5.2634	5.342	4.153	7.063	87.285
5700.0	5.2595	5.361	4.149	7.096	14.175
5995.9	5.0000	6.839	3.862	9.411	18.124
6633.8	4.5192	10.723	3.356	13.699	28.556
6635.8	4.5178	10.737	3.355	12.413	28.594
6999.9	4.2828	13.366	3.110	15.378	35.693
7438.7	4.0302	16.901	2.846	18.565	45.284
7441.1	4.0289	16.921	2.845	18.036	45.339
7848.0	3.8200	20.532	2.627	20.926	55.182
7850.4	3.8188	20.555	2.626	14.142	55.244
8193.3	3.6590	23.821	2.463	16.043	64.187
8196.2	3.6577	23.850	2.462	8.038	64.266
8196.7	3.6575	23.854	2.462	8.040	64.278
8198.5	3.6567	23.872	2.461	5.516	64.326
8207.0	3.6529	23.956	2.457	5.534	64.556
8209.0	3.6520	3.418	2.456	5.538	21.786
8265.4	3.6271	3.499	2.433	5.655	22.305
8268.1	3.6259	3.503	2.432	3.400	22.330
8499.9	3.5270	3.845	2.338	3.846	24.517
9000.1	3.3310	4.624	2.152	4.729	29.498
9500.0	3.1557	5.450	1.986	5.519	34.796
10000.1	2.9979	6.319	1.836	6.230	40.364

Table 4-9

Coefficients for Continuous Absorption (Continued)

in units of 10^{-40} ergs cm^3 s^{-1} Hz^{-1} T_ε = 10,000°K

λ(A)	$\nu \times 10^{-14}$	γ_H	γ_{2q}	$\gamma_{He\ I}$	$\gamma_{He\ II}$
1000.0	29.9790	0.001	0.000	0.046	0.082
1200.0	24.9830	0.009	0.000	0.079	0.880
1300.0	23.0610	0.023	5.611	0.126	2.196
1400.0	21.4140	0.051	8.236	0.210	4.800
1500.0	19.9860	0.100	9.210	0.400	9.448
1600.0	18.7370	0.181	9.478	0.869	17.069
1800.0	16.6550	0.486	9.196	1.650	45.649
2051.0	14.6170	1.272	8.433	3.895	119.142
2053.0	14.6030	1.280	8.426	3.926	2.996
2200.0	13.6270	2.026	7.947	6.087	4.760
2400.1	12.4910	3.453	7.314	9.014	8.150
2599.4	11.5330	5.404	6.747	11.861	12.817
2600.8	11.5270	5.419	6.743	7.156	12.853
2725.4	11.0000	6.928	6.423	9.976	16.400
2855.2	10.5000	8.740	6.096	12.651	20.857
2997.9	10.0000	11.020	5.770	15.326	26.387
3121.4	9.6044	13.231	5.508	17.443	31.775
3122.0	9.6026	13.242	5.507	15.873	31.802
3331.0	9.0000	17.477	5.106	20.976	42.181
3421.4	8.7623	19.490	4.947	22.989	47.142
3422.0	8.7607	19.505	4.946	4.321	47.177
3527.0	8.5000	21.976	4.771	4.902	53.289
3642.0	8.2315	24.841	4.588	5.501	60.400
3648.0	8.2180	1.387	4.579	5.532	10.742
3679.2	8.1483	1.434	4.532	5.687	11.102
3679.9	8.1467	1.435	4.531	1.450	11.110
4000.0	7.4948	1.950	4.088	2.070	15.107
4282.8	7.0000	2.461	3.746	2.807	19.067
4499.9	6.6622	2.883	3.512	3.310	22.344
4996.5	6.0000	3.929	3.055	4.943	30.465
5096.0	5.8829	4.150	2.975	5.231	32.177
5450.8	5.5000	4.959	2.715	6.176	38.463
5695.8	5.2634	5.534	2.554	6.759	42.934
5700.0	5.2595	5.544	2.552	6.771	17.101
5995.9	5.0000	6.252	2.375	7.583	19.321
6633.8	4.5192	7.801	2.064	9.086	24.210
6635.8	4.5178	7.806	2.063	8.661	24.226
6999.9	4.2828	8.693	1.913	9.391	27.042
7438.7	4.0302	9.754	1.751	10.176	30.426
7441.1	4.0289	9.760	1.750	10.000	30.445
7848.0	3.8200	10.729	1.616	10.607	33.555
7850.4	3.8188	10.735	1.615	8.081	33.574
8193.3	3.6590	11.537	1.515	8.497	36.161
8196.2	3.6577	11.544	1.514	5.880	36.183
8196.7	3.6575	11.545	1.514	5.880	36.186
8198.5	3.6567	11.549	1.514	5.041	36.200
8207.0	3.6529	11.569	1.511	5.048	36.263
8209.0	3.6520	4.317	1.511	5.050	21.162
8265.4	3.6271	4.368	1.496	5.098	21.411
8268.1	3.6259	4.371	1.496	4.360	21.423
8499.9	3.5270	4.578	1.438	4.596	22.443
9000.1	3.3310	5.020	1.324	5.065	24.609
9500.0	3.1557	5.449	1.221	5.483	26.718
10000.1	2.9979	5.867	1.129	5.860	28.766

Table 4-9

Coefficients for Continuous Absorption (Continued)

in units of 10^{-40} ergs cm^3 s^{-1} Hz^{-1} T_ε = 15,000°K

λ(A)	$\nu \times 10^{-14}$	γ_H	γ_{2q}	$\gamma_{He\ I}$	$\gamma_{He\ II}$
1000.0	29.9790	0.016	0.000	0.212	0.546
1200.0	24.9830	0.080	0.000	0.366	2.655
1300.0	23.0610	0.147	4.212	0.488	4.869
1400.0	21.4140	0.248	6.182	0.690	8.181
1500.0	19.9860	0.389	6.913	0.990	12.817
1600.0	18.7370	0.576	7.115	1.509	18.966
1800.0	16.6550	1.107	6.903	2.375	36.369
2051.0	14.6170	2.093	6.330	4.281	68.548
2053.0	14.6030	2.102	6.325	4.304	5.382
2200.0	13.6270	2.848	5.965	5.891	7.317
2400.1	12.4910	4.049	5.490	7.702	10.454
2599.4	11.5330	5.441	5.065	9.195	14.112
2600.8	11.5270	5.451	5.062	6.608	14.139
2725.4	11.0000	6.408	4.821	8.038	16.670
2855.2	10.5000	7.467	4.576	9.395	19.484
2997.9	10.0000	8.695	4.331	10.753	22.766
3121.4	9.6044	9.804	4.135	11.826	25.743
3122.0	9.6026	9.809	4.134	10.961	25.757
3331.0	9.0000	11.766	3.833	13.048	31.045
3421.4	8.7623	12.636	3.714	13.870	33.411
3422.0	8.7607	12.642	3.713	3.645	33.428
3527.0	8.5000	13.667	3.581	3.957	36.227
3642.0	8.2315	14.805	3.444	4.279	39.350
3648.0	8.2180	2.028	3.437	4.295	12.304
3679.2	8.1463	2.073	3.402	4.376	12.575
3679.9	8.1467	2.074	3.401	2.052	12.582
4000.0	7.4948	2.544	3.068	2.620	15.430
4282.8	7.0000	2.969	2.812	3.217	18.007
4499.9	6.6622	3.298	2.636	3.625	20.005
4996.5	6.0000	4.052	2.293	4.706	24.571
5096.0	5.8829	4.202	2.233	4.898	25.478
5450.8	5.5000	4.730	2.038	5.523	28.676
5695.8	5.2634	5.088	1.917	5.909	30.842
5700.0	5.2595	5.094	1.915	5.916	16.784
5995.9	5.0000	5.517	1.783	6.355	18.206
6633.8	4.5192	6.390	1.550	7.168	21.159
6635.8	4.5178	6.393	1.549	6.935	21.169
6999.9	4.2828	6.867	1.436	7.291	22.778
7438.7	4.0302	7.412	1.314	7.673	24.641
7441.1	4.0289	7.415	1.313	7.575	24.651
7848.0	3.8200	7.896	1.213	7.854	26.302
7850.4	3.8188	7.899	1.212	6.435	26.312
8193.3	3.6590	8.285	1.137	6.644	27.648
8196.2	3.6577	8.289	1.137	5.205	27.659
8196.7	3.6575	8.289	1.137	5.205	27.660
8198.5	3.6567	8.291	1.136	4.740	27.667
8207.0	3.6529	8.301	1.134	4.746	27.700
8209.0	3.6520	4.355	1.134	4.747	19.483
8265.4	3.6271	4.389	1.123	4.783	19.637
8268.1	3.6259	4.391	1.123	4.380	19.644
8499.9	3.5270	4.531	1.079	4.534	20.267
9000.1	3.3310	4.821	0.994	4.838	21.560
9500.0	3.1557	5.097	0.917	5.110	22.784
10000.1	2.9979	5.358	0.848	5.355	23.945

Table 4-9

Coefficients for Continuous Absorption (Continued)

in units of 10^{-40} ergs cm^3 s^{-1} Hz^{-1} T_ε = 20,000°K

λ(A)	$\nu \times 10^{-14}$	γ_H	γ_{2q}	$\gamma_{He\ I}$	$\gamma_{He\ II}$
1000.0	29.9790	0.067	0.000	0.378	1.299
1200.0	24.9830	0.218	0.000	0.653	4.236
1300.0	23.0610	0.344	3.383	0.860	6.664
1400.0	21.4140	0.507	4.966	1.170	9.817
1500.0	19.9860	0.710	5.552	1.580	13.724
1600.0	18.7370	0.952	5.714	2.150	18.382
1800.0	16.6550	1.551	5.544	3.100	29.860
2051.0	14.6170	2.494	5.084	4.667	47.852
2053.0	14.6030	2.502	5.080	4.682	6.845
2200.0	13.6270	3.137	4.791	5.695	8.612
2400.1	12.4910	4.077	4.410	6.769	11.243
2599.4	11.5330	5.078	4.068	7.577	14.068
2600.8	11.5270	5.085	4.066	5.915	14.087
2725.4	11.0000	5.734	3.872	6.760	15.931
2855.2	10.5000	6.424	3.675	7.562	17.898
2997.9	10.0000	7.192	3.479	8.363	20.102
3121.4	9.6044	7.861	3.321	8.998	22.032
3122.0	9.6026	7.864	3.320	8.441	22.041
3331.0	9.0000	8.997	3.078	9.529	25.334
3421.4	8.7623	9.484	2.983	9.958	26.760
3422.0	8.7607	9.487	2.982	3.370	26.769
3527.0	8.5000	10.050	2.876	3.583	28.423
3642.0	8.2315	10.661	2.766	3.801	30.229
3648.0	8.2180	2.359	2.761	3.812	12.658
3679.2	8.1483	2.398	2.732	3.869	12.867
3679.9	8.1467	2.399	2.732	2.370	12.871
4000.0	7.4948	2.797	2.464	2.850	14.998
4282.8	7.0000	3.141	2.258	3.331	16.838
4499.9	6.6622	3.400	2.117	3.660	18.219
4996.5	6.0000	3.969	1.842	4.436	21.253
5096.0	5.8829	4.079	1.794	4.573	21.838
5450.8	5.5000	4.459	1.637	5.022	23.861
5695.8	5.2634	4.711	1.540	5.300	25.200
5700.0	5.2595	4.715	1.538	5.304	16.070
5995.9	5.0000	5.007	1.432	5.586	17.086
6633.8	4.5192	5.594	1.245	6.109	19.137
6635.8	4.5178	5.596	1.244	5.960	19.144
6999.9	4.2828	5.906	1.153	6.176	20.233
7438.7	4.0302	6.257	1.055	6.409	21.471
7441.1	4.0289	6.258	1.055	6.350	21.477
7848.0	3.8200	6.563	0.974	6.519	22.557
7850.4	3.8188	6.565	0.974	5.590	22.563
8193.3	3.6590	6.807	0.913	5.729	23.425
8196.2	3.6577	6.809	0.913	4.810	23.432
8196.7	3.6575	6.809	0.913	4.810	23.433
8198.5	3.6567	6.810	0.913	4.500	23.437
8207.0	3.6529	6.816	0.911	4.505	23.458
8209.0	3.6520	4.254	0.911	4.506	18.123
8265.4	3.6271	4.280	0.902	4.539	18.231
8268.1	3.6259	4.281	0.902	4.270	18.237
8499.9	3.5270	4.385	0.867	4.382	18.675
9000.1	3.3310	4.599	0.798	4.604	19.576
9500.0	3.1557	4.800	0.736	4.802	20.418
10000.1	2.9979	4.988	0.681	4.980	21.209

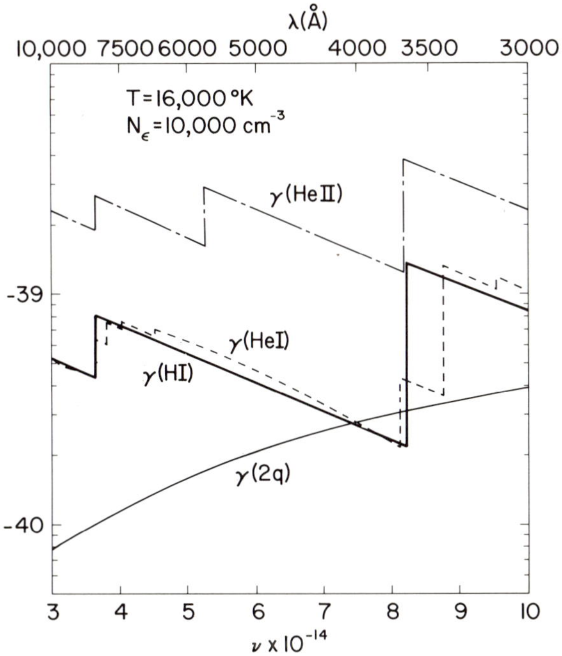

Figure 4-5

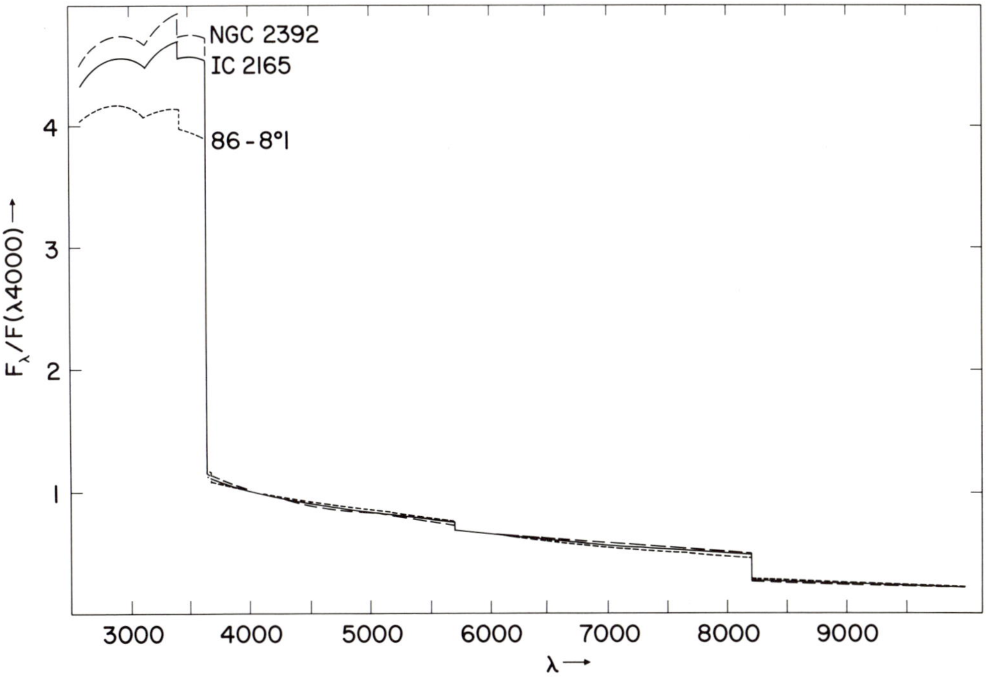

Figure 4-6

Helium is the only other element sufficiently abundant to contribute to nebular continua. The He I continuum is somewhat more complicated than that of H I because the levels are nonhydrogenic; i.e., you must consider the ℓ degeneracy. Brown and Mathews (1969) write:

$$\gamma(\text{He I}) = \sum_{n=n_o}^{\infty} \sum_{\ell=o}^{n-1} \gamma_{n\ell}(\nu) + \gamma_{ff}(\nu) . \tag{73}$$

For each (n,ℓ) level, one employs an expression of the form:

$$\gamma_{n\ell} = \text{const} \frac{\omega(n,\ell)}{T_\varepsilon^{3/2}} \nu^3 e^{X_{n\ell}} \alpha_{n\ell}(\nu) e^{-h\nu/kT_\varepsilon} , \tag{74}$$

where $\alpha_{n\ell}(\nu)$ is the atomic absorption coefficient that can be evaluated by equations given by Peach (1967). Table 4-9 gives the values of γ(H I), γ_o(2q,H), γ(He I), and γ(He II) for T_ε = 7500°K, 10,000, 15,000 and 20,000 in units of 10^{-40} erg cm^3 s^{-1} (Hz)$^{-1}$. Figure 4-5 displays the frequency variation of the separate continuous emission coefficients: γ(H I), γ(He II), and γ_o(2q,H) for T_ε = 16,000°K and N_ε = 10,000 cm^{-3}. Figure 4-6 shows the predicted energy fluxes in the optical region for several high-excitation planetaries.

In summary, we have described how the continuum extending from the near infrared to the ultraviolet, and arising from atomic processes, may be calculated as a function of N_ε, T_ε, $n(He^+)/n(H^+)$, $n(He^{++})/n(H^+)$. The shape of the continuum and its absolute flux gives independent diagnostics on N_ε and T_ε. In the radio frequency range, Sec. 4D, the emission is dominated by free-free emission but in the far infrared, 2 to 100 μm, atomic processes are often overwhelmed by thermal emission from solid grains. In diffuse nebulae, a background continuum may arise from scattered and reflected starlight superposed on a continuum arising from the above-described atomic processes. The contribution of the latter may be estimated from the flux in nebular Hβ since both are produced in the same volume element and are proportional to $N(H^+)N_\varepsilon$. Similar considerations may be employed in small objects where a nebula and its exciting star cannot be resolved. We calculate the nebular continuum from the Balmer line intensities and subtract it from the total flux to obtain the contribution of the underlying star. Frequently, the nebular continuum dominates in the optical region, the stellar in the UV.

4F. Hydrogen Plasmas at Moderately High Densities and Optical Thicknesses

If an emitting plasma is sufficiently optically thick, one might expect that the energy density in Lyα could build up enough by repeated scattering that the population in the 2p level would become appreciable and absorptions from this level could become significant. We can envisage such a situation occurring in an H II region surrounded

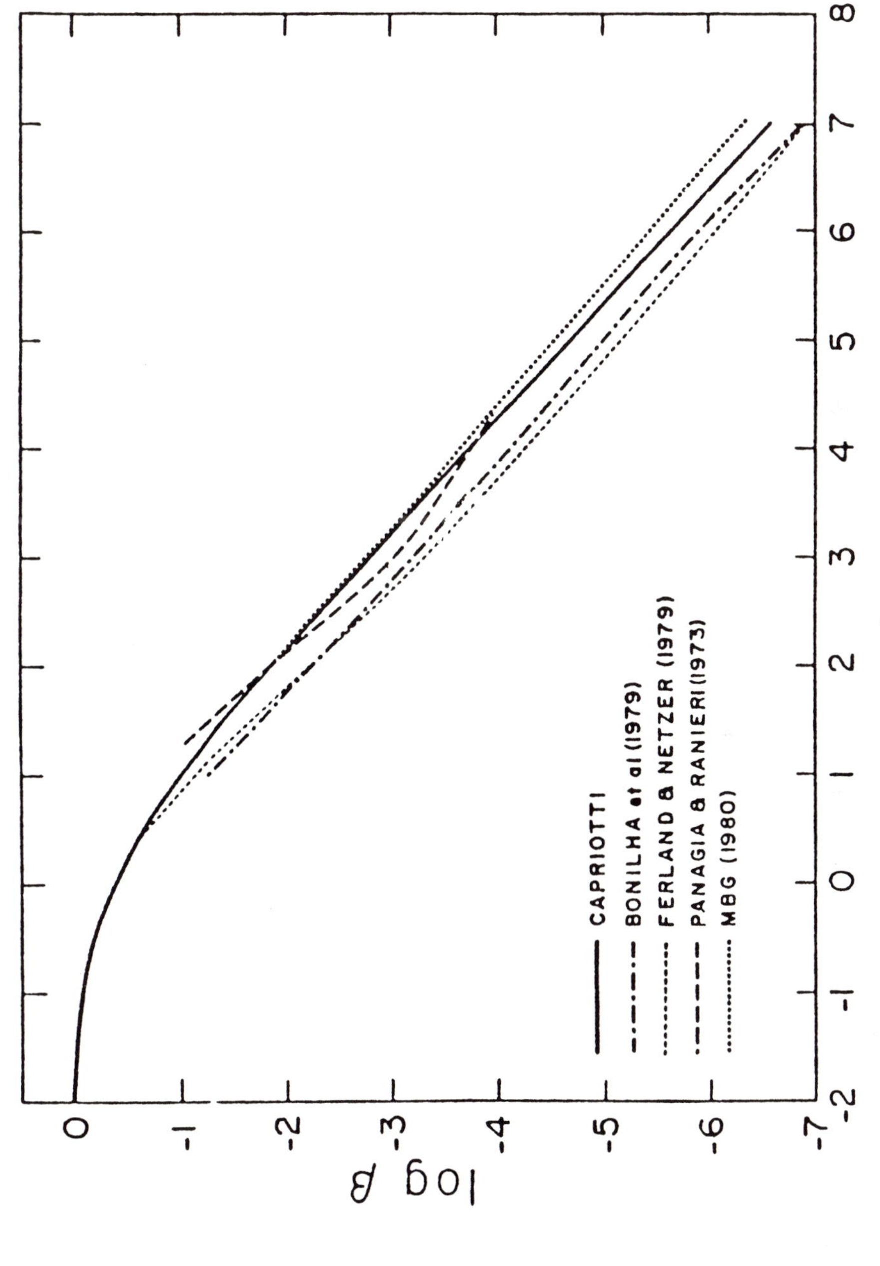
CAPRIOTTI
BONILHA et al (1979)
FERLAND & NETZER (1979)
PANAGIA & RANIERI (1973)
MBG (1980)
log τ
log β
-2
-1
0
1
2
3
4
5
6
7
8
0
-1
-2
-3
-4
-5
-6
-7

Figure 4-7

by a thick shell of neutral atomic hydrogen. Actually, the density of Lyα quanta is limited by the possibility of their escape, by dust absorption and by degradation into two-Photon emission. Atoms in the 2p level may undergo collisional transitions to the 2s from which they escape to the ground level with continuum radiation.

Consider first the escape of a quantum of Lyman α. A consistent procedure would be to set up appropriate equations of transfer for line and continua, employ proper line profiles and realistic assumptions about coherent versus noncoherent scattering, and solve for a spherically symmetrical or plane-parallel slab geometry. Differential motions within the shell of geometrical irregularities would complicate the calculations badly.

Another and more practical approach is to calculate the escape probability β that a photon can get out of the nebula without further scattering or absorption. We would also like to obtain the relation between β and the optical depth τ at the center of the line. The results depend on the geometry and whether there are differential motions. For example, the escape probability from a sphere is about three times that from a slab of the same τ value, provided τ is very large.

The shape of the line profile plays a decisive role. Suppose it were square! In order to escape, a photon would have to proceed by a random walk. If R was the radius of the sphere and τ_o the optical depth, the photon would move a distance $s = (R/\tau_o)$ between scatterings and would require $(R/s)^2 = \tau_o^2$ scatterings in order to escape! Actually, the line shape is Gaussian with a possible Lorentz wing at very high densities. Random walk processes are unimportant; what matters are single scattering events in which a photon finds itself displaced from ν_o by a frequency shift, $\Delta\nu$, so large that $\tau(x_1) = \tau_o \exp\left(-x_1^2\right) < 1$; here $x = \Delta\nu/\Delta\nu_o$. Then the photon escapes. We may neglect the spatial excursions up to the point of liberation. We assume a complete redistribution of frequency in each scattering. If we suppose that the photon escapes if $\tau(x_1) \leqslant 1$, otherwise not, then $x_1 = \sqrt{\ln \tau_o}$ defines the escape edge.

If $\phi(x)$ is the normalized profile function, the total probability of escape will be actually $\phi(x)$ multiplied by the escape probability, $\exp(-\tau[x])$. Thus, for a Gaussian profile:

$$\beta = \int \phi(x) \exp[-\tau(x)]\, dx = \frac{1}{\sqrt{\pi}} \int \exp\left(-x^2 - \tau_o e^{-x^2}\right) dx\,. \qquad (75)$$

Approximately, $$\beta \sim \frac{1}{1.9\sqrt{\ln \tau_o}} \frac{1}{\tau_o}\,. \qquad (76)$$

The number of scatterings before the photon escapes is $Q = 1/\beta = K_1\tau_o$ where $K_1 = 1.9\sqrt{\ln \tau_o} \sim 6$ if $\tau_c \sim 10^4$ to 10^5. In the same interval of time, random walk has moved the photon $\sim 3R/\sqrt{\tau_o}$. Between scattering events a photon moves a distance of the order R/τ_o; hence, in Q scatterings it moves a distance of the order of $K_1 \tau_o(R/\tau_o) \sim K_1 R$. The total flight time is about K_1R/c or six or seven times longer than it would have required to escape from the plasma directly. Under these circumstances, radiation pressure is not enhanced by a large factor. If $\tau_o \geqslant 10^5$, the Lorentz wings in the profile must be considered; the escape probability is thereby increased.

The relation between β and τ_o will depend on the geometry, any expansion of the plasma, and the assumed shape of the profile function. Since the pioneering calculations by Capriotti (1965) a number of investigators have examined this problem (see Fig. IV-7). Although the functional relationship is a smooth one, different geometries and assumptions may give for a specified τ_o half an order of magnitude spread in β! As a consequence, when the b_n factors are calculated for an optically thick slab, the results will depend on our choice of the $\beta - \tau_o$ relationship. For example, for $\tau(Ly\alpha) = 10^6$ Drake finds that b(2) and b(3) computed from the Mathews, Blumenthal and Grandi (1980) relationship may be 40% smaller than that found from Capriotti's original formulation. The effect on the $H\alpha/H\beta/H\gamma$ ratio is smaller.

At low densities, level populations tend to values fixed by the photoionization and recombination rates as indicated by the elementary theory and are independent of the precise form of the $\beta - \tau$ relation, while at high densities, collisional rather than radiative processes determine the level population. At intermediate densities, optical depth effects and collisional excitation processes combined to overpopulate the n = 2 and n = 3 levels strongly. The exact form of the $\beta - \tau$ relationship may be more significant here, but Drake finds that the dependence of b_n on N_ε is not strongly influenced by the precise $\beta - \tau$ relationship. For most purposes, one may use the Capriotti function.

Problem 1.

1) Let X_o be the fraction of recombinations to excited levels that directly or by cascade reach the 2^2S level. Show that if self-absorption occurs only from the ground level and all collisions are ignored except those between 2^2S and 2^2P, then X_o in Eq. (70) must be replaced by:

$$X = \frac{X_o}{1 + N_\varepsilon r} + \frac{QA_{2q} N_\varepsilon r(1 - X_o + N_\varepsilon r)}{3A_{21}(1 + N_\varepsilon r)^2} . \tag{77}$$

Here $A_{21} = 6.265 \times 10^8$ sec^{-1} is the probability for Lyα emission. Q, here assumed much larger than 1, is the mean number of scatterings for a Lyα photon; $r = q(T)/A_2$, where q(T) is the total rate coefficient for both electron and proton encounters for collisional transitions from $2^2S_{1/2}$ to 2^2P; also, $N_\varepsilon \ll 10^{12}$ cm^{-3} (Brown and Mathews, 1970, Ap. J., 160, 939).

Problem 2.

If radio frequency lines are widened by pure Doppler broadening, show that if the gas is in local thermodynamic equalibrium, the electron temperature T_ε^o may be found from:

$$\frac{\Delta\nu_L \, T_L}{T_c} = \frac{1.908 \times 10^4 \, \nu \, e^{X_n} \, N(H^+)/N_\varepsilon}{T_\varepsilon[1.5 \ln T^o_\varepsilon - \ln(20.18 \, \nu)]} \tag{78}$$

where:

$\Delta\nu_L$ is the full width of the line at half intensity (in equivalent radial velocity units);

T_c = Temperature from the continuum;

T_L = Temperature from the line.

(Shaver, McGee, Newton, Danks and Pottasch, 1983, M.N.R.A.S.)

Problem 3.

Show that in radio-frequency measurements corrections for nonLTE effects may be made by the relation:

$$\left(\frac{T_\varepsilon}{T^o_\varepsilon}\right)^{1.15} = b_n\left\{1 + \frac{1}{T_\varepsilon}\left[\frac{kT_\varepsilon}{h\nu}\frac{\Delta b}{b_n} - 1\right]\left[\frac{T_c}{2} + T_B + T^*_G\right]\right. \tag{79}$$

where $\Delta b = b_{n+1} - b_n$.

T_c = true continuum brightness temperature of the H II

T_B = 2.67°K = cosmic microwave background brightness temperature. Assume that all galactic nonthermal radiation whose brightness temperature = T^*_G originates behind the source (cf. Shaver et al. 1983).

References to Chapter 4

Early investigations of the Balmer decrement by Plaskett (1928), Carroll (1930), and Cillie (1932) culminated in the studies by:

Menzel, D.H., and Baker, J.B. 1937, Ap. J., 86, 70; 1938, 88, 52.

Further contributions were made by:

Seaton, M.J. 1959, M.N.R.A.S., 119, 90; 1964, 127, 177.
Pengelly, R.M., and Seaton, M.J. 1964, M.N.R.A.S., 127, 145.

The standard theory of Balmer decrement for H and He II is that by:

Brocklehurst, M. 1971, M.N.R.A.S., 153, 471.

whose results are very similar to those derived by W. Clarke (1965), which are described in Stars and Stellar Systems, 7, Nebulae and Interstellar Matter, 504, ed. B. Middlehurst and L. Aller, Chicago, Univ. of Chicago Press. See also:

References to Chapter 4 (Continued)

Brocklehurst, M., and Seaton, M.J. 1972, M.N.R.A.S., 157, 179.
Gerola, H., and Panagia, N. 1968, Astrophys. Space Sci., 2, 285; 1970, 8, 120.

Theoretical calculations for helium are given by:

Brocklehurst, M. 1970, M.N.R.A.S., 157, 211.
Robbins, R.R. 1968, Ap. J., 151, 497, 511; 1970, 160, 519.
Robbins, R.R., and Bernat, A.P. 1974, Ap. J., 188, 309.

Balmer decrement measurements have been made by many observers. See, e.g.:

Miller, J.S. 1971, Ap. J. Lett., 165, L101.
Lee, P., et al. 1969, Ap. J., 155, 853.

The theory of continuous thermal radio-frequency radiation from H II regions and gaseous nebulae is given by many workers. Some representative papers are:

Terzian, Y. 1974, Vistas in Astronomy, 16, 279.
Mezger, P., and Henderson, A.O. 1967, Ap. J., 147, 471.
Schraml, J., and Mezger, P.G. 1969, Ap. J., 156, 269.

The fundamental theory for formation of radio-frequency lines taking non-LTE effects into account was due to:

Goldberg, L. 1966, Ap. J., 144, 1225; see also general review by:
Brown, R.L., Lockman, F.J., and Knapp, G.R. 1978, Ann. Rev. Astron. Astrophys., 16, 445.

Atomic physics processes relavant to the radio frequency region:

Oster, L.F. 1961, Rev. Mod. Phys., 33, 525 (free-free processes).
Menzel, D.H. 1969, Ap. J. Suppl., 18, 221 (f-values for H).
Seaton, M.J. 1972, Comments on Atomic and Molecular Physics, 3, 107 (density broadening of r.f. lines).
Brocklehurst, M., and Leeman, S. 1971, Astrophys. Lett., 9, 35.

Some illustrative applications to determinations of T_ε in Orion are given by:

Mills, B.Y., and Shaver, P. 1967, Australian Jl. Phys., 21, 95.
Chaisson, E.J., and Dopita, M.A. 1977, Astron. Astrophys., 56, 385.
Pauls, T., and Wilson, T.L. 1977, Astron. Astrophys., 60, L31.
Lockman, F.J., and Brown, R.L. 1976, Ap. J., 207, 336; see also 1975, Ap. J., 201, 134.

References to Chapter 4 (Continued)

The optical continuum of gaseous nebulae was investigated by a number of workers; the importance of the two-photon emission was first pointed out by:

Spitzer, L., and Greenstein, J.L. 1951, Ap. J., 114, 407.

The basic reference on nebular optical continua is:

Brown, R.L., and Mathews, W.G. 1970, Ap. J., 160, 939.

See also: Peach, G. 1967, Mem. R.A.S., 71, 13; Cox, D.P., and Mathews, W.G. 1969, Ap. J., 155, 859, and Drake, S.A., and Ulrich, R.K., 1981, Ap. J., 248, 380.

An observational check on lines and continuum in NGC 7027 is given by:

Miller, J.S., and Mathews, W.G. 1972, Ap. J., 172, 593.

The theory of the emission line spectrum of an optically thick nebula at moderate-to-high densities is discussed, e.g., by:

Capriotti, E.R. 1965, Ap. J., 142, 1101.
Panagia, N., and Ranieri, M. 1973, Astron. Astrophys., 24, 219.
Ferland, G., and Netzer, H. 1979, Ap. J., 229, 274.
Mathews, W.G., Blumenthal, G.R., and Grandi, S.A. 1980, Ap. J., 235, 971.
Drake, S.A., and Ulrich, R.K. 1980, Ap. J. Suppl., 42, 351.

List of Illustrations

Fig. 4-1. Dependence of the b_n factor on n and type of approximation. The lower part of the scale gives the displacement of the level $\Delta\sigma = -109737/n^2$ cm^{-1} from the series limit. The values of b_n are calculated with allowance for:

I. Radiative processes only;
II. Radiative processes, collisional ionization and three-body recombination;
III. Radiative processes and collisional $n \to n \pm 1$ transitions;
IV. Radiative processes, collisional ionization and three-body recombination, and collisional $n \to n \pm 1$ transitions.

(Seaton, M.J., 1964, M.N.R.A.S., 127, 182). More detailed calculations have been made by Brocklehurst and Seaton (1972).

List of Illustrations for Chapter 4 (Continued)

Fig. 4-2. A Comparison of Observed and Theoretical Decrements in the Planetary Nebulae, NGC 6210 and NGC 6543. The intensities of the lines on the scale I(Hβ) = 100, as measured by photographic photometry, are plotted against the n-value of the upper level of the transition for NGC 6210 and against wavelength for NGC 6543. H8, λ3889 is blended with He I and H 14, λ3722 is blended with [S III]. The discordance between theory and observation for $n > 18$ for NGC 6210 may arise from difficulties in allowing for the position of the underlying continuum. The theoretical values, due to W. Clarke (1965) are very similar to those obtained by Brocklehurst (1971). Comparison of theoretical and observed Balmer decrements shows no sigificant interstellar extinction.

Courtesy, University of Chicago Press (NGC 6210: 1970, Czyzak, Aller, and Buerger, Ap. J., 162, 783; NGC 6543: 1968, Czyzak, Aller, and Kaler, Ap. J., 154, 543.

Fig. 4-3. Relation Between Emergent Intensity (or Flux Density) and Frequency ν for a Nebular Plasma. We give the schematic representation for a series of nebulae of differing optical thicknesses. Since the r.f. absorption coefficient goes as ν^{-2}, the optical thickness decreases as ν increases. At low frequencies where the plasma is optically thick, the flux density follows a Planckian curve (Rayleigh-Jeans law in this region). We choose T_ε = 10,000°K here. As ν increases, τ_ν decreases and ultimately the flux density approaches $B_\nu\tau_\nu$ as the curve flattens and slowly declines. For a given telescope aperture, the angular resolution improves as ν increases; conversely, as ν decreases, so also does the angular resolution. Not only does the source become blurred out, there also may occur confusion with other sources, particularly nonthermal ones that are brighter at low ν. The lettered curves correspond to gaseous nebulae of the same size (larger than the antenna pattern) but of differing optical thicknesses. For example, a corresponds to an optically thin nebula, (e.g, NGC 346 in the SMC), b corresponds to a nebula of medium optical thickness such as IC 418 and d to a very thick nebula such as NGC 7027. All planetary nebulae and H II regions become optically thin at sufficiently high frequencies.

Courtesy Sky and Telescope magazine.

Fig. 4-4. Determination of Optical Thickness and Temperature of an H II Region From Observations at Two frequencies. Suppose that a transverse scan is taken across a relatively dense nebula such as Orion and further that it is observed at a high frequency where it is optically thin and also at a low frequency where it is optically thick. At the high frequency, ν_2, $I(\nu_2) = \tau(\nu_2)\ B_{\nu 2}(T_\varepsilon)$; whereas at the low frequency, $I(\nu_1) = B_{\nu 1}(T_\varepsilon)\ (1 - \exp[-\tau\{\nu_1\}])$ approaches $B_{\nu 1}(T_\varepsilon)$ as $\tau(\nu_1)$ increases. Note that $B_\nu(T_\varepsilon) = 2kT_\varepsilon\ \nu^2/c^2$. $I(\nu 1)$ is indicated by the dotted curve. The solid curve depicts $I(\nu_2)\ \tau(\nu_1)/\tau(\nu_2) = \tau(\nu 1)\ B_{\nu 2}(T_\varepsilon) = \tau(\nu_1)\ (\nu_2/\nu_1)^2\ B_{\nu 1}(T_\varepsilon)$. The ratio $\tau(\nu_1)/\tau(\nu_2)$ can be

List of Illustrations for Chapter 4 (Continued)

found readily for a given T_ε. Mills and Shaver (1967) compared observations of the Orion nebula at 10 GHz with their own 408 MHz data. They constructed cross sectional profiles from the 10 GHz data and convolved the Mills Cross 408 MHz data with the antenna pattern to give the same angular resolution. Then they were able to construct profiles such as those depicted schematically above and derive τ_ν and T_ε. Thus they found T_ε = 8000°K for the Orion nebula. Not only does τ_ν vary across the object, but also T_ε varies.

Fig. 4-5. Frequency Variation of Contributions to the Continuous Absorption Coefficient. We plot the logarithm of the emission coefficient (ergs cm^{-3} sec^{-1} Hz^{-1}) against the frequency ν over the normal ground-based range from 3000 Å to 10,000 Å for H I, He I, He II, and the two-photon emission. The parameters chosen are appropriate to a high-excitation planetary such as (86 - 8°1). Numerical calculations are based on Brown and Mathews, 1970, Ap. J., 160, 934. Compare Figure 6.

Fig. 4-6. Predicted Continuous Energy Distributions for Three High-Excitation Planetaries. We plot:

$$\frac{F(\lambda)}{F(\lambda_{4000})} = \frac{\gamma_T(\nu)}{\gamma_T(\nu_o)} \left(\frac{4000}{\lambda}\right)^2 , \text{ where } \nu_o = 7.5 \times 10^{14} ,$$

against λ for the three high-excitation planetaries, IC 2165 (T_ε = 13,000°K; N_ε = 4400 cm^{-3}), NGC 2392 (T_ε = 13,000; N_ε = 3600) and (86 - 8°1) (T_ε = 16,000; N_ε = 10,000).

Courtesy P. Etzel

Fig. 4-7. Relation Between the Escape Probability and Optical Depth τ_o. In this figure, prepared by S. Drake, Capriotti's escape probability relation for a slab geometry is compared with similar calculations by Ferland and Netzer (1979), Bonilha et al. (1979) and Mathews, Blumenthal and Grandi (MBG) (1980). Also shown are results of more accurate Monte Carlo-type calculations by Panagia and Ramieri (1973).

Chapter 5

COLLISIONALLY EXCITED LINES AND PLASMA DIAGNOSTICS

5A(i) Collisional Excitation of Hydrogen Lines

To this point we have discussed lines which are excited by radiative processes, mostly photoionization followed by cascade. Collisional excitation can play an important role in many instances. If mechanical energy or magnetic energy from hydromagnetic waves are dissipated in a gas, the temperature may be raised and atomic levels excited by collisions with electrons. Collisional excitation of hydrogen, in particular, could become important. In a low-density gas which is ionized by radiation, collisional excitation of hydrogen may be less important, but that of low-lying levels of many other atoms and ions can become significant.

First, we shall briefly examine the collisional excitation of H. For the moment let us consider the low-density nebular situation where all collisional and radiative excitations occur from the ground level. The number of collisional excitations to level n from the ground level per unit time will be:

$$\mathcal{F}_{1n} = N_1(H^0)\, N_\varepsilon \int_{v_n}^{\infty} \sigma_{1n}(v) v\, f(v)\, dv = N_i(H^0)\, N_\varepsilon\, q_{1n}(T_\varepsilon) \ , \quad (1)$$

which defines $q_{1n}(T_\varepsilon)$. Here σ_{1n} is the target area for collisional excitation from the ground level to level n; f(v) is the Maxwellian velocity distribution (Eq. 3-32) and v_n is given by:

$$\frac{1}{2} m v_n^2 = \chi_n = \chi_o - \frac{hR}{n^2},$$

where χ_o = hR is the ionization energy. Analogously, the number of collisional ionizations will be:

$$\int_{\nu_1}^{\infty} \mathcal{F}_{1\kappa}\, d\nu = N_1(H^0)\, N_\varepsilon \int_{v_o}^{\infty} \sigma_c(v)\, f(v)\, dv = N(H^0)\, N_\varepsilon\, q_{1c}(T_\varepsilon) \ . \quad (2)$$

The σ's have been calculated by quantum mechanics and checked by experiments.

Equations for radiative equilibrium for discrete levels and the continuum have to be modified. Thus, Eq. 4-7 now becomes:

$$\sum_{n''=n+1}^{\infty} F_{n''n} + \int_{\nu_n}^{\infty} F_{\kappa n}\, d\nu + F_{1n} + \mathcal{F}_{1n} = \sum_{n'=n_o}^{n-1} F_{nn'} \ . \quad (3)$$

If ionizations are all produced by collisions (as occurs for ions in the solar corona, for example):

$$\int_{\nu_o}^{\infty} \mathcal{F}_{1\kappa}\, d\nu = \sum_{n=n_o}^{\infty} \int_{\nu_n}^{\infty} F_{\kappa n}\, d\nu \ . \quad (4)$$

Here n_o = 1,2 for Menzel and Baker's Cases A and B, respectively.

Table 5-1 gives the collisional rate coefficient $q_{1n}(T_\varepsilon)$ for H, n = 2 → 20, and also the ionization rate coefficient $q_{1c}(T_\varepsilon)$ as defined by Eq. (2), in units of $cm^{-3}\ sec^{-1}$. We also tabulate the ratios $q(1s-n\ell)/q(1,n)$ for the individual (n,ℓ) levels.

If b_n^c denotes the b_n factors for a purely collisional situation as described above and b_n^R the corresponding factors for the purely radiative case, Chamberlain showed that for a mixture of collisional and photoionization processes we could write:

$$b_n = (1 - \mathcal{H})b_n^r + b_n^c . \tag{5}$$

Here:

$$\mathcal{H} = 1 - \frac{[c^2/2h]e^{X_1}}{G(T_\varepsilon)} b_1 \int J(\nu)g(\nu)\frac{d\nu}{\nu^4} = \frac{q_{1c}(T_\varepsilon)\ T_\varepsilon^{3/2}\ N_i(H^o)}{G(T_\varepsilon)\ KN_\varepsilon} , \tag{6}$$

where $J(\nu)$ is the energy distribution in the radiation field. We choose $G_A(T_\varepsilon)$ for case A and $G_B(T_\varepsilon)$ for Case B. Thus, for any given external radiation field and ionic concentration, $N(H^+)$ we can solve for $(1 - \mathcal{H})/\mathcal{H}$ and compute the importance of collisional excitation. Collisional excitation of helium levels can occur only in extremely hot plasmas where T_ε exceeds about 40,000°K.

5A(ii) Collisional Excitation of Lines of Heavier Elements

Low-lying levels of ions such as C^{++}, C^{3+}, N^{++}, N^{3+}, N^{4+}, O^{++}, O^{3+}, O^{4+}, Si^{++} and Si^{3+} can be excited by electron collisions.* As these ions return to their ground levels, they emit permitted lines that fall in the ultraviolet domain, observable with satellite telescopes. Atomic term structure, elemental abundances, and ionic concentrations are such that in gaseous nebulae very few collisionally excited permitted lines are observed in optical spectral regions. One example is Mg I $3^1S - 3^3P_1$ λ4571, which is detected in the low-excitation planetary NGC 40 and several other objects with zones of very low ionization. In denser plasmas such as that associated with η Carinae, numerous collisionally-excited, permitted as well as forbidden lines of ionized iron are observed.

The most frequently observed lines excited by impacts with electrons are the so-called forbidden ones that violate the Laporte parity rule. For them we need to know transition probabilities and target areas for collisional excitation of metastable levels. These lines cannot be observed in the laboratory. Hence, we have to rely entirely on theory to determine the absolute A-values and collision strengths. Observational checks on certain line intensity ratios are sometimes possible, but for most of the transitions on which we depend for nebular plasma diagnostics, rigorous observational or experimental checks are not available.

*Spectra of neutral atoms such as N^o or O^o and of ions such as C^+, O^{++}, Ne^{4+} are denoted as N I, O I, C II, C III, and Ne V, respectively. Some writers, however, often use the spectroscopic notation also to denote the atom or ion as well as its spectrum.

Table 5-1. Collisional Rate Coefficients q_{1n} for H (cm^{-3} s^{-1})

1	5000	7500	10,000	12,500	15,000	17,500	20,000
2	2.23-18	5.15-15	2.42-13	2.43-12	1.12-11	3.35-11	7.62-11
3	5.25-21	5.27-15	5.19-15	8.10-14	5.05-13	1.87-12	4.99-12
4	4.13-22	6.92-18	8.83-16	1.61-14	1.11-13	4.44-13	1.25-12
5	9.74-23	2.07-18	2.98-16	5.83-15	4.23-14	1.74-13	5.05-13
6	3.70-23	8.95-19	1.37-16	2.80-15	2.08-14	8.75-14	2.57-13
7	1.81-23	4.73-19	7.54-17	1.57-15	1.19-14	5.06-14	1.50-13
8	1.03-23	2.83-19	4.62-17	9.79-16	7.49-15	3.20-14	9.54-14
9	6.45-24	1.84-19	3.06-17	6.54-16	5.04-15	2.17-14	6.47-14
10	4.34-24	1.27-19	2.13-17	4.60-16	3.56-15	1.54-14	4.60-14
11	3.07-24	9.13-20	1.55-17	3.60-16	2.61-15	1.13-14	3.39-14
12	2.26-24	6.81-20	1.17-17	2.54-16	1.98-15	8.57-15	2.58-14
13	1.72-24	5.23-20	9.00-18	1.97-16	1.54-15	6.66-15	2.01-14
14	1.34-24	4.11-20	7.10-18	1.56-16	1.22-15	5.29-15	1.59-14
15	1.06-24	3.29-20	5.70-18	1.25-16	9.80-16	4.26-15	1.29-14
16	8.59-25	2.67-20	4.65-18	1.02-16	8.02-16	3.49-15	1.05-14
17	7.05-25	2.21-20	3.85-18	8.47-17	6.65-16	2.90-15	8.74-15
18	5.86-25	1.84-20	3.22-18	7.09-17	5.57-16	2.43-15	7.34-15
19	4.93-25	1.55-20	2.72-18	6.00-17	4.72-16	2.06-15	6.22-15
20	4.19-25	1.32-20	2.32-18	5.13-17	4.03-16	1.76-15	5.32-15
q_{1c}	4.22-23	2.05-18	4.87-15	1.35-14	1.28-13	6.50-13	2.23-12

$q(1s \rightarrow n\ell)/q(1,n)$

	5000	10,000	20,000
1s-2s	0.3444	0.339	0.323
2p	0.655	0.661	0.677
3s	0.229	0.314	0.296
3p	0.739	0.611	0.601
3d	0.032	0.073	0.103
4s	0.208	0.202	0.190
4p	0.685	0.689	0.702
4d	0.097	0.096	0.098
4f	0.011	0.013	0.011

Courtesy Stephen A. Drake

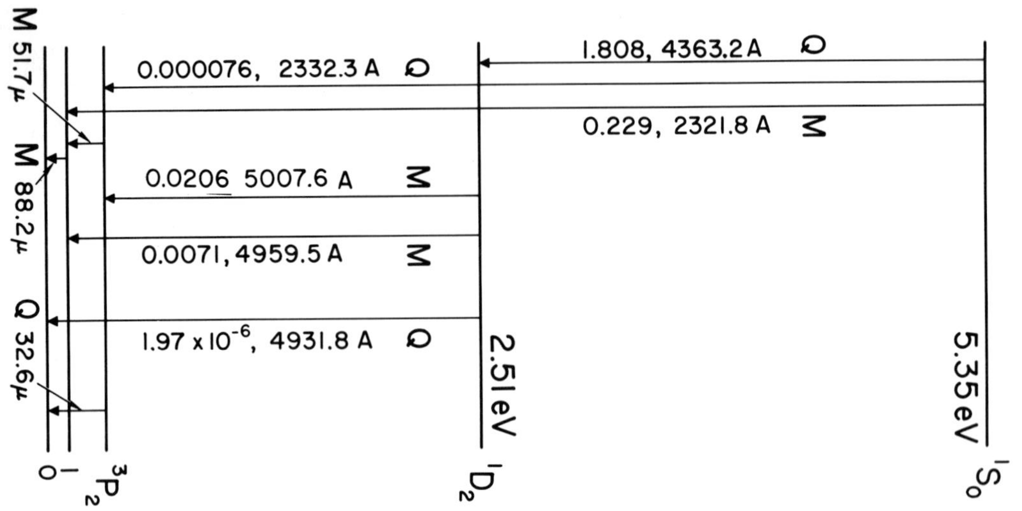

Figure 5.1.a

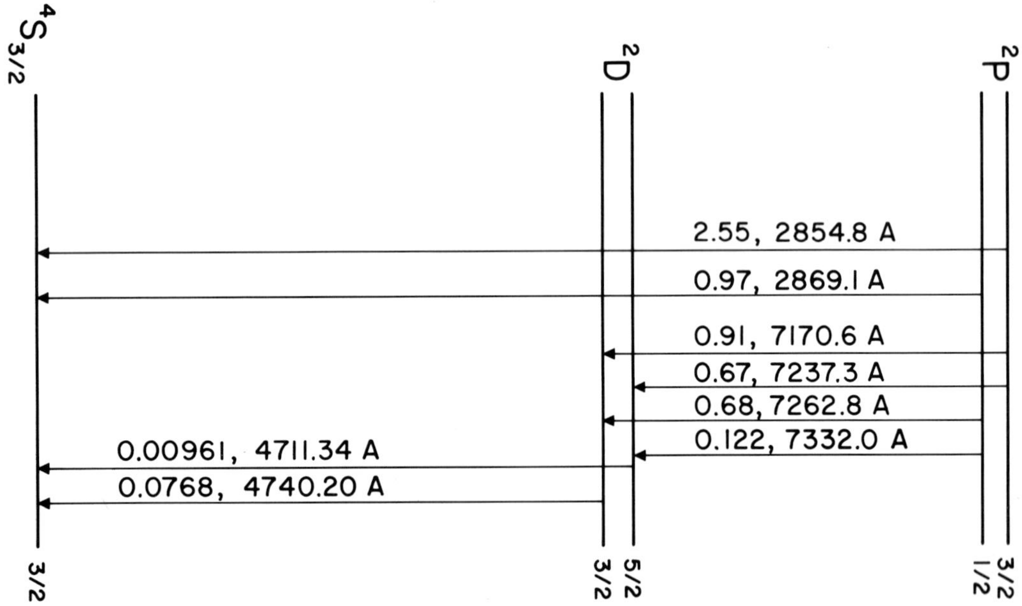

Figure 5.1.b

5A(iii) Transition Probabilities for Magnetic Dipole and Electric Quadrupole Transition: a Quantum Mechanical Digression

The coefficient for spontaneous emission for a magnetic dipole transition may be written as:

$$A_m(LJ;\ L'J') = 35320\ \left(\frac{\nu}{R}\right)^3 \frac{S_m(LJ;\ L'J')}{2J+1}\ \sec^{-1}\ , \tag{7}$$

where the matrix elements are given in atomic units and ν is expressed in terms of R (frequency of Lyman limit = $3.29 \times 10^{15}\ \sec^{-1}$), L and L' define the spectroscopic terms involved. S_m is the strength of the transition $J \rightarrow J'$, which is obtained by summing the matrix elements over individual Zeeman states, thus:

$$S_m(\alpha J;\ \alpha' J') = \sum_{M,M'} \left|(\alpha JM\ |\vec{M}|\ \alpha' J' M')\right|\ , \tag{8}$$

where:

$$(\alpha JM\ |\vec{M}|\ \alpha' J' M') = \int \Psi(\alpha JM)\ \vec{M}\ \Psi^*(\alpha' J' M')\ d\tau\ , \tag{9}$$

and:

$$\vec{M} = -\frac{\varepsilon}{2mc}\ (\vec{L} + 2\vec{S})\ . \tag{10}$$

Here α and α' represent the configuration and term of the upper and lower levels, respectively, e.g., α denotes $2s^2\ 2p^2\ {}^1D$ and α' denotes $2s^2\ 2p^2\ {}^3P$ for an [O III] nebular-type transition.

Magnetic dipole radiation occurs only as a consequence of deviations from LS coupling. In pure LS coupling, the terms of a $2p^2$ configuration, for example, would be separated from one another because of electrostatic repulsions between the electrons, but the three levels of the ^3P term would be coincident. Splitting of the ^3P term into $^3P_2$, $^3P_1$ and $^3P_0$ levels is produced by magnetic spin-orbit interactions in the atom. See Fig. 5-1a. These magnetic effects cause a deviation from LS coupling. When a breakdown from LS coupling occurs, J remains a "good" quantum number while L and S do not. As a consequence, within a given configuration, levels of the same J will mix. The unprimed wave functions are in pure LS coupling, while we use a prime (') to denote the actual perturbed functions. The wave functions for the actual levels may be expressed in terms of the pure LS levels. Thus:

$$\Psi({}^1D_2') = a\Psi({}^1D_2) + b\Psi({}^3P_2),\ \Psi({}^3P_2') = a\Psi({}^3P_2) - b\Psi({}^1D_2),\ \text{etc.} \tag{11}$$

where the coefficient $a \rightarrow 1$ and $b \rightarrow 0$ as the ion approaches perfect LS coupling. Condon and Shortley (1935) define a parameter, χ, which expresses the deviation from LS coupling. It is essentially a measurement of the ratio of the magnetic splitting of the levels of a term to the term separation, which is an index of the electrostatic effects. Individual magnetic dipole line strengths may then be given in terms of χ for each type of transition. When χ is small, we may use asymptotic expressions, for example:

$$S_m(^1D_2', \ ^3P_2') = \frac{15}{2} a^2b^2 = \frac{375}{155} \chi^2[1 + \frac{5}{6} \chi + \ldots \text{etc.}] \ . \qquad (12)$$

For [O III], $\chi = 0.012$ (good LS coupling) but for Pb ($6p^2$), $\chi = 1.583$, i.e., there is no semblance of LS coupling at all!

A determination of χ is not always simple. For example, the positions of individual terms may be affected by configuration interaction. We try to allow for this by using observed values of the energies in the equations. In p^3 configurations, one must take into account second-order spin-orbit interaction and also the interaction of the spin of one electron with the orbit of another. These effects determine the splitting of the ^2D and ^2P terms. See Fig. 5-1b. An error in χ markedly affects the value of S_m. For the fine structure transitions:

$$S_m(^3P_2' - \ ^3P_1) = \frac{5}{2}(1 - \frac{25}{72} \chi^2 + \ldots,$$

$$S_m(^3P_1 - \ ^3P_0') = 2(1 - \frac{2}{9} \chi^2 + \ldots, \qquad (13)$$

but $A_m(J, J') \to 0$ as $\chi \to 0$, since $\nu(J, J') \to 0$.

Electric quadrupole transitions can occur in pure LS coupling. We have:

$$A_q(\alpha J; \alpha' J') = \frac{2648}{2J+1} \left(\frac{\nu}{R}\right)^5 S_q(\alpha J; \alpha' J') \text{ sec}^{-1} \ . \qquad (14)$$

Here the strength S_q consists of two factors: a) a coefficient involving J, L, S, ℓ, etc., which can be evaluated for each forbidden line; and b) a factor which depends on radial quantum wave functions:

$$s_q = \frac{2}{5} \varepsilon^2\{\int R(n\ell) r^2 R(n'\ell') r^2 dr\} \ . \qquad (15)$$

The strengths of the λ4363 and transauroral lines are given by:

$$S_q(^1S_0' \ ; \ ^1D_2') = \frac{20}{3} s_q^2(1 - \frac{7}{24} \chi^2 + \ldots \ . \qquad (16)$$

$$S_q(^1S_0' \ ; \ ^3P_2') = \frac{5}{6} s_q^2 \chi^2(1 + \frac{11}{6} \chi + \ldots \ . \qquad (17)$$

Notice that in the first approximation the strength of λ4363 doesn't depend on χ^2 at all!

5B. Collisional Excitation of Atomic Levels

5B(i) General Principles

In order to calculate theoretical intensities for collisionally-excited lines, be they permitted or forbidden, we must know not only the relevant A-values, but also the collisional cross section $\sigma(v)$ for excitation from a level α'L'S'J' to a level αLSJ. For forbidden lines, the configurations α' and α will be the same.

Consider an electron colliding with an ion with an impact parameter, R: R is defined as the distance of closest approach between the colliding particles if no attraction existed between them. The corresponding angular momentum is:

$$\vec{L} = mv\vec{R} \quad . \tag{18}$$

In the quantum theory we would write $L = h/2\pi \sqrt{\ell(\ell + 1)}$. We find:

$$R_\ell = \sqrt{\ell(\ell + 1)}/k \text{ with } k = mv/(h/2\pi) \quad . \tag{19}$$

Rarely would an electron encounter a target with precisely the right combination of R and ℓ to fulfill this condition. Hence, a quantum mechanical description of the impinging electron involves a summation of waves of differing angular momentum: an s-wave, a p-wave, a d-wave, etc. The electron must have sufficient energy to excite the ion from the lower level J' to the upper level J. On the other hand, the ℓ-value cannot be too high; classically, such an electron with its high angular momentum would simply miss the atom.

If $P(J' \to J, R)$ is the probability for an excitation from level J' to J, the classical cross section.

$$\sigma(J' \to J) = \int_0^\infty P(J' \to J; R)\, 2\pi R dR, \qquad P(J' \to J, R) \leqslant 1 \ , \tag{20}$$

is replaced by the quantum mechanical sum:

$$\sigma(J' \to J) = \frac{\pi}{k^2} \sum_{\ell=0}^{\infty} (2\ell + 1)\, P(J' \to J; R_\ell) \ , \tag{21}$$

since $2R\,\delta R = (2\ell + 1)\,\delta\ell/k^2$ with $\delta\ell = 1$ in the collision. In their pioneering investigations, Hebb and Menzel (1940) noted that $\sigma(v)$ varied almost as v^{-2}. Hence, they introduced a quantity, Ω, defined as the <u>collision</u> <u>strength</u>.

$$\sigma_v(J' \to J) = \frac{\pi}{k^2}\, \Omega(J',\, J) = \frac{h^2}{4\pi m^2}\, \frac{\Omega(J',J)}{v^2} \ . \tag{22}$$

Actually, each $\Omega(J'\ J)$ has to be obtained by calculating the Ω's for combinations of the individual Zeeman levels, i.e., $\Omega(J'\ M_{J'},\ J\ M_J)$, summing them up and dividing by $2J' + 1$. We get finally for the total rate of collisional excitations per unit volume per second:

$$\mathcal{F}_{J'J} = N_{J'} N_\varepsilon \int_{v_0}^\infty v\sigma_v(J',J)\, f(v)\, dv = N_{J'} N_\varepsilon\, q(J'J) \ , \tag{23}$$

where f(v) is the Maxwellian distribution of velocities, and q(J'J) is called the activation coefficient. The integration is carried out over all velocities greater than v_0, given by:

$$1/2\ mv_0^2 = \chi_J - \chi_{J'} = \chi_{JJ'} \ ,$$

the JJ' energy interval.

We establish a deactivation coefficient $q(J,J')$ by noticing that under conditions of thermodynamic equilibrium the number of collisional excitations from level J' to J must exactly equal the number of collisional deexcitations from J to J'. Since $N_J/N_{J'}$ is given by Boltzmann's equation, we have that:

$$q(J'J) = q(JJ') \frac{2J + 1}{2J' + 1} e^{-\chi_{JJ'}/kT} . \tag{24}$$

Further, we can show that $\Omega(J',J) = \Omega(J,J')$. (25)

Thus, collision strengths are analogous to line strengths which had the property that $S(\alpha J;\alpha' J') = S(\alpha' J';\alpha J)$, for all types of transitions. Now, putting in numerical values:

$$q(J'J) = 8.63 \times 10^{-6} \frac{\Omega(J'J)}{(2J' + 1)} T_\varepsilon^{-1/2} \exp\left(- \frac{\chi_{J'J}}{kT_\varepsilon}\right) \tag{26}$$

and

$$q(JJ') = 8.63 \times 10^{-6} \frac{\Omega(J'J)}{(2J + 1)} T_\varepsilon^{-1/2} , \tag{27}$$

as long as the approximation $\sigma \cong \Omega v^{-2}$ is valid.

A useful relation holds for collision strengths between a term consisting of a single level and one consisting of several levels, e.g., between $^1D_2$ and ^3P in a p^2 or p^4 configuration. Thus:

$$\Omega(SLJ: S'L'J') = \frac{2J' + 1}{(2S' + 1)(2L' + 1)} \Omega(SL, S'L') . \tag{28}$$

Here $(2S' + 1)(2L' + 1)$ equals the statistical weight of the term while $(2J' + 1)$ is the statistical weight of the level under consideration. If we denote by $\Omega(SL, S'L')$ the total collision strength for $^3P \rightarrow {}^1D$:

$$\Omega(^1D_2, {}^3P_2) = \left(\tfrac{5}{9}\right) \Omega(^3P, {}^1D_2) \text{ since } 2J' + 1 = 5,\ 2S' + 1 = 3,$$
$$(2L' + 1) = 3.$$

5B(ii) Comments on Calculation of Collision Strengths

The quantum mechanical calculation of collision strengths turned out to be very difficult. Early efforts which were based on what is called the Born-Oppenheimer approximation violate the Mott-Bohr-Peierls-Placzek conservation theorem which requires that:

$$\sum_{n'} \frac{\Omega(n,n')}{2J_n + 1} \leqslant (\ell_o + 1) ,$$

where ℓ_o is the azimuthal quantum number of the partial wave that contributes the greatest share to the cross section.

The problem of calculating reliable cross sections was first solved successfully by Seaton who devised two procedures known as the "Exact Resonance Method: and the "Distorted Wave Method." A third method known as the "Close-Coupling (CC) Approximation" also has been developed.

Numerous collision strength calculations have been carried out and applied in evaluations of electron temperatures, densities and ionic concentrations in gaseous nebulae and extended stellar envelopes. How accurate are such calculations? Unfortunately, direct experimental checks are not possible. At the moment, we have to rely on consistency arguments and while these are necessary conditions they are not sufficient. A set of collision strengths that give the same T_ε and N_ε for a filament in a gaseous nebula probably is to be preferred over a set that gives inconsistent values, sometimes for different lines observed in the same ion, e.g., [S II] or [Ar IV]. On the other hand, the grossly erroneous Ω-values calculated by the Born approximation led to electron temperatures that were consistent with those then found from other evidence, such as line broadening and from a comparison of energy distribution in the Balmer continuum with Hβ.

First, different collision strengths result from the exact resonance method (which is best for the p-wave interacting with p^n electrons) than from the distorted wave method. The proper choice of which method to use is fairly well indicated, normally. Choice of wave functions may have an effect. For example, one can proceed using entirely calculated energies, or one can use the observed energy levels to fix parameters in the wave functions. Different collision strengths result! It is best to calculate wave functions taking superposition of configurations (configuration interaction) into account to obtain an agreement between observed and calculated energy levels.

The assumption that $\sigma \sim \Omega v^{-2}$ often is inaccurate. For example, it fails for neutral atoms where Ω depends on v and approaches zero at the threshold energy. Seaton writes the deactivation coefficient as:

$$q_{JJ'} = \frac{8.63 \times 10^{-6}}{\omega_J \sqrt{T_\varepsilon}} \left[\int_0^\infty \Omega(J',J;\ v) \exp\left(-\frac{E}{kT}\right) \frac{dE}{kT}\right] \text{cm}^3 \text{ sec}^{-1} , \quad (29)$$

where ω_j is the statistical weight of the upper level and $E = 1/2\ mv^2$ is the kinetic energy of the free electron. Thus, we have to tabulate $q_{JJ'}(T_\varepsilon)$ for atoms such as O I, N I, S I, etc.

Seaton noted that the Ω = constant approximation fails also for ions such as O^{++}. Consider a p-electron interacting with an O^{++} ion in its ground ^3P level (see Fig. 2). The addition of such an electron to an O^{++} ion corresponds to an O^+ ion in an unbound state. Further, certain discrete energies can correspond to O^+ Auger levels. The normal ground configuration of O+ is $2s^2\ 2p^3$ and ordinary excitations to levels below the first ionization potential correspond to excitations of one of the 2p electrons. Excitation of a 2s electron to a 3s configuration, however, can lead to a populating of levels that can lie above the first ionization potential. In O^+ there exist two levels:

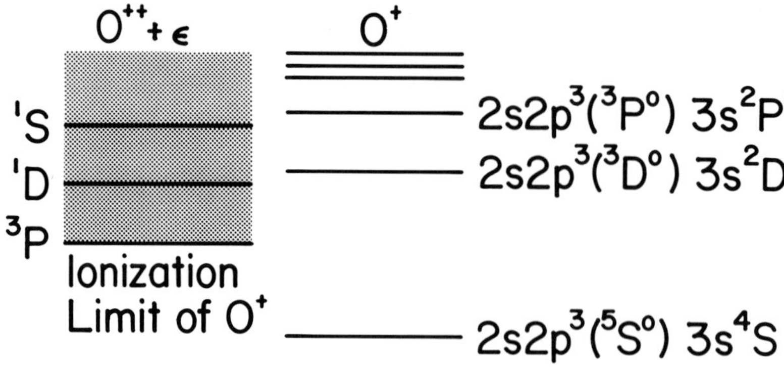

Figure 5-2

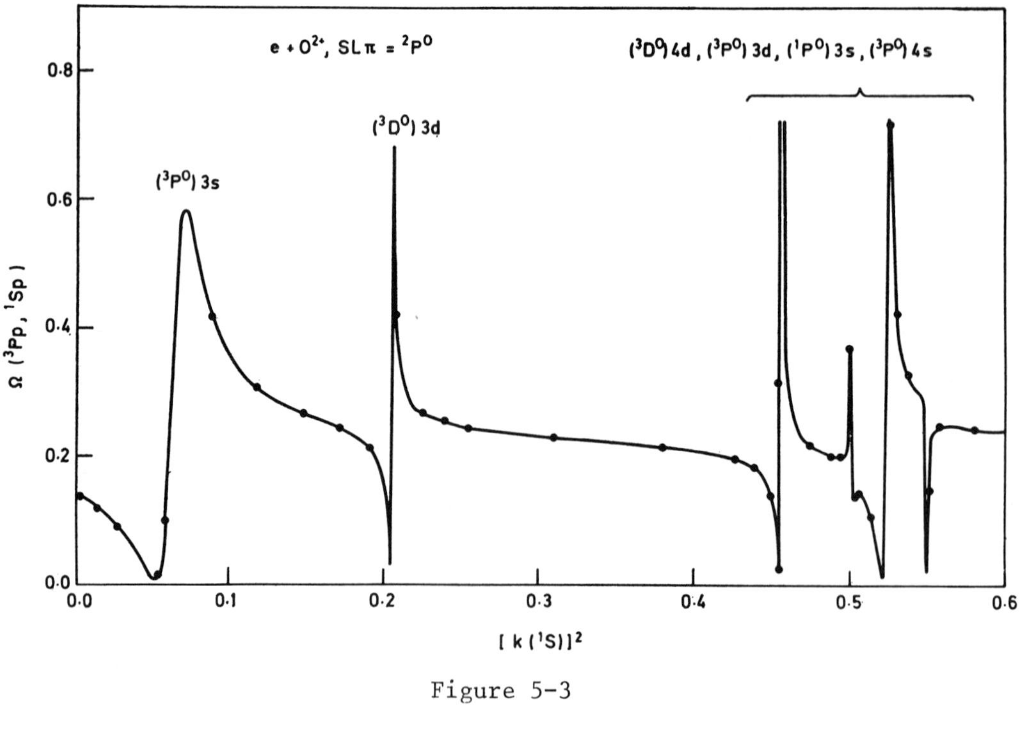

Figure 5-3

Table 5-2. Data for Calculation of Ionic Concentrations, $N^*(ion)/N(H+)$
p^2 and p^4 configurations

	λ	A_{BA} (s^{-1})	<A/λ(μ)	γAB	t	a	b	d	Neb. Ratio	Notes
N II	6584 *	3.94 (-3)	0.0060	2.64	0.5	1.753(-6)	0.116	0.951	0.746	(1,4)
				2.68	1	1.711(-6)	0.118			
				2.72	1.5	1.663(-6)	0.119			
				2.73	2	1.635(-6)	0.120			
O III	5007,4959	0.0277	0.0554	2.02	0.5	1.294(-6)	1.3 (-2)	1.254	(2.88^2)	1,4
				2.17	1	1.201(-6)	1.35(-2)			
				2.30	1.5	1.125(-6)	1.44(-2)			
				2.39	2	1.073(-6)	1.50(-2)			
F IV	4062,3997	0.135	0.333	1.93		1.10 (-6)	2.5 (-3)	1.55	(2.83)	3
Ne III	3868 *	0.225	0.578	1.34	1	1.99 (-6)	1.0 (-3)	1.59	0.764	1
Ne V	3343,3426	0.512	1.504	1.78	1	1.00 (-6)	6.0 (-4)	1.846	(2.75)	1,4
				1.85	1.5	9.67 (-7)	6.2 (-4)			
				1.92	2	9.22 (-7)	6.5 (-4)			
Na IV	3362,3241	0.796	2.44	1.17	1	1.47 (-6)	2.4 (-4)	1.91	(0.294)	3
S III	9069 *	0.0788	0.0234	8.39	1	2.11 (-6)	1.83(-2)	0.673	0.297	1
Cℓ II	8579 *	0.133	0.153	3.86	1	1.49 (-6)	5.0 (-3)	0.707	0.791	2
Cℓ IV	8045,7530	0.250	0.318	5.42	1	7.64 (-7)	3.7 (-3)	0.803	(2.30)	1
				5.88	1.5	7.05 (-7)	4.0 (-3)			
Ar III	7135 *	0.396	0.546	5.10	1	9.27 (-7)	2.2 (-3)	0.841	0.806	3
Ar V	7005,6435	0.680	0.996	3.72	1	9.63 (-7)	9.4 (-4)	0.932	(2.14)	2
K IV	6101 *	1.012	1.626	1.90	1	2.10 (-6)	3.2 (-4)	0.968	0.822	2
K VI	6228,5603	1.63	2.71	0.81	1	3.90 (-6)	8.6 (-5)	1.06	(1.87)	3
Ca V	5309 *	2.325	4.28	1.12	1	3.04 (-6)	8.3 (-5)	1.10	0.855	3

(1) The collision strengths depend on t; see Mendoza's compilation (1983).
(2) See Mendoza (1983).
(3) Collision strengths, Ω, are from Czyzak.
(4) The A-values are from Czyzak and Poirier.

The first column gives the ion; the second column gives the nebular-type transitions used. In some instances, denoted by *, we give the parameter, a, for only one of the nebular lines because the other transition is either blended or rather weak.
$A_{BA} = A(^1D_2-^3P_1) + A(^1D_2-^3P_2)$ and $< A/\lambda(\mu)> = A_1/\lambda_1(\mu) + A_2/\lambda_2(\mu)$ where the A-values are given in sec^{-1} and λ in μm. Then we tabulate the collision strength $\gamma_{BA} = \gamma(^3P-^1D)$. For several ions where the collision strength is temperature-dependent, we list several values of

Table 5-2. (Continued)

t for which the constants are computed. The next columns give the constants a, b and d = 0.504 χ_{BA} as defined in Eq. (39). The numbers in parenthesis, e.g., (-6), (-2) indicate the power of ten by which the coefficient is to be multiplied. The column headed "Neb. ratio" is to be interpreted as follows: When only one of the two nebular transitions is employed, we give the ratio of the intensity of that line to that of the pair. If both lines are tabulated, the entry is enclosed in parentheses ( ) and then expresses the intensity ratio of the nebular transition of longer wavelength to that of shorter wavelength.

Table 5-3. Data for Calculation of Ionic Concentrations From UV Lines

Ion	λ	Transition	χ(J'J)	t	$A_{\lambda,i}$	d
C II	2329	$2s^2\,2p\,^2P - 2s\,2p^2\,^4P$	5.33	1.0	2.57 (-7)	2.69
C III	1906 1909	$2s^2\,^1S - 2s\,2p\,^3P^o$	6.50	1.0	1.112 (-7)	3.276
C IV	1548 1550	$2s\,^2S_{1/2} - 2p\,^2P_{1/2,3/2}$	8.00	1.3	2.04 (-8)	4.032
N III]	1747 -1754	$2s^2\,2p\,^2P - 2s\,2p^2\,^4P$	7.08	1.0	2.99 (-7)	3.568
N IV]	1487	$2s^2\,^1S_o - 2s\,2p\,^3P^o_1$	8.34	1.5	1.064 (-7)	4.203
N V	1239 1243	$2s\,^2S - 2p\,^2P^o$	10.0	1.5	2.17 (-8)	5.04
O III]	1661 1666	$2s^2\,2p^2\,^3P - 2s\,2p^3\,^5S_2$	7.45	1.0	7.29 (-7)	3.75
O IV	1403 1409	$2s^2\,2p\,^2P^o - 2s\,2p^2\,^4P$	8.82	1.5	3.88 (-7)	4.45
O V	1218	$2s^2\,^1S - 2s2p\,^3P$	10.18	1.5	1.056 (-7)	5.13
O VI	1032 1038	$2s\,^2S - 2p\,^2P$	11.97	1.5	2.41 (-8)	6.03
Mg II	2800	$3s\,^2S - 3p\,^2P$	4.43	1.0	1.98 (-8)	2.22
Si III	1892	$3s\,^1S - 3s\,3p\,^3P$	6.55	1.0	2.04 (-8)	3.301
Si IV	1391	$3s\,^2S - 3p\,^2P$	8.86	1.5	9.62 (-9)	4.465

In some ions Ω varies with temperature (usually rather slowly). The column headed t gives the value of t = $T_\varepsilon/10{,}000$ for which the coefficient $A_{\lambda,i}$ defined by Eq. (40) is calculated.

$$2s\ 2p^3\ (^3D^o)\ 3s\ ^2D^o \quad \text{and} \quad 2s\ 2p^3\ (^3P^o)\ 3s\ ^2P^o ,$$

a short distance above the ionization limit; in fact, they lie just above the thresholds for excitation of ^1D and ^1S levels in O III. The positions of these levels can be calculated accurately from an extrapolation of isoelectronic sequences. Accordingly, for the calculation of excitation or deexcitation rates, one must use instead of a constant Ω, the parameter:

$$\gamma_{J'J}(T_\varepsilon) = \int_0^\infty \Omega(J'J) \exp\left[- E_i/kT_\varepsilon\right] dE_i/kT_\varepsilon , \tag{30}$$

where $E_i = 1/2\ mv_i^2$ is the initial kinetic energy of the colliding electron and $\Omega(J'J)$ is the collision strength.

Considering the wild fluctuations of Ω in the neighborhood of resonances, as shown in Fig. 3, it is truly remarkable that the resultant Ω values are so insensitive to T_ε. The explanation seems to be that in a Maxwellian distribution there is such a wide range of velocities that peaks and valleys tend to be smoothed out when Ω is convolved with f(v). For example, for Ne^{++}, the variation is very small indeed.

Nevertheless, the huge Ω fluctuations cannot be ignored. We can anticipate that the effects may be troublesome in some ions of the $3p^3$ configuration, for example Ar^{+++}, where observational data do not seem to harmonize well with theoretical predictions.

5C. Level Populations for Collisionally-Excited Lines

We can easily write down the formal general equations of statistical equilibrium for collisionally-excited lines. Noting that the only possible radiative transitions are downward ones, we get for the ith level:

$$\sum_{j\neq i} \mathcal{F}_{ji} + \sum_{j''>i} F_{j''i} = \sum_{j'<i} F_{ij'} + \sum_{j\neq i} \mathcal{F}_{ij} , \tag{31}$$

where levels denoted by j" lie above level i; those denoted by j' lie below level i. The rates of radiative transitions are denoted by $F_{j''i}$ and $F_{ij'}$ while the rate of collisional transitions i-j is denoted as $\mathcal{F}_{ij}$. We can write Eq. (31) as:

$$N_\varepsilon \sum_{j'<i}^{i-1} N_{j'}\, q_{j'i} + \sum_{j''>i} N_{j''}\left(A_{j''i} + N_\varepsilon\, q_{j''i}\right)$$

$$= N_i\left\{ \sum_{i>j'} A_{ij'} + N_\varepsilon \sum_{i\neq j} q_{ij}\right\} \tag{32}$$

where the summation is taken over all m levels involved in the scheme. For p^2, p^3 and p^4 configurations, m is 5, but if higher terms are involved as in considerations of many UV line intensities, m is often of the order of 15.

Substituting in Eq. (26) and (27), and introducing explicit values of the A's and Ω's, we obtain m equations, only (m-1) of which are independent, however, so we can solve only for the ratios $n_i = N_i/N_1$, i = 1 . . . m. Consider, for example, a p^3 configuration for which the levels are $^4S_{3/2}(i=1)$, $^2D_{3/2}(i=2)$, $^2D_{5/2}(i=3)$, $^2P_{1/2}(i=4)$, and $^2P_{3/2}(i=5)$. We follow Seaton's notation and let

$$x = 10^{-2}\, N_\varepsilon/\sqrt{T_\varepsilon},\quad t = T_\varepsilon/10{,}000. \tag{33}$$

We can use either (x,t) or (N_ε, T_ε) as diagnostic parameters. We have:

$$-\left[\frac{4635}{x} A_{21} + \left\{(\Omega_{12} + \Omega_{23}) + (\Omega_{24} + \Omega_{25}) \exp\left(-\frac{\chi_{24}}{kT_\varepsilon}\right)\right\} n_2 + \frac{2}{3}\Omega_{23}\, n_3\right.$$

$$+\left[\frac{4635}{x} A_{42} + 2\Omega_{42}\right] n_4 + \left[\frac{4635}{x} A_{52} + \Omega_{52}\right] n_5 + \Omega_{12} \exp\left(-\frac{\chi_{12}}{kT_\varepsilon}\right) = 0 \tag{34}$$

with analogous expressions for levels (3, 4, and 5). Radiative transitions between doublet levels are neglected. Although these equations are readily solved on a computer, it is instructive to consider some limiting cases:

5C(i) <u>"Low Density" Domain for p^2 and p^4 Configurations.</u>

At densities $N_i < \sim 10^5$ ions/cm^3, we can usually make the approximation that the excitation from the ground ^3P term to ^1D and ^1S terms are much more important than collisional ^1D to ^1S excitations. Further, as reference to Eq. (26) and (28) will show, we need not be concerned with details of distributions of ions within the levels of the ^3P ground term. In effect, the five-level problem reduces to a three-level problem. We denote the P, D, and S terms by A, B, and C, respectively. At the low densities postulated, we can neglect cascades and deactivation (downward) transitions from the ^1S term. We will be making an error of a magnitude determined by taking the ratio of the auroral to the nebular type transition, e.g., the λ4363 to (4959 + 5007) ratio in [O III]. Thus, the number of atoms entering $^1D_2$ by collisional excitation equals the number escaping by radiative plus collisional deexcitations:

$$N_A N_\varepsilon q_{AB} = N_B N_\varepsilon q_{BA} + N_B A_{BA}\ . \tag{35}$$

Using Eq. (26) and (27), noting $(2J_A + 1) = 9$, $(2J_B + 1) = 5$, we have:

$$\frac{N_B}{N_A} = \frac{b_B}{b_A}\left(\frac{N_B}{N_A}\right)^o \ ;\quad \frac{b_A}{b_B} = \left[1 + \frac{A_{BA}}{N_\varepsilon q_{BA}}\right] = \left[1 + \frac{5794}{x\,\Omega_{AB}} A_{BA}\right]\ , \tag{36}$$

where $(N_B/N_A)^o$ is the rate appropriate to thermodynamic equilibrium at T_ε. For the nebular-type transition, $(^1D - {}^3P)$:

$$E_{neb} = N(^1D_2)\ \{A(^1D_2 - {}^3P_2)\ h\nu(^1D_2 - {}^3P_2)$$

$$+ A(^1D_2 - {}^3P_1)\}\ h\nu(^1D_2 - {}^3P_1)\} = N_B h \langle A\nu\rangle_{BA}\ . \tag{37}$$

Suppose that the Hβ emission (see Eq. 4-15) originates in the same volume:

$$E(H\beta) = N(H^+)N_\varepsilon\ \alpha_B(H\beta)\ h\nu = H(H^+)N_\varepsilon\ E^o_{4,2} \times 10^{-25}\ (\text{erg cm}^{-3}\ \text{s}^{-1})$$

Let $N^*(\text{ion}) = N(\text{ground term}) \lesssim N(\text{ion})$ at nebular densities. Then:

$$\frac{N^*(\text{ion})}{N(H^+)} = 0.906 \times 10^{-9} \left\{\frac{x + 5794A_{BA}/\Omega_{BA}}{\langle A/\lambda(\mu)\rangle}\right\} 10^{0.504\chi_{AB}/t} \sqrt{t}\ E^o_{4,2}\ \frac{I_{neb}}{I(H\beta)} \qquad (38)$$

which (see Table 5-2) can be expressed in the form:

$$\frac{N^*(\text{ion})}{N(H^+)} = a(1 + bx)\sqrt{t}\ E^o_{4,2}\ 10^{d/t}\ \frac{I(\text{neb})}{I(H\beta)} \qquad (39)$$

The amount by which the (1 + bx) factor differs from 1.0 is a good criterion for a "low density," for if it appreciably exceeds 1, collisions between ^1D and ^1S can become important and we may no longer accept the low-density approximation but will have to solve the complete set of equations. Under these circumstances $N(\text{ion}) = \Sigma N_J$ (where the summation over all levels, J, of low-lying configurations) may substantially exceed N(ground term).

5D(i) Collisionally-Excited Lines in the Near-Ultraviolet Region.

In the ultraviolet, the strongest collisionally-excited lines in many planetaries are permitted lines such as C IV λ1548, 1550, SI IV λ1394 and 1403, although in high-excitation nebulae the [Ne IV] lines are also strong. Collision strengths for permitted lines are sometimes estimated by approximate formulae which invoke the approximation that target areas for electron impact excitation are proportional to the f-values for the corresponding optical transition.

Collisional deexcitation can be neglected. We simply equate the number of collisional excitations to the number of quanta emitted. If J',J denote ground and excited levels, respectively, $\Omega_{J'J}$ is the collision strength and $\chi_{J'J}$ the excitation potential, then the rate of emission from ions $N(X^i)$ in the $J \rightarrow J'$ transition per unit volume is $E(JJ') = N_J(X^i)\ A(J,J')h\nu = N_{J'}(X^i)\ N_\varepsilon\ q(J'J)\ h\nu$. We find:

$$\frac{N_{J'}(X_i)}{N(H^+)} = 5.841 \times 10^{-11}\ \lambda(\text{Å})\ \frac{2J'+1}{\Omega(J',J)}\ e^{+\chi(JJ')/kT}\ E^o_{4,2}\sqrt{t}\ \frac{I(\lambda,X_i)}{I(H\beta)}$$

$$= A_{\lambda,i}\ 10^{d/t}\ E^o_{4,2}\sqrt{t}\ \frac{I(\lambda,X_i)}{I(H\beta)} \qquad (40)$$

(See Table 5-3.) Several complications can occur with UV lines: a) the calculation of N_{ion} is sensitive to T_ε in the zone; hence, its value must be estimated accurately. Theoretical models (see Chapter 7) are sometimes helpful here; b) levels can be populated by recombination, particularly dielectronic recombination so the pertinent cascade effects have to be taken into account; c) for resonance lines, e.g., 1548/1550 of C IV, we must consider the transfer problem which may become complicated, particularly if scattering and absorption by dust are involved.

5D(ii) Infrared Transitions

In many gaseous nebulae, some of the most heavily populated ionization stages of abundant elements like C, N, Ne, and S supply no optical region lines, although important transitions may be found not only in the UV but also in the infrared. For example, infrared lines of [Ne II], [S IV], and [Ar III] corresponding to transitions within ground configurations of $2p^5$, 3p, and $3p^4$ at 12.8 μm, 10.5 μm, and 9.0 μm, have been observed in a number of nebulae from the earth's surface. Note that for the ^2P terms of the p and p^5 configurations, we can use Eq. (35) where now A refers to the lower and B to the upper of the doublet levels. The influence of higher levels can be neglected. For p^2 and p^4 configurations we should solve the appropriate five-level equations, although to some extent we can treat the levels of the ^3P terms as constituting an isolated group of states. At nebular densities, trickling down from ^1D and ^1S levels is less important than the shuffling between levels of the same term. Table 5-4 gives data for calculation of ground term concentrations of ions Ne^+, S^{+3}, and Ar^{++} from their infrared lines.

For infrared lines, very small energy differences are involved. Uncertainties in T_ε have a minimal effect on abundance estimates from these transitions. Generally, because of the ν^3 factor in the expression for the A-value, transition probabilities between closely adjacent levels are small. At moderate densities, long before such effects become critical for the higher levels of the ground configuration, collisions will dominate populating and depopulating processes and occupation numbers may approach the ratios of statistical weights. Also, the collision strengths can be calculated for transitions between levels of the same term more accurately than they can be found for transitions between different terms. Further, interstellar extinction is small so errors in the extinction factor, C, will have a minimal effect. For all these reasons, infrared lines are becoming increasingly important for abundance and plasma diagnostic studies.

With the Kuiper Airborne Observatory it has been possible to extend observations to the spectral regions, 5 - 8 m and 16 - 100 m. In addition to the transitions mentioned above, Ms. Dinerstein lists the following lines as having been measured in planetary nebulae by various observers: [Mg IV] 4.49 μm, [Mg V] 5.61 μm, [Ni II] 6.62 μm, [Ar II] 6.98 μm, [Cℓ IV] 11.76 μm, [S III] 18.71 μm,, [Ne V] 24.28 μm, [O IV] 25.87 μm, [O III] 51.81, 88.36 μm, (see Fig. 5-1), [N III] 57.33 μm, and [O I] 63.17 μm. Additional transitions will certainly be added to this list, but a few transitions such as [Ne IIIl 15.4 μm will require observations from a space platform.

5D(iii) Ionic Concentrations for p^3 Configurations

For the calculation of $N(X^1)/N(H^+)$ from auroral and nebular-type transitions in $2p^3$ or $3p^3$ configurations simple equations such as (39) or (40) are not available. We can replace (1 + bx) by a function $L_A(x,t)$ or $L_N(x,t)$ which can be tabulated. (See Table 5-5.)

Table 5-4.

Data for Calculation of Ionic Concentrations From Some Fine Structure Lines.

$$\frac{N(X^i)}{N(H^+)} = a(1 + bx)\sqrt{t}\, E^{o}_{4,2}\, 10^{d/t} \frac{I(\lambda)}{I(H\beta)}$$

where $N(X^i)$ is the total population in the ground term of the configuration. (The Ar^{++} data are valid for T_ε in the neighborhood of 10,000°K, $N_\varepsilon < 10^5$.)

Ion	Transition	λ	A	Ω	a	b	d
[Ne II]	$^2P_{1/2} - {}^2P_{3/2}$	12.8μm	8.55(-3)s^{-1}	0.368	8.13(-5)	0.0278	0.0489
[S IV]	$^2P_{1/2} - {}^2P_{3/2}$	10.5	7.73(-3)	6.42	1.91(-6)	0.490	0.0595
[Ar III]	$^2P_1 - {}^2P_2$	8.99	3.08(-2)	2.24	9.54(-6)	0.034	0.0695

Table 5-5a.

Data for Calculation of Ionic Concentrations in p^3 Configurations.

$$\frac{N(X^i)}{N(H^+)} = a\, L_j(x,t)\sqrt{t}\, E^{o}_{4,2}\, 10^{d/t} \frac{I(\lambda)}{I(H\beta)}$$

We here label the individual energy levels by the scheme:

$$^4S_{3/2} = [1],\ ^2D_{5/2} = [2],\ ^2D_{3/2} = [3],\ ^2P_{3/2} = [4],\ ^2P_{1/2} = [5]$$

Here $d = 0.5040\ \chi(\mathrm{ev})$, $t = T_\varepsilon/10{,}000$

Nebular transitions N

$L_i = L_N = x[(N_3A_{31} + N_2A_{21})\exp(\chi_{13}/kT_\varepsilon)]^{-1}$, [O II, [Ne IV], [Cℓ III], [S II]

$L_i = L_N = x[N_2A_{21})\exp(\chi_{13}/kT_\varepsilon)]^{-1}$

Auroral transitions A

$L_j = L_A = x[N_5(A_{53} + A_{52})\exp(\chi_{52}/kT_\varepsilon) + N_4(A_{42} + A_{43})\exp(\chi_{42}/kT_\varepsilon)]^{-1}$

$L_j = L_A = x[N_5A_{53}\exp(\chi_{51}/kT_\varepsilon) + N_4A_{43}\exp(\chi_{41}/kT_\varepsilon)]^{-1}$ [Ne IV]

Ion	λ		a	d
O II	3726, 3729	N	1.88 (-10)	1.676
	7319, 7330	A	3.68 (-10)	2.529
Ne IV	2422, 2424	N	1.22 (-10)	2.573
	4725, 4727	A	2.38 (-10)	3.90
S II	6717, 6730	N	3.39 (-10)	0.930
Cℓ III	5517, 5537	N	2.78 (-10)	1.130
Ar IV	4740	N	2.39 (-10)	1.322

Table 5-5b. Auxiliary Quantities L_N and L_A for Calculation of Ionic Concentrations in p^3 Configurations

[O II]

logx =	-2	-1.5	-1	-0.5	0.0	+0.5	+1.0	+1.5
5000(N)	3514	3656	4050	5122	8262	18000	48800	146000
(A)	13220	11530	8807	6034	4157	3269	2959	2931
10000(N)	3426	3575	3988	5112	8411	18700	51000	153400
(A)	13210	11510	8811	6086	4256	3397	3101	3081
15000(N)	3339	3507	3972	5238	8964	20600	57200	173000
(A)	13220	11550	8929	6309	4565	3754	3481	3482
20000(N)	3276	3468	3998	5442	9701	23000	64800	198000
(A)	13250	11630	9111	6615	4965	4204	3956	3982

[Ne IV]

10000(N)	3288	3299	3335	3436	3688	4229	5495	9133
(A)	37000	36400	34900	31300	25400	19100	13700	9987
15000(N)	3254	3266	3305	3415	3688	4272	5646	9606
(A)	37000	36400	34900	31400	25600	19400	14100	10500
20000(N)	3217	3231	3274	3397	3699	4346	5873	10300
(A)	37000	36400	35000	31500	25800	19800	14700	11200

[S II] Nebular Transitions only

5000	602	625	690	878	1447	3231	8864	26640
10000	580	610	695	937	1680	4007	11350	34540
15000	566	603	708	1013	1941	4857	14080	43340
20000	556	600	722	1076	2156	5553	16320	50660

[Cℓ III] Nebular Transitions Only

5000	1454	1462	1487	1556	1729	2161	3383	7159
10000	1399	1409	1441	1529	1748	2296	3860	8695
15000	1349	1362	1402	1512	1785	2470	4441	10550
20000	1314	1329	1376	1505	1822	2624	4943	12150

[Ar IV] Nebular Transitions Only

5000	7640	7618	7552	7361	6899	6157	5646	6155
10000	7512	7490	7423	7233	6789	6128	5808	6703
15000	7283	7262	7199	7021	6620	6081	5994	7351
20000	7097	7077	7018	6853	6494	6058	6162	7912

5E(i) Collisionally-Excited Lines as Tools for Plasma Diagnostics

Electron Temperatures

How electron temperatures can be established from forbidden line ratios may be understood easily by the following elementary argument. Consider a nebula with a moderately low density and an ion of a p^2 or p^4 configuration. Denote the ^3P, ^2D and ^1S terms by A, B, and C, respectively.

Observationally, it is well-established that at low densities, the intensity of the auroral transition, C-B, is very much smaller than that of the nebular-type transition, B-A. Since A_{CB} is normally much larger than A_{BA}, this fact means that collisional deexcitations of level C then are unimportant and that in setting up Eq. (35) we were also justified in assuming that the number entering level B by radiative cascade from C can be neglected. Only at high densities will the number of collisional decays to level B from C be important and then we must use the full equations. In some important ions such as N^+, collision deexcitation of the $^1D_2$ term is significant at normal nebular densities. Under the conditions described, Eq. (35) suffices to determine the N_B/N_A ratio. For level C we can suppose that collisional excitation is followed only by radiative decay. Thus:

$$N_A N_\varepsilon q_{AC} = N_C(A_{CB} + A_{CA}) , \tag{41}$$

whence, using Eq. (26), (27), (33), (35), and (41), we would find:

$$\frac{N_C}{N_B} = \frac{\Omega_{AC}}{\Omega_{AB}} \frac{A_{BA}}{A_{CB} + A_{CA}} [1 + 1.73 \times 10^{-4} \frac{x\,\Omega_{AB}}{A_{BA}}] e^{-\chi_{BC}/kT_\varepsilon} , \tag{42}$$

Since I(auroral) = $N_C A_{CB} h\nu_{CB}$ and I(nebular) = $N_B A_{BA} h\nu_{BA}$, the intensity ratio of the auroral and nebular-type lines can give the temperature. Thus, the coefficient of the Boltzmann factor depends on density, the A-values, but most important, on the ratio of the collisional strengths. (At higher densities, one must employ the complete expressions for b_C/b_B.) Using the results of Problem 3:

$$\frac{I(aur)}{I(neb)} = \frac{\Omega_{AC}}{\Omega_{AB}} \frac{A_{CB}}{A_{CB} + A_{CA}} \frac{\lambda_{BA}}{\lambda_{CB}} [\frac{1 + a_2 x}{1 + a_1 x}] e^{-\chi_{BC}/kT_\varepsilon} . \tag{43}$$

To be rigorous we must replace Ω_{AB} and Ω_{AC} by Υ_{AB} and Υ_{AC} and find T_ε by an iterative process. One can write the expression in the general form as:

$$\frac{I(neb.\ type)}{I(aur.\ type)} = C_T [\frac{1 + a_1 x}{1 + a_2 x}] 10^{d(2,3)/t} . \tag{44}$$

Table 5-6 lists for a number of ions the auroral and nebular-type transitions, the constants C_T, a_1, a_2 and d(2,3).

The most popular line ratio that has been used for a determination of T_ε is [I(4959) + I(5007)]/I(4363). For the first approximation (for O III), we can use Seaton's (1975) formula:

$$\frac{I(4959) + I(5007)}{I(4363)} = R(O^{++}) = 7.7 \left[\frac{1 + 0.0004x}{1 + 0.044x}\right] 10^{1.432/t} , \qquad (45)$$

as modified for choices of more recent atomic constants.

For both [O III] and [N II], C_T, a_1 and a_2 change slowly with temperature (Seaton 1975). We can neglect the temperature variation of a_1. For most applications, these variations actually enter as second order effects. In practice, we use the values of the constants tabulated for T_ε = 10,000°K and then proceed by iteration.

In principle, line ratios of other ions can be used, but the difficulty is that except for [O III] and [N II] and occasionally [S III], the auroral-type transition is weak and difficult to measure. For example, the [Ne III] auroral line is not only faint but is blended with λ3343 of [Cℓ III] and possibly other lines, unless observations are secured with high dispersion. Similarly, although nebular-type transitions of [Ar III], [Ar V], and [Cℓ IV] are often well observed, the auroral-type transitions lie near the practical limit of detection. In practice, electron temperatures of gaseous nebulae are generally derived from [O III] and [N II] auroral and nebular transitions. Other ionic lines serve merely to supply corroborative evidence. With the advent of reticon systems for the near infrared, [S III] offers advantages. The λ6312 auroral transition is easily observed; the nebular λ9069 line is often quite strong. Although λ9532 is blended with a Paschen line at low dispersion, the 9069/9532 ratio is well known.

5E(ii) Collisionally-Excited Lines as Tools for Nebular Diagnostics: Electron Densities.

Ions of p^3 configurations can also be employed for temperature determinations, and more generally for determination of both temperature and density. The same principle, comparison of auroral with nebular transitions, is involved, but here the influence of the density plays a dominant role. In a p^3 configuration there is a ground $^4S_{3/2}$ level and $^2D_{5/2,\ 3/2}$ and $^2P_{3/2,\ 1/2}$ excited levels. Ions are collisionally excited mainly from the ground level to excited levels, whence they cascade back with the emission of a pair of closely spaced nebular lines and four auroral-type transitions that usually fall close to one another.

Let us first examine a simplified situation in which we consider only the two nebular-type transitions $^4S_{3/2} - {}^2D_{5/2}$ (denoted as 1 - 2) and $^4S_{3/2} - {}^2D_{3/2}$ (1 - 3). Let us assume that processes involving the ^2P term have only a second-order influence on ^2D-level populations. Then we may write the equations of statistical equilibrium, taking into account collisional excitation from the ground level to levels 2 and 3 and radiative and collisional deexcitation therefrom. Since the excitation difference between levels 2 and 3 is extremely small, we can set $\chi_{12} = \chi_{13}$. Then:

Table 5-6. Coefficients for Temperature Determinations

Ion	t	Auroral Type Transition	Nebular Type Transition	C_T	See Eq. (44) a_1	a_2	d(2,3)
N II	0.5	5755	6548	5.02*	5.2 (-4)	0.249	1.086
	1.0		6584*	5.15*	5.3 (-4)	0.256	
	1.5			5.19*	5.3 (-4)	0.261	
	2.0			5.20*	5.4 (-4)	0.265	
O III	0.5	4363	4959	8.01	3.2 (-4)	3.88 (-2)	1.432
	1.0		5007	7.77	3.8 (-4)	4.37 (-2)	
	1.5			7.66	4.0 (-4)	4.54 (-2)	
	2.0			7.62	4.0 (-4)	4.55 (-2)	
Ne III	1.0	3342	3969 3869*	10.4 *	7.0 (-5)	2.6 (-3)	1.87
Ne V	1.0	2972	3426	15.29	9.3 (-5)	1.83 (-3)	2.107
	1.5		3346	16.52	9.6 (-5)	2.06 (-3)	
	2.0			17.2	1.01(-4)	2.30 (-3)	
S III	1.0	6312	9069 9532	6.45	8.8 (-4)	4.7 (-2)	0.99
Cℓ IV	1.0	5323	7530	3.09	5.9 (-4)	6.2 (-3)	1.174
	1.5		8045	3.23	6.8 (-4)	7.3 (-3)	
Ar III	1.0	5192	7751 7135*	10.17*	2.0 (-4)	4.6 (-3)	1.203
Ar V	1.0	4625	6435 7005	6.08	2.1 (-4)	1.8 (-3)	1.351

Values of C_T indicated by a * apply to intensity measurement ratios of only the stronger nebular-type transition to the auroral line, e.g., I(6584)/I5755) [N II], I(3869)/I(3342) [Ne III] and I(7135)/I(5192) [Ar III]. If both transitions are used, the C_T values are to be multiplied by $I(neb)/I(\lambda_*)$ = 1.340, [N II]; 1.30, [Ne III], and 1.24, [Ar III].

$$N_1 N_\varepsilon q_{12} + N_3 N_\varepsilon q_{32} = N_2 A_{21} + N_2 N_\varepsilon(q_{23} + q_{21}) \ . \quad (46)$$

$$N_1 N_\varepsilon q_{13} + N_2 N_\varepsilon q_{23} = N_3 A_{31} + N_3 N_\varepsilon(q_{32} + q_{31}) \ . \quad (47)$$

Radiative transitions between the two ^{2}D levels are negligible. Let $C = 8.63 \times 10^{-4}x$ and use Eq. (26) and (27). We then obtain:

$$\frac{N_3}{N_2} = \frac{N(^2D_{5/2})}{N(^2D_{3/2})} = \frac{\omega_3}{\omega_2} \; \frac{\dfrac{\Omega_{23}}{\Omega_{13}} + \dfrac{\Omega_{23} + \Omega_{12}}{\Omega_{12}} + \dfrac{A_{21}\,\omega_2}{C\,\Omega_{12}}}{\dfrac{\Omega_{23}}{\Omega_{12}} + \dfrac{\Omega_{23} + \Omega_{13}}{\Omega_{13}} + \dfrac{A_{31}\,\omega_3}{C\,\Omega_{13}}} \ . \quad (48)$$

Consider first the limiting case of high density (large x). The term involving the A's drops out. Collisions alone determine the population of the levels. Accordingly, the relative populations of the levels is simply the ratio of their statistical weights, $N_3/N_2 = \omega_3/\omega_2$, whence the expression for the intensity ratio of the two nebular lines is:

$$\frac{I_{31}}{I_{21}} = \frac{\omega_3 A_{31}}{\omega_2 A_{21}} \frac{\nu_{31}}{\nu_{21}} = \frac{\omega_3 A_{31}}{\omega_2 A_{21}} \ , \quad (49)$$

since $\nu_{31} = \nu_{21}$ for a close doublet. Note that the collision strengths do not enter at all It is sufficient to know that the population ratio is determined by collisions and therefore is given by Boltzmann's law, and that $I = NAh\nu$. In the limiting case of low density, the C factor becomes small, so the A term dominates. Thus, $N_3/N_2 = (A_{21}\,\Omega_{13})/(A_{31}\,\Omega_{12})$ and $I_{31}/I_{21} = N_3A_{31}\Omega_{31}/N_2A_{21}\Omega_{21}$ now depends on the Ω ratio:

$$\frac{I_{31}}{I_{21}} = \frac{\Omega_{13}}{\Omega_{12}} \ . \quad (50)$$

Under these conditions, every atom that reaches ^{2}D escapes by radiation. N_2 and N_3 are determined by the number of collisional excitations. The A-values are not involved at all. In between these two extremes (which depend on the Ω's and A's involved), the ratio changes rapidly with density, while outside this range the ratio remains constant at one or the other of its extreme values.

In actual practice, the term separations, the sizes of the A-values and of the collision strengths are such that the ^{2}D - ^{4}S doublet ratio does depend slightly on T_ε because populations in the ^{2}D levels are influenced by interchanges with the ^{2}P levels. The temperature effect is relatively minor. Several line ratios involving p^3 configurations are of practical importance for establishing electron densities in gaseous nebulae. These are: [N I] I(5200)/I(5198),

[O II] I(λ3729)/I(3726), $2p^3$; [S II] I(λ6717)/I(λ6730), $3p^3$
[Cℓ III] I(λ5517)/I(λ5537), $3p^3$; [Ar IV] I(λ4711)/I(λ4740), $3p^3$

and the UV ratio: [Ne IV] I(λ2422)/I(λ2425), $2p^3$.

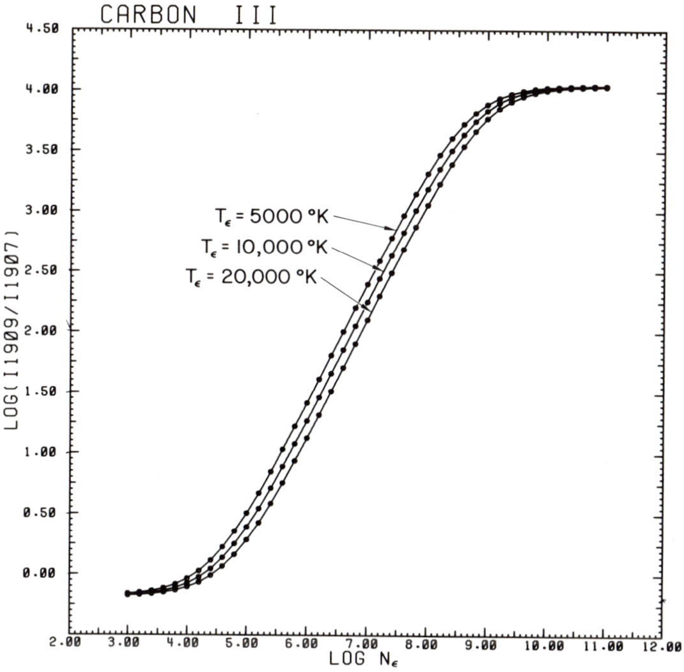

Figure 5-4

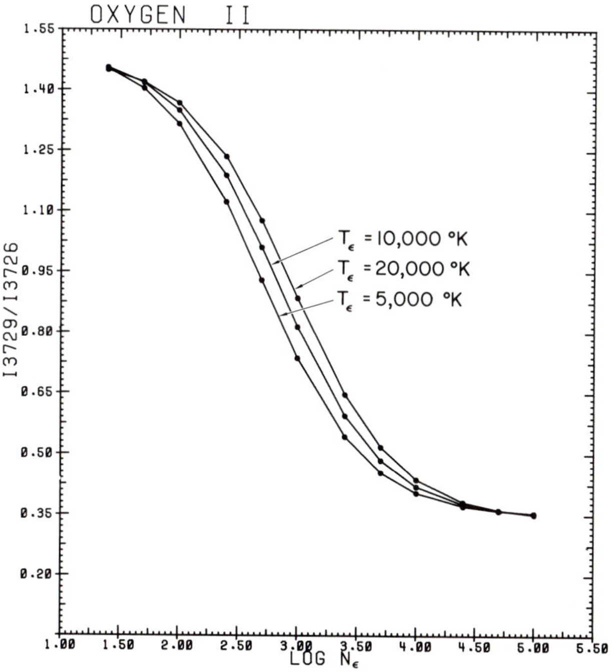

Figure 5-5

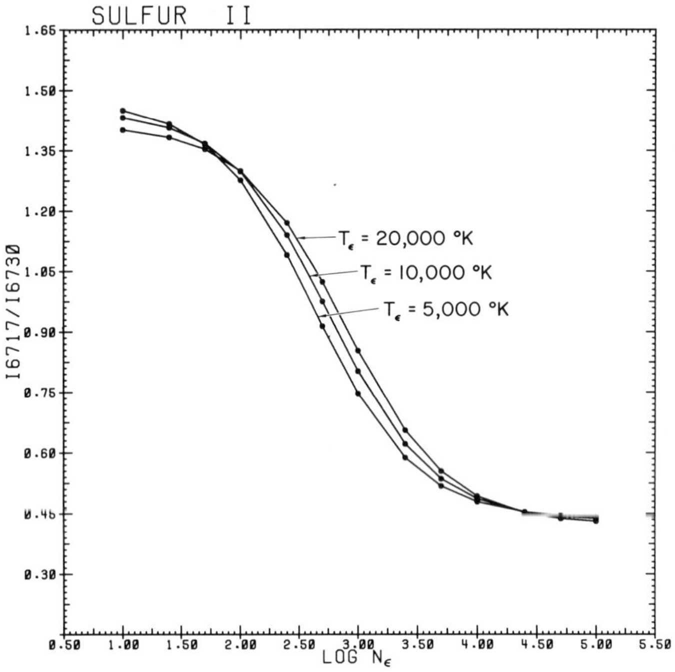

Figure 5-6

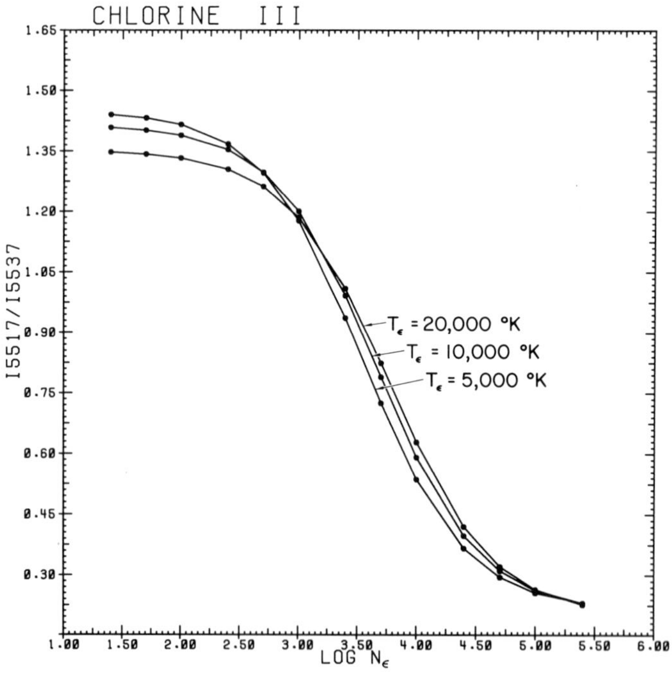

Figure 5-7

The C III] intercombination line ratio $2s^2\ ^1S_0 - 2s2p\ ^3P^o_1$ (λ1908.73) to $2s^2\ ^1S_o - 2s2p\ ^3P^o_2$ (λ1906.68) provides a useful density discriminant for the approximate range, $3.5 < \log N_\varepsilon < 5.5$. Diagnostic curves have been published by Nussbaumer and Schild (1979); Feibelman finds that this C III] ratio tends to give N_ε values lower than those found by other methods. See Figure 5-4. High-resolution ultraviolet spectra enable us to use ratios such as N III I(1748.7)/I(1752.2) for $\log N_\varepsilon > 5.0$; N IV I(1486)/I(1483) for $\log N_\varepsilon > 5.00$, and Si III I(1892/1894) for $\log N_\varepsilon > 5.00$.

The classical and possibly the most accurate ratio for obtaining N_ε in gaseous nebulae is λ3729/λ3726 [O II], which depends slowly on T_ε but is sensitive to N_ε over an important range in x. The great advantage of line ratios of this type is that we do not have to worry about wavelength-dependent factors in photometry or extinction. The [S II] ratio is a good density indicator over an important range characteristic of dense filaments. The lines are well separated in wavelength, are often strong and can be measured accurately. See Figures 5-5 and 5-6. The [Cℓ III] ratio is more difficult to measure as these lines are often very weak. See Figure 5-7. The λ4711 [Ar IV] transition is often blended with λ4713 He I at moderate dispersions. We can estimate the intensity of the latter from λ4471 He I to obtain a corrected value of I(λ4711). In high-excitation objects, the He I contribution is often negligible. See Figure 5-8. The [N I] line ratio is useful in exploring regions where H is probably only partly ionized, as in H I - H II region interfaces and high-density filaments.

5E(iii) Line Ratios That Depend on Both Temperature and Density

In addition to forbidden line doublets one may compare auroral and nebular transitions for p^3 ions of [O II], [S II], [Ar IV], [Cℓ III] and [Ne IV]. Ultraviolet and optical region ratios may be compared. Some examples are: [N II] [I(2140) + I(2144)]/[I(6584) + I(6548)], versus I(5755)/[I(6584) + I(6548)]. O II: I(2470)/I(3727) versus I(3729)/I(3726). See Fig. 5-9. O III: [I(2321) + I(2330)]/[I(1660) + I(1666)] or [I(1160) + I(1666))]/[I(4959) + I(5007)] plotted against I(4363)/[I(4959) + I(5007)].

There exists, however, one severe practical difficulty. Since the lines involved fall far apart in wavelength, any wavelength-dependent factor, such as interstellar extinction or an error in a stellar energy distribution used to calibrate the nebular measurements, will be reflected in the results. Also, the numerical values of the ratios are dependent on both T_ε and N_ε or t and x. In a nebula with variable density and temperature, the derived N_ε and T_ε-values will be weighted in different ways.

In the optical region, we may compare the red 7319 and 7330 lines with 3727 (see Fig. 5-10) or we may compare the transauroral [S II] 4068 line with the nebular-type 6717, 6730 transitions (see Fig. 5-11). One of the most useful ratios is I(4726)/I(2423) in [Ne IV] since it enables us to probe conditions in the hotter interior zones of planetaries (see Fig. 5-12).

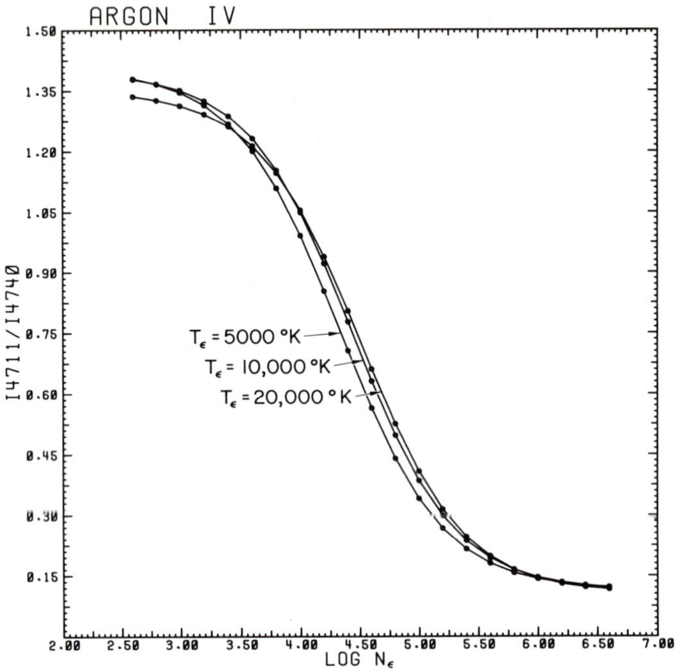

Figure 5-8

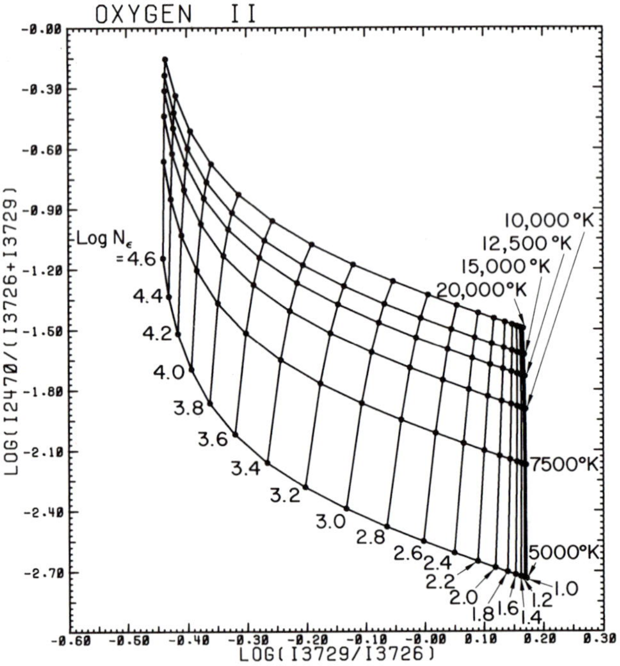

Figure 5-9

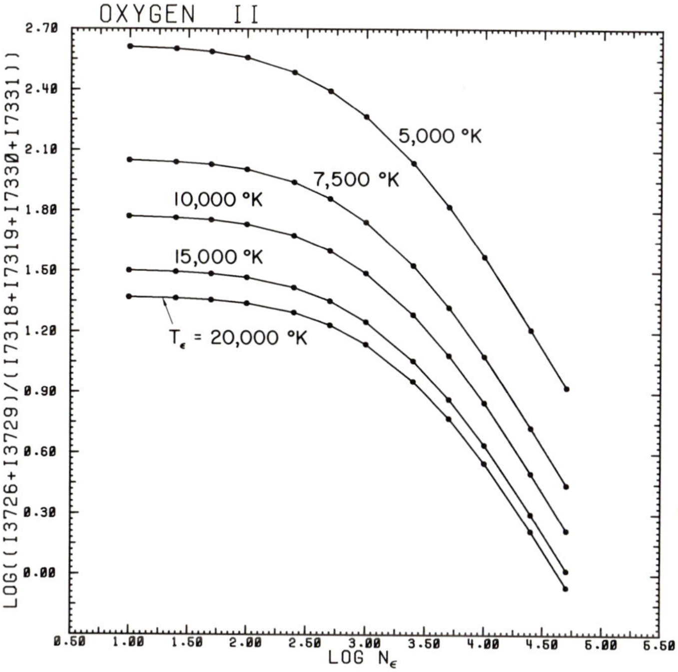

Figure 5-10

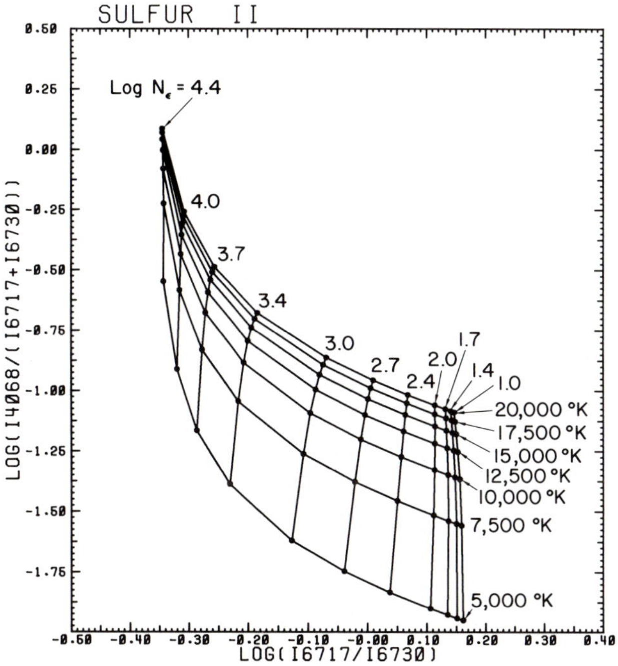

Figure 5.11

For any nebula we combine various diagnostic criteria to obtain estimates of N_ε and T_ε. Note that each observed value of a ratio, 3729/3726, 6717/6730, 6583/5755, etc., defines a curve in the (N_ε, T_ε) plane. If the nebular plasma were of uniform density and temperature and all atomic parameters were accurate, all these curves would intersect at a point. Figure 5-13 shows a diagnostic diagram for NGC 6741. Note that p^3 nebular forbidden doublet ratios give curves running almost vertically, while those for [O III] and [N II] run roughly horizontally at low densities and drop downwards as N_ε increases. The auroral/nebular ratio 3727/7325 for [O II] runs diagonally. The curves do not intersect at a point, indicating that the nebular plasma is not of uniform density and temperature, or there are errors in the atomic parameters, observational errors, or perhaps all three effects are operative. The temperature difference indicated by the [O III] and [N II] lines is somewhat larger than that suggested by theoretical model calculations for this nebula. See Chapter 7. The 3727/7325 curve may be affected by extinction, [S II] and [O II] may originate in different strata or may reflect effects of theoretical and observational uncertainties.

Combinations of infrared and optical data for ions of p^2 and p^4 configurations offer some valuable plasma diagnostics. For example, one may compare the ratio of the nebular 5007A [O III] transition to the fine structure 88 μm line with the familiar ratio λ4363/λ5007 which depends primarily on T_ε. The λ5007A/88 μm ratio, however, is sensitive to both T_ε and N_ε. See Fig. 5-14. By combining optical and infrared measurements at 88 μm, Ms. Dinerstein was able to find both N_ε and T_ε and show that [O III] densities were similar to those found from [O II] lines. The same principle can be applied for [Ar III] and [S III].

5F. Observational Checks on Accuracy of Theoretical Parameters

A fundamental difficulty inherent in nebular plasma diagnostics is the scarcity of suitable observational checks on the theory. We are restricted mostly to a comparison of observed and predicted line ratios. The easiest and most convenient check is provided by the intensity ratio of two nebular-type transitions in p^2 and p^4 configurations. Early photoelectric measurements (Liller and Aller 1954) gave I(5007)/(I(4959) = 3.03 ± 0.11 versus a predicted value of 2.93 (Garstang 1951). Subsequent investigations employing both improved observational material and theoretical calculations have shown here a satisfactory agreement between theory and observations.

Different calculations of A's and Ω's yield distressingly different results. Presumably, a set of atomic parameters that gave consistent values of T_ε for regions of comparable excitation levels, e.g., [O III], [Ar III], and [Ne III] would be preferred over those that gave discordant results. Unfortunately, until accurate intensity measurements of very weak lines are available, this criterion is of limited usefulness.

The p^3 configurations offer greater difficulties. We saw that the limiting intensity ratio for a nebular-type doublet $I(3 \rightarrow 1)/I(2 \rightarrow 1)$ for very high densities would be $\omega_3 A_{31}/\omega_2 A_{21}$. For [O II], 3729/3726 the limiting value of the ratio is 0.35. Until

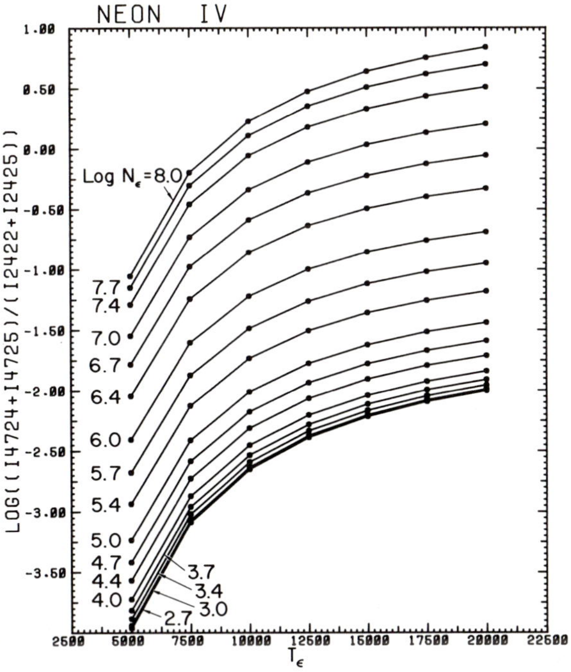

Figure 5.12

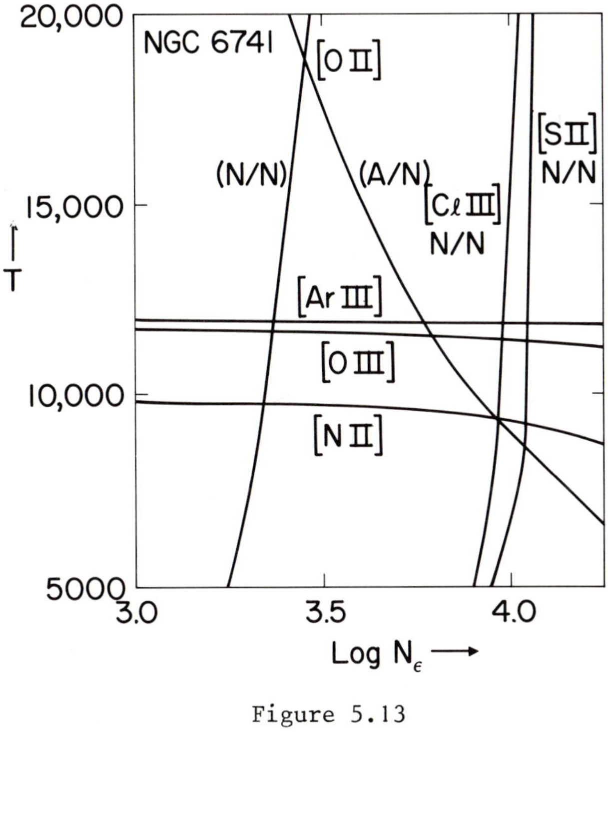

Figure 5.13

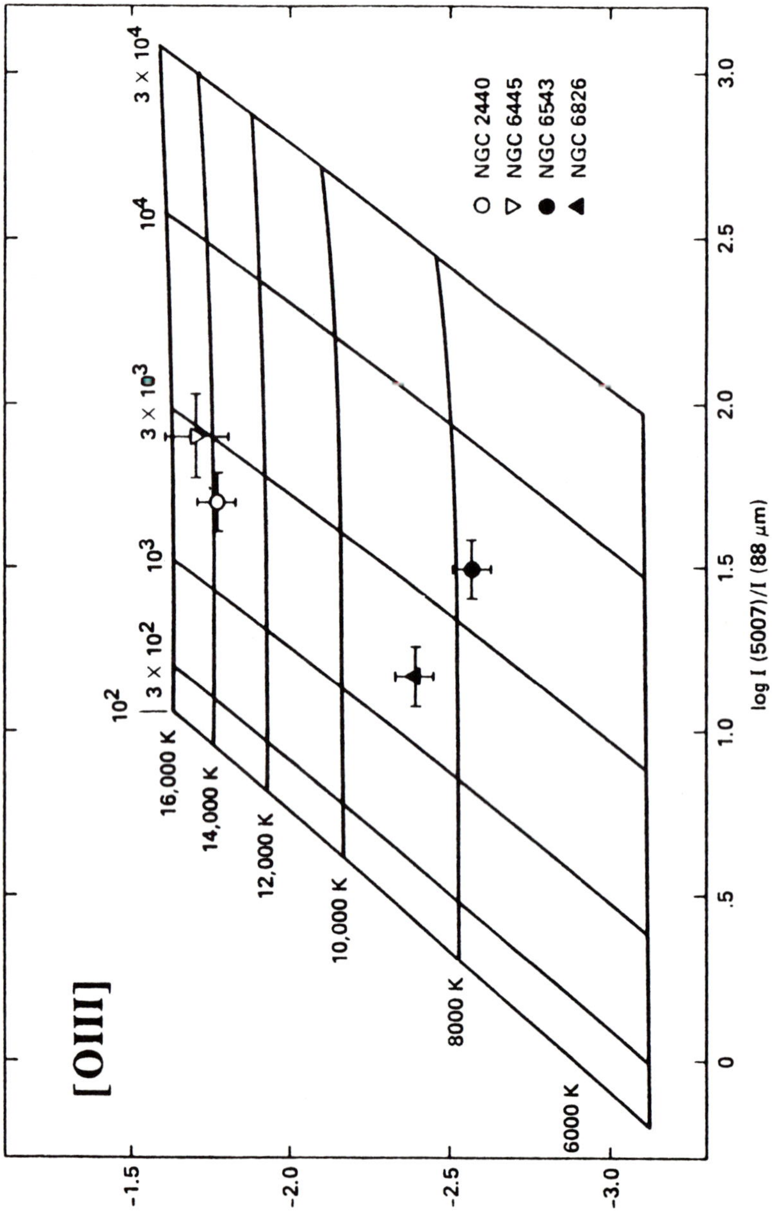
[OIII]
6000 K
8000 K
10,000 K
12,000 K
14,000 K
16,000 K
10^2
3×10^2
10^3
3×10^3
10^4
3×10^4
O NGC 2440
▽ NGC 6445
● NGC 6543
▲ NGC 6826
log I (4363)/I (5007)
-1.5
-2.0
-2.5
-3.0
log I (5007)/I (88 μm)
0
.5
1.0
1.5
2.0
2.5
3.0

Figure 5.14

Table 5-7. Plasma Diagnostics for a Number of Planetary Nebulae

(1)	(2) [O II]	(3) [Cℓ III]	(4) [S II]	(5) O^+	(6) $C\ell^{++}$	(7) S^+	(8) [O III]	(9) t	(10) log x	(11) [N II]	(12) t	(13) log x	(14)
NGC 40		1.03	0.81		- 0.48	- 0.83				114	0.80	- 0.8	2
Hu 1-1		0.97	0.75		- 0.40	- 0.70	124	1.15	- 0.7	60.4	1.0	- 0.7	6
IC 1747	0.58	1.03	0.60	- 0.64	- 0.48	- 0.40	152	0.98	- 0.4	58.6	0.98	- 0.4	6
IC 351	0.49	1.0	0.71	- 0.26	- 0.44	- 0.64	113	1.19	- 0.5				8
IC 2003		0.83	0.53		- 0.20	- 0.21	168	1.03	- 0.20	30.7	1.30	- 0.20	8
IC 418		0.72	0.44		- 0.03	- 0.10	367	0.83	+ 0.05	74.3	0.87	0.05	3
NGC 2022	0.92	1.16	0.76	- 1.11	- 0.70	- 0.73	78.7	1.37	- 0.9				10
IC 2165	0.62	0.87	0.61	- 0.59	- 0.25	- 0.43	79.6	1.36	- 0.5	43.2	1.12	- 0.5	9
J 900		0.82	0.61		- 0.18	- 0.43	114.4	1.18	- 0.4	49.1	1.06	- 0.4	7
NGC 3242	0.5	0.88	0.65	- 0.30	- 0.26	- 0.54	142	1.09	- 0.3	53.2	1.01	- 0.3	7
IC 4593	0.7		0.76	- 0.74		- 0.73	238	0.94	- 0.74				4
NGC 6210	0.45	0.83		- 0.12	- 0.20		230	0.94	- 0.10				5
IC 4634		0.78	0.51		- 0.12	- 0.15	219	0.95	- 0.15	37.0	1.11	- 0.15	5
NGC 6309	0.56	0.71	0.59	- 0.48		- 0.38	142	1.10	- 0.5	60	0.98	- 0.5	8
NGC 6543	0.54	0.68	0.61	- 0.44	+ 0.2	- 0.45	417	0.80	- 0.4	62.3	0.96	- 0.4	5
IC 4776		0.62	0.49		+ 0.12	- 0.07	340	0.84	- 0.1	23.6	(1.4)	- 0.1	6
Hu 2-1		0.89	0.43		0.24	0.16	272	0.875	+ 0.25				
NGC 6741	0.64	0.61	0.49	- 0.63	+ 0.14	- 0.06	120	1.17	- 0.6	61.9	0.93	- 0.15	8
NGC 6751		1.36	0.61		- 1.26	- 0.43	192	1.00	- 1.0	1.14	0.80	- 1.0	7
IC 4846	0.44	0.58	0.58	- 0.07	+ 0.16	- 0.35	202	0.97	- 0.05	28:	1.3:	- 0.05	5
NGC 6778	0.77	1.01	0.92	- 0.86	- 0.46	- 1.0	490:	0.78	- 0.90	109	0.80	- 0.9	5
NGC 6790	0.41	0.67	0.46	+ 0.10	+ 0.04	+ 0.05	126	1.13	0.0				6
MI-74	0.48	0.39	0.47	- 0.22	+ 0.6	0.0	221	0.93	0.25	27:	1.3:	0.25	
NGC 6818	0.62	0.94	0.74	- 0.60	- 0.35	- 0.79	100	1.25	- 0.6	44.5	1.12	0.6	9
NGC 6826	0.70	0.88	0.67	- 0.73	- 0.25	- 0.57	173	1.03	- 0.6				5
NGC 6884	0.45	0.73	0.54	- 0.11	- 0.06	- 0.25	164	1.06	- 0.1	37.7	1.12	0.1	6
NGC 6886	0.48	0.54	0.51	- 0.24	+ 0.3	- 0.32	92.5	1.27	- 0.2	50	1.05	0.2	8
NGC 7026		0.76	0.51		- 0.10	- 0.14	288	0.88	- 0.10	60	0.92	0.10	6
IC 5117		0.3	0.45		(+ 1)	+ 0.07	99	1.18	+ 0.5	18.8	1.25	+ 0.5	6
Hu 1-2		0.69	0.52		+ 0.02	- 0.14	52.5	1.62	- 0.10	38	1.12	- 0.10	10
IC 5217	0.45	0.76	0.50	- 0.12	- 0.10	- 0.10	143	1.09	- 0.10	39	1.11	- 0.10	6
NGC 7662	0.57	0.87	0.65	- 0.5	- 0.25	- 0.53	86.5	1.31	- 0.50	61	1.0	- 0.10	8

We update a discussion by Aller and Czyzak (1979) Astrophys. Space Sci., 62, 397, who employed diagnostic relationships expressed in terms of (x, t) rather than (N_ε, T_ε). Here $x = 0.01\, N_\varepsilon/\sqrt{T_\varepsilon}$ and $t = T_\varepsilon/10{,}000$. Column (1) gives the usual name of the planetary nebula. Columns (2), (3), and (4) give respectively the ratios 3729/3726 [O II], 5517/5537 [Cℓ III], and 6717/6730 [S II], while Columns (5), (6), and (7) give the values of log x obtained from the ratios in Columns (2), (3), and (4) for T_ε = 10,000°K. Column (8) gives [I(4959) + I(5007)]/I(4363), [O III]. Column (9) gives t, where the log x values in Column (10) are adopted. Column (11) gives I(6583)/I(5755), [N II]. Column (12) gives the corresponding t values where the log x values assumed for the N+ zone are taken from Column (13). Here Eq. (44) with Seaton's values of the constants was employed for [O III] and [N II]. Column (14) gives the excitation class of the nebula (cf. Aller: Gaseous Nebulae, 1956, p. 66). The [O II] ratios are from various sources (for references, see Aller and Czyzak 1979). The [Cℓ III] and [S II] ratios were all measured with the image tube scanner at Lick Observatory. Ratios of [O III] and [N II] lines are from various sources (references are cited in Aller and Czyzak, 1983, Ap. J. Suppl., 51, 211).

recently, predicted values were always larger. Eissner and Zeippen (1981) showed that this discordance could be removed by improving the theory of magnetic dipole transitions and including relativistic terms. Zeippen's limiting value for the corresponding [N I] doublet is 0.54, in agreement with observations by Czyzak and Aller. In nebulae of the highest density, the [S II] $(^2D_{5/2} - {}^4S_{3/2})/(^2D_{3/2} - {}^4S_{3/2})$ ratio seems to approach a value near 0.43; the limiting asymptotic value $r(x \rightarrow \infty)$ is perhaps a little lower, perhaps near 0.40. The theoretical predicted limiting values of the ratio are 0.38 (Czyzak and Krueger 1963) and 0.44 (Mendoza and Zeippen 1982), respectively. The curves in Fig. 5-6 were based on the Mendoza-Zeippen values. Note that because of uncertainties in the values of the transition probabilities, estimates of log N_ε are likely to be in error at high densities. In view of Eq. (28), the limiting value of the ratio $r(N_\varepsilon \rightarrow \infty) \sim 1.5$ gives us no useful information on the Ω's.

In [Ar IV], where several distinct auroral-type transitions can be compared with the nebular-type doublet, a more exacting test can be applied (Czyzak et al. 1980) (see Fig. 5-1). Consider the auroral line ratio I(7262)/I(4740) which is a function of N_ε and T_ε. For typical high-excitation nebulae such as NGC 7662, 86-8°1, and IC 2165, this ratio gives log x ~ 1.3, 1.2, and 1.9, respectively, while other diagnostic criteria give log x ~ -0.5, -1, and -0.5. Thus, densities derived from this [Ar IV] ratio are two orders of magnitude too high! The discordance cannot be removed by juggling T_ε. Other [Ar IV] lines such as λ7237 and (at dispersions normally employed), λ4711 are markedly affected by blends.

In summary, the higher the density of the object, the greater the effect of the errors in the A values. Uncertainties in the Ω's can affect the shape of the curve, while the absolute value of Ω defines the domain of log N_ε where one approaches the limiting ratio.

Table 5-7 gives diagnostics for a number of planetaries. The [Cℓ III] densities are often larger than the others, possibly because of a lack of improved Ω's. Except for objects showing or suspected of having pronounced filamentary structure, the densities generally are in harmony with those found from the surface brightness in Hβ and plausible nebular distances.

Problems

1. Noting that:

$$(2J'+1)\ \sigma_v(J',J) = \left(\pi/k^2\right) \sum_{M_{J'},M_J} \Omega\left(J'M_{J'};JM_J\right)$$

and applying the principle of detailed balancing to the energy interval dE' = dE or v'dv' = vdv, show:

$$(2J'+1)\ v'^2\ \sigma(J',J) = (2J+1)\ v^2\sigma(J,J')\ ,$$

and that:

$$\Omega(J',J) = \Omega(J,J')\ .$$

Problems (Continued)

2. Consider the term structures in ions like N^+, O^{++}, Ne^{++}, Ne^{+4}, S^{++}, Ar^{++}, Ar^{+4}. Under what condtions can we treat the ^3P term in the ground p^2 and p^4 configurations as a single "level" in equations of statistical equilibrium, so that we need solve only three simultaneous equations instead of five for typical gaseous nebulae?

3. Show that:

$$\frac{b_C}{b_B} = \frac{1 + \varepsilon_\kappa + \dfrac{5794\, A_{BA}}{\Omega_{AB}\left(1 + \Omega_{BC}/\Omega_{AC}\right) x}}{1 + \varepsilon_\kappa + \dfrac{1159\left[A_{CA} + A_{CB}\left(1 + (\Omega_{AC}/\Omega_{AB})\exp\left(-\chi_{BC}/kT_\varepsilon\right)\right)\right]}{\Omega_{CA}\left(1 + \Omega_{BC}/\Omega_{AC}\right)x}}$$

where:

$$\varepsilon_\kappa^{-1} = \left[\Omega_{AB}/\Omega_{BC} + \Omega_{AB}/\Omega_{AC}\right] \exp\left(\chi_{BC}/kT_\varepsilon\right)$$

so that if we can neglect ε_κ and the exponential term in the denominator, we get in Eq. (43):

$$\left[\frac{1+a_2 x}{1+a_1 x}\right] \cong \frac{1 + \dfrac{1.73 \times 10^{-4}\, \Omega_{AB}\left(1 + \Omega_{BC}/\Omega_{AC}\right)}{A_{BA}}\, x}{1 + \dfrac{8.63 \times 10^{-4}\, (\Omega_{AC} + \Omega_{BC})}{(A_{CA} + A_{CB})}\, x}$$

References

1. Collisional Excitation and Ionization Rates for H are given, e.g., by:

Johnson, L.C. 1972, Ap. J., 174, 227, see also:
Drake, S., and Ulrich, R.K. 1980, Ap. J. Suppl., 42, 351, and references therein cited.

2. A-Values and Collision Strengths

See I.A.U. Symposia No. 76 (1978) and No. 103 (1983), also see references in Appendix. A review article summarizing developments up to 1967:

References (Continued)

Czyzak, S.J. 1968, Stars and Stellar Systems, **7**, 403; Nebulae and Interstellar Matter, ed. B. Middlehurst and L.H. Aller, Chicago, Univ. Chicago Press.

For early references not otherwise cited, see Menzel, D.H. (ed.) Physical Processes in Ionized Plasmas, 1962, New York, Dover.

The literature is so extensive that we quote here only a few recent illustrative papers. Thus, for A-values:

Mendoza, C., and Zeippen, C.J. 1982, M.N.R.A.S., **198**, 127; **199**, 1025.
Nussbaumer, H., and Rusca, C. 1979, Astron. Astrophys., **72**, 129.
Eissner, W., and Zeippen, C.J. 1981, J. Phys. B., At. Mol. Phys., **14**, 2125.

Ω-Values

Krueger, T.K., Czyzak, S.J. 1970, Proc. R. Soc. London A, **318**, 531.
Seaton, M.J. 1975, M.N.R.A.S., **170**, 476.
Pradhan, A.K. 1978, M.N.R.A.S., **183**, 89P.
Baluja, K.L., Burke, P.G., and Kingston, A.I. 1981, J. Phys. B., **13**, 4675.

3. Applications to Nebular Diagnostics

Early determinations of T_ε employing [O III] lines (p^2 configuration) were made by Menzel, Aller and Hebb (1941) using Ω-values computed by Hebb and Menzel (1940), A-values from Pasternack (1940), and from Shortley *et al*. (1941). Although the absolute values of these Ω's, which were obtained from an inadequate theory, were substantially in error, the ratios were much less inaccurate, such that electron temperatures found by this method were of the order of 10% lower than the best modern values. They sufficed to show that electron temperatures of gaseous nebulae were in the neighborhood of 10,000°K.

Since forbidden line intensity ratios depend on both electron density and temperature, if we suppose that lines of two ionic species, e.g., O^{++} and N^+, arise in the same strata, we can use the two sets of ratios to obtain N_ε and T_ε (Aller and White 1949, A.J., **54**, 181). This is the principle applied in diagnostic diagrams (Fig. 13). The method requires accurate Ω-values, which were not available in 1949. Seaton's breakthrough (see e.g., Proc. Roy. Soc. London, **A218**, 400, 1953; **A231**, 37, 1955; Phys. Soc. Proc., **68**, 457, 1955), in cross section theory which enables reliable collision strengths to be computed, made it possible to determine trustworthy temperatures and densities from nebular line ratios.

That forbidden line doublet ratios could depend on electron density was established by Aller, Ufford, and Van Vleck (1949) who

References (Continued)

measured the [O II] I(3727)/I(3726) ratio in a number of planetary nebulae and compared results with theoretical predictions. The observed intensity ratio was opposite to that predicted by the first-order theory. Improvements in the theory required taking into account the second-order spin-orbit interaction and magnetic interaction between the spin of one electron and the orbit of another! Even with these refinements all discordance was not removed. It was noted that the denser the nebula, the closer was the observed ratio to the predicted value. This variation was interpreted as a result of radiative and collisional processes that competed at different rates to populate and depopulate the ^2D levels. Thus, the 3729/3726 ratio could provide a clue to the density. It was not until after Seaton had developed an adequate theory of collisional excitation that Seaton and Osterbrock (1967, Ap. J., 125, 66) were able to give a satisfactory treatment of the problem. Applications of np^3 nebular line ratios of [S II], [Cl III], [Ar IV], and [K V] have been made by a number of workers, see, e.g., Krueger et al., 1980, Proc. Nat'l. Acad. Sci. USA, 66, 14, 282; Ap. J., 160, 921; Saraph and Seaton, 1970, M.N.R.A.S., 148, 367.

With modern computers, we can easily solve equations of statistical equilibrium for five to typically fifteen levels using the best available atomic parameters (see, e.g., the compilation by Mendoza in I.A.U. Symposium No. 103).

A useful compilation for forbidden lines of np^q configurations of atoms and ions of C, N, O, Ne, Mg, Si, S, and Fe in the temperatore range between 5000 and 2,000,000 degrees has been given by M. Kafatos and J.P. Lynch (1980, Ap. J. Suppl., 42, 611).

4. Evaluations of the accuracy of nebular line intensity predictions are difficult. Illustrative examples are given by Czyzak et al., 1980, Ap. J., 241, 719, for [Ar IV], and in Highlights of Astronomy, 1983, 6, 791, ed. R.M. West, Reidel Publ. Co. Much more work needs to be done on this problem.

5. Observational Data Pertinent to Nebular Plasma Diagnostics

Optical Region. A compilation of spectroscopic data on gaseous nebula up to 1975 is given by:

Kaler, J.B. 1976, Ap. J. Suppl., 31, 517.

More recent data obtained by photoelectric photometry, image tube scanners, etc., include:

Torres-Peimbert, S., and Peimbert, M. 1977, Rev. Mex. Astron. Astrofis., 2, 181.
Barker, T. 1978, Ap. J., 219, 914; 220, 193.
Kaler, J.B. 1978, Ap. J., 226, 947.

References (Continued)

Aller, L.H., and Czyzak, S.J. 1982, Ap. J. Suppl., 51, 211, and references cited therein.

Ultraviolet Data for individual nebulae appear in many papers in contemporary periodicals. For a good starter see I.A.U. Symposium No. 103 and references therein, also: Universe at Ultraviolet Wavelengths, 1981, ed. R.D. Chapman, NASA Conference Publication No. 2171.
Infrared. For summaries and references for work on planetaries, see IAU Symposia No. 76, 1978; No. 103, 1983. See also:

Aitken, D.K., Roche, P.F., Spenser, P.M., Jones, B. 1979, Ap. J., 233, 925.
Aitken, D.K., and Roche, P.F. 1982, M.N.R.A.S., 200, 217.
Dinerstein, H. 1983, I.A.U. Symposium No. 103, 79.
Zeilik, M. 1977, Ap. J., 218, 118 (H II regions).
Grasdalen, G. 1979, Ap. J., 229, 587.
Beck, S.C., Lacy, J.H., Townes, C.H., Aller, L.H., Geballe, T.R., and Baas, F. 1981, Ap. J., 249, 592.
Pottasch, S., et al. 1984, Ap. J. Letters (IRAS results).

List of Illustrations for Chapter 5

Fig. 5-1. Term Schemes for p^2 and p^3 Configurations.
(Left) The $2p^2$ Configuration of [O III]. The ground ^3P term is not drawn to scale. M denotes magnetic dipole transitions which obey the selection rules, $\Delta n = 0$, $\Delta \ell = 0$, $\Delta J = 0, \pm 1$, 0 ($0 \nrightarrow 0$ excluded). Q denotes electric quadrupole transitions for which the selection rules are Δn = arbitrary, $\Delta \ell = 0, \pm 2$; $\Delta J = 0, \pm 1, \pm 2$ (0 − 0, 1/2 −1/2, $1 \nleftrightarrow 0$ excluded). Theoretical transition probabilities according to Czyzak are given for each optical line. Transitions within the ground-term all fall in the infrared. Nussbaumer and Rusca give the following transition probabilities:

$$A(^3P_2 - {}^3P_1) = 9.7 \times 10^{-5},\ A(^3P_1 - {}^3P_0) = 2.6 \times 10^{-5} \text{ and}$$
$$A(^3P_2 - {}^3P_0) = 3.2 \times 10^{-11}.$$

The quadrupole contributions to λ5007 and λ4959 are completely negligible. The λ4931 $^1D_2 - {}^3P_0$ transitions can be seen in the spectra of several planetaries (e.g., Fig. 2-2, NGC 2440).
(Right) The $3p^3$ Configuration of [Ar IV]. The ^2P and ^2D level separations are not drawn to scale. For each line we give the wavelength in A and the transition probability.

Fig. 5-2. Energy Scheme for an Encounter Between a Ground Level O^{++} Ion and a Free p-Electron ($\ell = 1$). The doubly-excited 2s $2p^3$ ($^3D^o$) 3s $^2D^o$ and 2s $2p^3$ ($^3P^o$) 3s $^2P^o$ levels of O^+ lie just above the energies of an O^{++} ion and in a ^1D and a ^1S level, respectively.

List of Illustrations for Chapter 5 (Continued)

Thus, an incoming p-electron with an energy slightly above that needed to excite the $^1D_2$ level of O^{++} may match the energy of the doubly-excited 2s $2p^3$ ($^3D^o$) 3s $^2D^o$ level of O^+; hence, its cross section for capture will get a profound enhancement. Similarly, the capture cross section of an incoming p-electron whose energy matches the 2s $2p^3$ ($^3P^o$) 3s $^2P^o$ level will be profoundly perturbed. The presence of such resonances greatly complicates the calculation of collision strengths.

List of Illustrations for Chapter 5 (Continued)

Chapter 6

RADIATIVE EQUILIBRIUM OF A GASEOUS NEBULA

For radiatively ionized plasmas with densities exceeding about 10^2 electrons/cm^3, the relaxation times of relevant physical processes are such that we can usually regard a given volume element as being in a steady state. If we can neglect mechanical effects, particularly energy gained or lost by compression or expansion, we can impose the condition of radiative equilibrium or energy balance. The amount of energy absorbed in each volume element then equals the amount emitted.

6A. The Energy Balance of a Gaseous Nebula

We follow the treatment given by Menzel and his coworkers (1938), who set up the equation of energy conservation first for a pure hydrogen nebula in which photoionization, free-free and bound-free transitions and later collisional effects were taken into account. This analysis showed very clearly the thermostatic action of collisional line excitation. Although the exciting star temperature, T_1, might attain very high values, e.g., 100,000°K or more, the gas kinetic temperature, T_ε, would not rise above about 20,000°K. Most of the energy loss occurred in the excitation of low-lying levels of ions of the first two rows of the periodic table, a situation that had already been envisaged by Zanstra (1931) and applied by him in a method for getting central star temperatures. Let us consider a nebula that consists of hydrogen and some coolants, such as carbon, oxygen and nitrogen, the ions of which have low-lying "normal" and metastable levels. Collisional excitation of these levels gives rise to strong lines in the optical region and satellite ultraviolet. Also, at such very low densities as prevail in some gaseous nebulae, H II regions and the interstellar medium, excitation of fine structural levels can produce a substantial contribution to cooling.

In the notation of Chapter 3, we can write down the equation of conservation of energy for the continuum:

$$\int_{\nu_1}^{\infty} F_{1\kappa}\, h\nu_{1\kappa}\, d\nu = \int_0^{\infty}\int_0^{\nu} F_{\kappa\kappa'} h\nu_{\kappa\kappa'} d\nu_{\kappa'} d\nu_{\kappa} + \sum_{n_0}^{\infty} \int_{\nu_n}^{\infty} F_{\kappa n} h(\nu_{\kappa n} + \nu_{n1}) d\nu + E_{coll} \,. \qquad (1)$$

Energy absorbed by photo-ionization	=	Free-free emission	+	Energy liberated in capture and subsequent events	+	Energy emitted in collisionally excited lines

To find the terms in Eq. (1), we first recall Eq. 3-65,

$$\iint F_{\kappa\kappa'}(T_\varepsilon)h\nu_{\kappa\kappa'}d\nu_\kappa d\nu_{\kappa'} = N(H^+)\,N_\varepsilon \frac{KZ^4}{T_\varepsilon^{3/2}} \left\{\frac{g_{III}\,k^2\,T_\varepsilon^2}{2h\,RZ^2}\right\} ; \tag{2}$$

Then from Eq. 3-38 we obtain:

$$\sum_{n_o}^{\infty} \int_{\nu_n}^{\infty} F_{\kappa n}\,h\nu_{\kappa n}\,d\nu = N(N^+)\,N_\varepsilon \frac{KZ^4}{T_\varepsilon^{3/2}} \left\{kT_\varepsilon \sum_{n_o}^{\infty} \frac{\bar{g}}{n^3}\right\} , \tag{3}$$

Similarly:

$$\sum_{n_o}^{\infty} \int_{\nu_n}^{\infty} h\nu_{n1}\,F_{\kappa n}\,d\nu = \sum_{n_o}^{\infty} hRZ^2\left(1 - \frac{1}{n^2}\right) \int_{\nu_n}^{\infty} F_{\kappa n}\,d\nu$$

$$= N(H^+)\,N_\varepsilon \frac{KZ^4}{T_\varepsilon^{3/2}}\, hRZ^2\left(G[T_\varepsilon] - \sum_{n_o}^{\infty} \frac{S_n}{n^5}\right) , \tag{4}$$

where

$$S_n = e^{X_n} \int \exp(-\,h\nu/kT)\,d\nu/\nu \sim \bar{g}\,e^{X_n}\,E_2(X_n) \tag{5}$$

(cf. Eq. 3-39) and $G(T_\varepsilon)$ is defined by Eq. 3-50. Here $n_o = 1$ for Case A and 2 for Case B. Noting $E_{1\kappa} = F_{1\kappa}h\nu_{1\kappa}$ and using Eq. 3-30, 46 and the definitions of D and K (Eq. 3-39):

$$\int_{\nu_1}^{\infty} F_{1\kappa}\,h\nu_{1\kappa}\,d\nu = 4\pi\,N_1 \int_{\nu_1}^{\infty} J_\nu\,a_1(\nu)\,d\nu$$

$$= N(H^+)\,N_\varepsilon \frac{KZ^4}{T_\varepsilon^{3/2}} \frac{c^2}{2}\,b_1\,e^{X_1} \int J_\nu\,g_{1\kappa}\,\nu^{-3}\,d\nu\,. \tag{6}$$

Using Eq. 3-51, we find:

$$\int_{\nu_1}^{\infty} F_{1\kappa}\,h\nu_{1\kappa}\,d\nu = N(N^+)\,N_\varepsilon \frac{KZ^4}{T_\varepsilon^{3/2}}\, hG(T_\varepsilon) \frac{\int_{\nu_n}^{\infty} J(\nu)\,g_{1\kappa}\,\nu^{-3}\,d\nu}{\int_{\nu_1}^{\infty} J(\nu)\,g_{1\kappa}\,\nu^{-4}\,d\nu} . \tag{7}$$

Finally,

$$E_{coll} = \sum_i \sum_{JJ'} N_J(X^i)\,A(J,J')\,h\nu(JJ') , \tag{8}$$

where N(J) equals the number of atoms in a collisionally-excited level, J.

Let $\mathcal{F}_{J'J}$ represent the number of collisional excitations from a lower level J to an upper level J per unit volume of time and $p_{s,i}$ represents the probability it will be deexcited by a superelastic collision. Then the emission per unit volume arising from collisionally-excited lines will be:

$$E_{coll} = \sum_{i} \sum_{J,J'} \left(1 - p_{s,i}\right) \mathcal{F}_{J'J}\, h\nu(JJ') \,. \tag{9}$$

We may now substitute Eq. (2), (3), (4), (7), and (9) into Eq. (1) and 5-(26) into (9) and divide by N(H+) $N_\varepsilon K Z^4 T_\varepsilon^{-3/2}$ and $G(T_\varepsilon)$ to obtain:

$$\frac{h\int_{\nu_1}^{\infty} J(\nu) g_{1\kappa} \nu^{-3} d\nu}{\int_{\nu_1}^{\infty} J(\nu) g_{1\kappa} \nu^{-4} d\nu} = \frac{g_{III}\, k^2}{2hRZ^2} T_\varepsilon^2 + kT_\varepsilon \sum_{n_o}^{\infty} \frac{g}{n^3} + hRZ^2 \left[G\left(T_\varepsilon\right) - \sum_{n_o}^{\infty} \frac{S_n}{n^5}\right]$$

$$+ 2.65\, T_\varepsilon \sum_{i} \sum_{JJ'} \left(1 - p_{s,i}\right) \frac{\Omega_i (J'J)\, N_{J'}\left(X^i\right)}{\omega_{i,J'}\; N\left(H^+\right)} e^{-\chi_{(i,J)}/kT_\varepsilon} \frac{1}{G\left(T_\varepsilon\right)} \,. \tag{10}$$

Note that Eq. (10) may be written in the form:

$$f_1(J[\nu]) = f_2(T_\varepsilon) + C(T_\varepsilon) = f_2'(T_\varepsilon) \tag{11}$$

where the left-hand side depends only on the radiation field and the right-hand side depends only on the electron temperature, the ionic concentration and the electron density.*

Let us look at the physical situation from the most elementary point of view. Consider first the idealization where we have a fully ionized pure hydrogen nebula, and where we can also neglect collisional excitation of hydrogen. Suppose further than the radiation field can be taken as Planckian at a temperature T_1 multiplied by a dilution factor W. Then $f_1 = f_1(T_1)$. In the simplest approximation, we set g = 1. Then for an optically thin nebula, Menzel et al. (1938) found for the solution of Eq. (11):

T_1 =	10,000	20,000	40,000	80,000
T_ε =	9,500	18,000	34,000	57,000

*If neutral H is present and if T_ε approaches about 20,000°K, a substantial amount of energy can be dissipated by the collisional excitation of the second and higher levels. Balmer, Paschen, etc., quanta produced by radiative transitions from these levels escape from the nebula (see, e.g., Aller and Liller, 1968, Stars and Stellar Systems, 7, 521.)

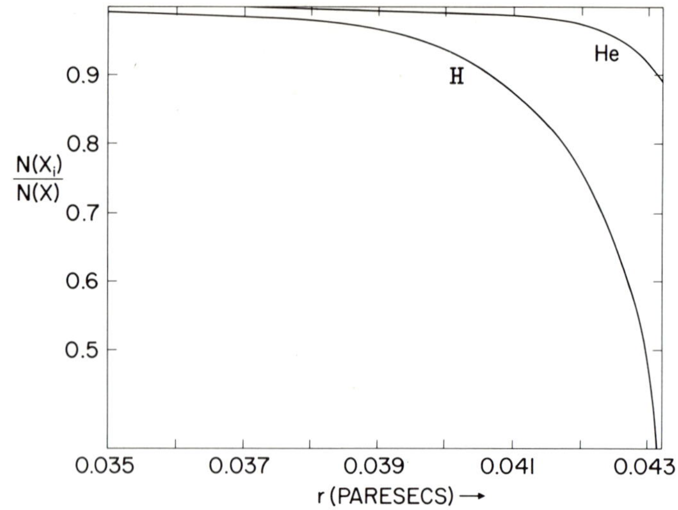

Figure 6-1

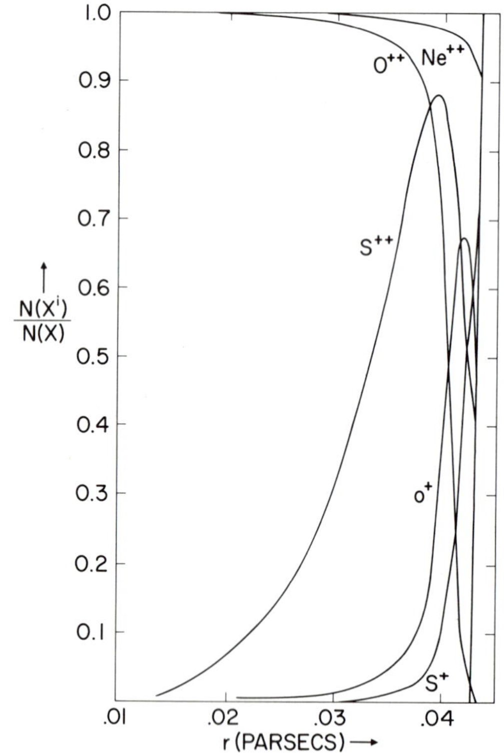

Figure 6-2

In this idealization of zero collisional dissipation, T_ε drops appreciably below T_1 only at relatively high temperatures where bremsstrahlung becomes important. Now, consider what happens when an "impurity" with a concentration N_p is added. Although the concentration, N_p, of cooling ions such as O^{++} may be four orders of magnitude below that of H^+, T_ε is depressed markedly through the medium of collisional excitation. Note further that as $N(H^+) = N_\varepsilon$ is increased, the cooling effect decreases for a fixed value of N_p.

In an optically thick nebula where Case B is appropriate, we should take $n_o = 2$ to allow for recaptures on the first level. These recaptures will modify the radiation field $J(\nu)$, since they take place from an assembly of electrons characterized by a gas kinetic temperature for which 3 kT/2 is less than the average energy of the photoejected electron. Furthermore, electrons moving with low speeds are picked off more rapidly than those that move with higher speeds. We must take this changing radiation field into account in order to calculate the ionization structure of the nebula. An early solution (1939) of the transfer equation for a pure hydrogen nebula of optical thickness at the Lyman limit, $\tau(\nu_1) = 3$, neglecting all collisional effects predicted that the "hardening" of the radiation field would cause a rise in T_ε as the outer boundary is approached. If collisional effects of "impurities" such as C, N, O, Ne, S, and Ar are taken into account, the effect is to lower T_ε. Figures 6-1 and 6-2 show the ionization structure of a theoretical model of IC 4846. Figure 6-3 predicts a rapid rise in T_ε as the outer boundary of the ionized hydrogen zone is approached, largely as a consequence of the "hardening" of the nebular radiation field as shown in Fig. 6-4. Unfortunately, definitive, quantitative observational checks on these predictions are not now available.

Later Spitzer (1948, 1949) derived an alternative form of the equation of energy conservation for the continuum alone, and used it to get a relation between radiation field and electron temperature. Spitzer's equation can easily be obtained from the expression given by Menzel and his coworkers.

From each side of Eq. (1) we subtract the total energy $h\nu_1 = h\nu_n + h\nu_{n1}$ multiplied by the number of photoionizations or recombinations, i.e.:

$$\int_{\nu_1}^{\infty} F_{1\kappa}\, h\nu_1\, d\nu = \sum_{n_o}^{\infty} \int F_{\kappa n}\left(h\nu_n + h\nu_{n1}\right) d\nu \,. \tag{12}$$

We thus obtain:

$$\int_{\nu_1}^{\infty} F_{1\kappa}\, h\left(\nu - \nu_1\right) d\nu$$

$$= \iint E_{\kappa\kappa'}\, d\nu\, d\nu' + \sum_{n_o}^{\infty} \int F_{\kappa n}\, h\left(\nu_{\kappa n} - \nu_n\right) d\nu + E_{coll} \,. \tag{13}$$

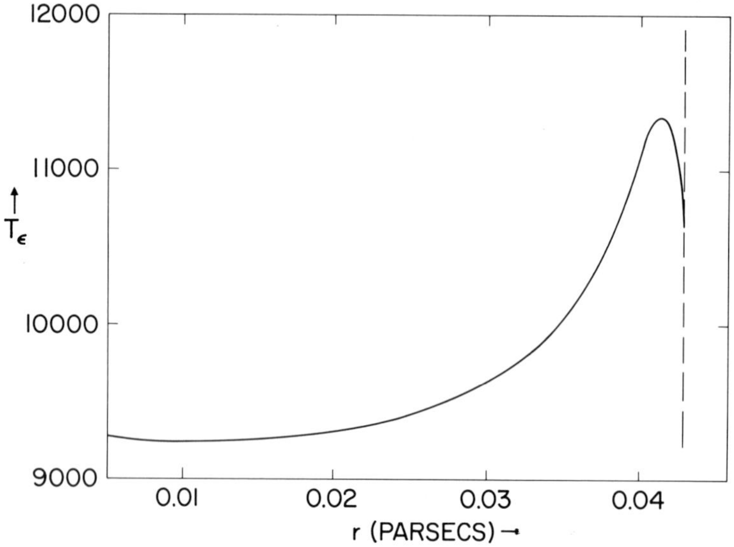

Figure 6-3

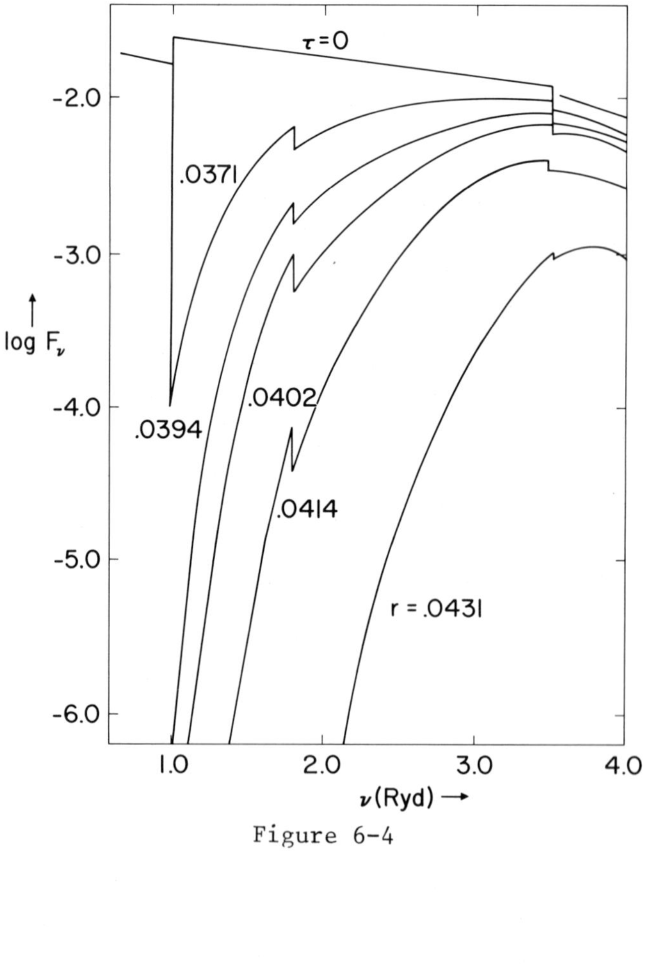

Figure 6-4

Also:

$$h\nu_{\kappa n} = \frac{1}{2} mv^2 + h\nu_n = \frac{1}{2} mv^2 + \frac{hRZ^2}{n^2} , \tag{14}$$

and

$$\sum_{n_o}^{\infty} \int_{\nu_n}^{\infty} F_{\kappa n}\, h(\nu - \nu_n)\, d\nu = \sum_{n_o}^{\infty} \int_{\nu_n}^{\infty} F_{\kappa n} \frac{1}{2} mv^2\, d\nu$$

$$= N(H^+)\, N_\varepsilon \frac{KZ^4}{T_\varepsilon^{3/2}} \sum_{n_o}^{\infty} \left(\frac{\bar{g}kT_\varepsilon}{n^3} - \frac{hRZ^2\, \bar{g}S_n}{n^5}\right)$$

$$= N(H^+)\, N_\varepsilon \frac{KZ^4}{T_\varepsilon^{3/2}} \mathcal{A}_H(T_\varepsilon) , \tag{15}$$

where $\mathcal{A}_H(T_\varepsilon)$ is given by:

(T°K)	Case A	Case B
2,500	0.1181 (-13)	0.0767 (-13)
5,000	0.4056 (-13)	0.241 (-13)
10,000	1.368 (-13)	0.733 (-13)
20,000	4.502 (-13)	2.195 (-13)

so the right-hand side of the equation becomes:

$$N(H^+)\, N_\varepsilon \frac{KZ^4}{T_\varepsilon^{3/2}} \left\{\frac{g_{III} k^2}{2hRZ^2} T_\varepsilon^2 + \mathcal{A}_H(T_\varepsilon)\right\} + E_{coll} . \tag{16}$$

Using Eq. (7), the left-hand side of the equation may be written as:

$$\int_{\nu_1}^{\infty} F_{1\kappa}\, h(\nu - \nu_1)\, d\nu = \int_{\nu_1}^{\infty} F_{1\kappa}\, h\nu\left(1 - \frac{\nu_1}{\nu}\right) d\nu$$

$$= N(H^+)\, N_\varepsilon \frac{hK}{T_\varepsilon^{3/2}} G(T_\varepsilon) \frac{\int_{\nu_1}^{\infty} J_\nu\, g(\nu)\, (1 - \nu_1/\nu)\, d\nu/\nu^3}{\int_{\nu_1}^{\infty} J(\nu)\, g(\nu)\, \nu^{-4} d\nu} \tag{17}$$

whence:

$$\frac{h\int_{\nu_1}^{\infty} J_\nu\, g(\nu)\,(1 - \nu_1/\nu)\, d\nu/\nu^3}{\int_{\nu_1}^{\infty} J(\nu)\, g(\nu)\, \nu^{-4}\, d\nu} = \left\{4.371 \times 10^{-22}\, T_\varepsilon^2/Z^4 + \mathcal{A}_H(T_\varepsilon) + 3.067 \times 10^5\, \frac{E_{coll}}{N(H^+)N_\varepsilon}\, \frac{T_\varepsilon^{3/2}}{Z^4}\right\} \frac{1}{G(T_\varepsilon)}, \quad (18)$$

which is analogous to Eq. (10).*

Note that the choice of $\mathcal{A}_H(T_\varepsilon)$ depends on whether we have an optically thick or thin nebula. The equations can be generalized to include additional elements. Helium is the only sufficiently abundant element to be important. For an account of some of the complexities to be anticipated. see Osterbrock (1974; AGN; Kaler (1976) reference, Chapter 3. Consider first a relatively low excitation nebula where helium is not doubly ionized. Shortward of the He I ionization limit, both H and He compete in absorbing photons. The quantitative behavior will depend on N(H°), N(He°) and the continuous absorption coefficients of H and He. When He^+ ions and electrons recombine, several types of photons capable of ionizing hydrogen are produced: First, direct recombinations to the helium ground level give ultraviolet photons analogous to Lyman continuum photons. Recaptures in singlet levels that result in cascade to 2^1P produce λ584 photons that can ionize hydrogen. Cascades to 2^1S produce two-photon emissions, a fraction of which can ionize H. Recaptures to triplet levels result in cascades that lead eventually to 2^3S which is strongly metastable (cf. Chap. 2). Helium atoms may escape from this level by collisions, although some may escape by photoionization by Ly α. Kaler concluded that in a typical situation, about 22 percent of helium recombinations lead to hydrogen photoionizations, but the photon energy is supplied in a markedly non-Planckian fashion. Detailed quantitative calculations which must take into account the interlocking between helium and hydrogen continua become quite intricate. In very high excitation nebulae where helium is doubly ionized, it can make a substantial contribution to the ionization, recombination and free-free terms.

6B. Persistence of a Maxwellian Velocity Distribution in a Photoionized Nebula

Although a gaseous nebula deviates far from thermodynamic equilibrium, it was assumed quite generally that the velocity distribution was closely Maxwellian. A rigorous examination of the problem was made by Bohm and Aller (1947). One must compare processes tending to destroy a Maxwellian distribution with those tending to restore it.

---

*In Spitzer's notation $KT_\varepsilon^{-3/2}\, \mathcal{A}_n(T_\varepsilon) = kT\, \beta(T)$.

Consider how a steady state is produced. Ions and atoms (mostly H) are ionized photoelectrically by ultraviolet quanta that may be traced to illuminating stars or a blue nonthermal continuum. Electrons are ejected with a kinetic energy equal to that of the photoionizing quantum minus the ionization energy. The number emitted in an energy range ΔE at E is determined by the frequency distribution of the impinging radiation and the continuous absorption coefficient as a function of energy. Thus, the energy distribution function of these newly arriving electrons is markedly non-Maxwellian.

Now, examine what happens to our newly liberated electron: It can experience elastic scattering by other electrons or ions (or perhaps even neutral atoms). It can collide with an atom or ion and excite it to an energy level within a few eV of the ground level or undergo a superelastic collision with an atom in a metastable level. It may lose energy in free-free transitions or it may be recaptured to a bound level with the emission of radiation, either directly or through dielectronic recombination. The balance of all these processes produces a steady state. Elastic collisions act to make a Maxwellian distribution; all other processes tend to destroy it. Therefore, the deviation from such a distribution will be related to the frequency of these other processes compared with that of collisional redistributions. In a steady state, the total rate of recaptures must equal the total rate of photoionizations. The target area for the recapture of an electron of velocity v on the nth hydrogen level is:

$$\sigma_{\kappa n} = \frac{2^4}{3\sqrt{3}} \frac{\varepsilon^2}{m^2} \frac{h}{c^3} \frac{1}{\left(\frac{1}{n'^2} + \frac{1}{\kappa^2}\right)} \frac{\kappa^2}{n'^3} g_{II} , \qquad (19)$$

where κ is defined by:

$$1/2\ mv^2 = hRZ^2/\kappa^2 . \qquad (20)$$

The total number of captures to a single ion is obtained by summing $\sigma_{\kappa n}$ over all levels of n and multiplying by velocity, v. The total number of captures per cm^3/sec is $N_i v\ \Sigma\ \sigma_{\kappa n}$. For $N_i = 10^4$ and an electron energy 1.36 eV ($v = 6.9 \times 10^7$ sec) $\Sigma\ \sigma\kappa n = 3.57 \times 10^{-21}/cm^2$. The recombination probability is about $2.5 \times 10^{-9}\ sec^{-1}$, so the electron will wander about for the order of 13 years before recapture. An electron of average energy 0.5 eV would last for eight years.

The target area for the collisional excitation of a level by an electron moving with a velocity v is:

$$\sigma = \frac{1}{\omega} \frac{h}{4\pi m^2} \frac{\Omega(A \rightarrow B)}{v^2} = \frac{4.17}{2J_A + 1} \frac{\Omega}{v^2} , \qquad (21)$$

provided $1/2\ mv^2 \geqslant \chi$, where χ is the excitation potential of level B. Hence, per electron of velocity v, the probability of energy loss by excitation is $v \sum N_j \sigma_{jk}$, where the summation is taken over all levels of every ion for the excitation of which the electron possessed enough energy. Typically, the mean time required for an energy loss to occur will be of the order of 10^7 seconds or about four months. Energy gains by superelastic collisions will be even less frequent.

For free-free transitions we can show that the mean rate of energy loss is:

$$\langle\frac{dE}{dt}\rangle = \frac{2^4}{3\sqrt{3}} \frac{\varepsilon^2}{m^2} \frac{hR}{c^3} g_{III} \frac{v}{2} N_i \ . \tag{22}$$

The time that would be required for an electron to dissipate an amount of energy equal to its initial supply E is:

$$\langle t_{ff}\rangle = \frac{E}{\langle dE/dt\rangle} = 1.2 \times 10^9 \sqrt{E(eV)} \text{ secs} \ . \tag{23}$$

For electron energies of about 1 eV, $\langle t\rangle_{ff}$ amounts to about 40 years.

Energy-rearranging collisions between charged particles may be described as coulombic scattering. The appropriate cross section σ for the scattering of a particle of energy E through an angle θ is given by the Rutherford formula:

$$\sigma = \frac{\varepsilon^4}{16 E^2 \sin^4 (\theta/2)} \ . \tag{24}$$

and the corresponding number of particles scattered into angles between θ and $\theta + d\theta$ will be $dn = 2\pi N_\varepsilon v\sigma \sin\theta d\theta$. Since the overwhelming majority of deflections are through small angles, these encounters will play the dominant role and we may replace $\sin\theta$ by θ. Then the total number of particles scattered into angles between θ and $\theta + d\theta$ will be:

$$dn = \frac{2\pi N_\varepsilon v \varepsilon^4}{E^2} \frac{d\theta}{\theta^3} \ . \tag{25}$$

We may neglect energy transfers from electrons to ions since the latter are so very much heavier. To calculate energy transfers in electron-electron encounters, note that the Rutherford formula applies to a frame of reference in which the center of gravity is at rest. Since the scattering angle in the fixed-coordinate system is one-half that in the center-of-mass system, and since the relative energy is one-half that of the incident electron, the above equation is unchanged if θ refers to the actual angle of scattering in the fixed system and E refers to the incident energy. In the first approximation, the energy gained by the scattering electron is:

$$\Delta E = E \sin^2 \theta \sim E \theta^2 \ . \tag{26}$$

The mean rate of energy transfer for θ deflections is:

$$\langle\frac{\Delta E}{\Delta t}\rangle_\theta \, dn = \frac{2\pi N_\varepsilon v\varepsilon^4}{E}\frac{d\theta}{\theta} . \tag{27}$$

The total energy transfer is estimated by integrating from some appropriate minimum value of θ, i.e., θ_{min} to θ_{max}, which is π. Then:

$$\langle\frac{dE}{dt}\rangle \sim \frac{2\pi N_\varepsilon v\varepsilon^4}{E} \ln\left(\frac{\pi}{\theta_{min}}\right) . \tag{28}$$

Here θ_{min} may be taken as the deflection corresponding to the maximum encounter distance h by the relation $\theta_{min} \sim \varepsilon^2/Eh$. In plasmas, we identify h with the Debye limit:

$$h = \sqrt{\frac{kT}{8\pi\ N_\varepsilon\ \varepsilon^2}} . \tag{29}$$

At $N_\varepsilon \sim 10^4/cm^3$ and $T_\varepsilon = 10{,}000°K$, h ~ 5 cm, while the mean interelectronic distance is of the order of 0.05 cm. The mean time required for an electron to exchange an amount of energy, $\sum \sqrt{\overline{\Delta E^2}} = E$ is:

$$\frac{E}{\langle dE/dt\rangle} \sim \frac{E^2}{2\pi N_\varepsilon v\varepsilon^4} \ln\left(\frac{\theta_{min}}{\pi}\right) . \tag{30}$$

For nebular densities and temperature $\ln(\pi/\theta_{min})$ will be of the order of 15. A three-volt electron will lose most of its energy in the order of four seconds, and since most electrons have less energy than this, they may reshuffle what they have in less time.

In summary then, in a nebular plasma at a density of 10^4 electrons/cm^3 at a temperature of 10,000°K, an electron will remain free for a period of the order of ten years before recapture. It will lose an amount of energy equal to its mean energy by inelastic collisions in an interval of weeks or months, and it will radiate but a small amount of its energy as free-free emission. On the other hand, if reshuffling by coulombic encounters occurs in the order of a second, the distribution must certainly approach the Maxwellian form. Hence, the high-velocity tail of the distribution function, which is responsible for the excitation of collisionally-produced lines in the optical and near ultraviolet portions of the spectrum, must result from the boiling up of low-energy electrons. All traces of the initial photoionizing spectral energy distribution will have been obliterated, except insofar as it has contributed to the total energy supply in a thermodynamic sense.

Under what circumstances, if any, would we expect appreciable deviations from a Maxwellian distribution? Such deviations would affect the high-energy end of the range rather than the low-energy range. Deviations might be expected to occur in the interstellar medium where electrons were ejected by cosmic rays rather than by photionization processes. More particularly, under strongly time-dependent circumstances such as shock wave phenomena, electrons and ions may achieve Maxwellian distributions corresponding to different temperatures. A considerably longer time would be needed for the electrons and ionic temperatures to achieve equality.

To calculate the deviations from a Maxwellian distribution, we must solve the appropriate Boltzmann integrodifferential equation that expresses the rate of change of the distribution function that arises from various processes such as elastic collisions, recombinations, free-free emissions, and inelastic collisions. The equations can be generalized to allow for diffusion and other time-dependent effects. One may start with an initial distribution function that is Maxwellian and apply a perturbation method. We express the deviation from the Maxwellian law as $b(E) = 1 + \delta(E)$ and evaluate $\delta(E)$ which turns out to be negligibly small for the steady-state situations described above. Further refinements in the mathematical formulation have been introduced by Gould and Thakur (1971) for low-density plasmas. In a neutral gas, the situation becomes more complicated. Thus, Gold and Levy (1976) find that deviations from a Maxwellian distribution may occur in regions of interstellar molecular hydrogen. In summary then, it is established that the gas-kinetic electron temperature, T_ε, is the appropriate one against which physical processes are to be compared.

Problem

The energy balance concept was utilized in Stoy's method (Chapter 3) for the determination of temperatures of planetary nebulae nuclei. The total radiative flux in collisionally excited lines is compared with the flux in Hβ.

Assuming that the star radiates as a black body at some temperature T^*:

$$\pi F_\nu = \pi B_\nu(T^*) = \frac{2h\nu^3}{c^2} \frac{1}{e^{h\nu/kT^*} - 1}$$

show that for a pure H nebula,

$$\frac{\sum F_N}{F(H\beta)} = \frac{\alpha_B}{\alpha(H\beta)h\nu_{h\beta}} \frac{\int_{\nu_1}^{\infty} B_\nu(T^*)\left(\frac{\nu-\nu_\varepsilon}{\nu}\right)d\nu}{\int_{\nu_1}^{\infty} \frac{B_\nu(T^*)}{h\nu}\,d\nu}$$

where $\sum F_N$ is the sum of the fluxes over all collisionally-excited lines in the IR, optical and UV regions and $h\nu_\varepsilon = h\nu_1 + (3/2)kT_\varepsilon$. What complications are introduced if the nebula contains helium in a ratio of about $n(He)/n(H) = 0.1$? Note that $\nu(\text{He I}) = 1.8\ \nu_1(H)$, $\nu_1(\text{He I}) = 4\nu_1(H)$. Consider particularly the role of He I λ584, and Lyα(He II).

For further information on the Stoy method see:

Kaler, J.B. 1976, Ap. J., 210, 843, and
Preite-Martinez, A., and Pottasch, S.R. 1983, Astron. Astrophys., 126, 31.

References

The idea that energy supplied by photoionization processes appears in the excitation of forbidden lines was first applied by Zanstra in one of his methods for obtaining temperatures of central stars of planetary nebula, viz.:

Zanstra, H. 1931, Zeits. f. Astrofis., 2, 1.

The pioneering papers on energy balance or thermal equilibrium in gaseous nebulae are:

Menzel, D.H., Aller, L.H., and Baker, J.G. 1938, Ap. J., 88, 313.
Baker, J.G., Menzel, D.H., and Aller, L.H. 1938, Ap. J., 88, 422.
Menzel, D.H., and Aller, L.H. 1941, Ap. J., 94, 30.

Figures 2 and 3 in the last cited of these papers show the pronounced lowering of T_ε by the presence of "impurities" and how for a fixed value of N("impurity"), the cooling effect decreases as $N(H^+) = N_\varepsilon$ increases. Further applications are given in Ap. J., 118, 574, 1953.

Spitzer's method is described by him in:

Spitzer, L. 1948, Ap. J., 107, 6; 1949, Ap. J., 109, 337, and by:
Spitzer, L., and Savedoff, M.P. 1950, Ap. J., 111, 593.

The persistence of a Maxwellian velocity distribution was discussed by:

Bohm, D., and Aller, L.H. 1947, Ap. J., 105, 1.

See also:

Gould, R.J., and Thakur, R.K. 1971, Phys. Fluids, 14, 1701.
Gould, R.J., and Levy, M. 1976, Astrophys. J., 206, 435.

List of Illustrations

Fig. 6-1. Detail of Ionization Structure of a Model Nebula Calculated to Represent IC 4846. We plot $N(H^+)/N(H)$ and $N(He^+)/n(He)$

as a function of distance in parsecs from the central star. Note that the concentration of neutral H atoms increases more abruptly than that of helium in spite of higher ionization potential of the latter.

Fig. 6-2. Ionization Structure of IC 4846. We plot ionization fraction $N(X^i)/N(X)$ for O^+, O^{++}, Ne^{++}, S^+ and S^{++} as a function of distance from the central star expressed in parsecs. N_H = 11,000, R_* = R(sun), $F_\nu(*)$ corresponds to a model calculated by Cassinelli for T_{eff} = 48,000°K.

Fig. 6-3. Variation of Theoretical Electron Temperature for Model in IC 4846. Notice the rapid temperature rise towards the outer nebulae boundary in spite of the increase in concentration of cooling ions of N^+, O^+, and S^+. It appears that the hardening of the radiation field as the nebular boundary is approached here has an even more important effect than the rise in the concentration of coolants. Such an increase in T_ε towards the outer nebular boundary was anticipated in theoretical work in the 1930s (see Ap. J., 90, 601, 1939, Fig. 4).

Fig. 6-4. Energy Distributions in Radiation Field. The top curve gives log $F_\nu(*)$, the stellar energy distribution. Energy distributions are also given for radii 0.0371 pcs (τ_H = 10.05), 0.0394 pcs (τ_H = 19.6), 0.0402 pcs (τ_H = 27.2), 0.0414 pcs (τ_H = 51), 0.0431 (τ_H = 142). Here τ_H refers to optical depth at 1 Rydberg. Note the pronounced "hardening" of the radiation field as the outer boundary of the nebula is approached.

Chapter 7

NEBULAR MODELS

Unlike theoretical stellar models which have certain enormous simplifications such as spherical symmetry and hydrostatic equilibrium, models of gaseous nebulae entail idealizations that are rarely encountered in nature. Few nebulae or H II regions are spherically symmetrical. The density distribution is seldom a simple function of distance from the exciting star. Nevertheless, it is possible to construct reasonably straightforward models that are often useful in the interpretation and assessment of observations. Thus, models have been calculated for H II regions, for planetary nebulae, as well as for QSOs and similar plasmas presumably excited by something like synchrotron radiation, where $F(\nu) \sim \nu^{-\alpha}$. The objectives of such investigations involved representation of observed nebular continua and line intensities, including such diagnostic ratios as yield electron densities and temperatures. In this discussion, we consider only radiative models. Some of the complex characteristics of observed nebulae are reviewed in Chapter 10, but we allude to some of these matters here to emphasize that one must be cautious in applying simple theories to intricate structures.

7A. Initial Assumptions

The simplest models usually postulate spherical symmetry, i.e., N_H = constant or a function of distance from the illuminating star r, but not of angular variables, θ and ϕ. We may postulate shells or a gaussian distribution about $r = r_0$. We can also choose whether the nebula is to be material-bounded (truncated) or radiation-bounded, in which case neutral material lies outside the ionized Strömgren sphere. The chemical composition is also an adjustable parameter; theoretical models are often used to establish nebular abundances. We may also select the radius of the exciting star and the emergent flux. For H II regions we choose illuminating stars with radii appropriate to population Type I. For planetary nuclei, radii comparable with that of the sun or even smaller are often used.

Choice of a proper emergent stellar flux, $F_\nu(*)$, poses one of the most perplexing problems. Ultraviolet observations secured with rockets and satellites are of limited help here since we cannot get direct information on the region shortward of λ912A. Many workers have favored black body energy distributions (see, e.g., Köppen 1980). Kurucz (1979) calculated atmospheric structures and emergent fluxes for a network of models extending up to T_{eff} = 30,000, 35,000, 40,000, and 45,000°K and standard compositions. These are line-blanketed, LTE models and appear to be good starting points for the computation of radiative fields for H II regions ionized by hot main sequence stars. For moderate-to-high excitation planetaries, a number of workers have used theoretical models by Hummer and Mihalas (1970), by Cassinelli (1971), or by Kunasz, Hummer, and Mihalas (1975). The Hummer-Mihalas models are computed for local thermodynamic equilibrium (LTE) and involve plane-parallel stratification, but they are extensive enough to permit interpolation in T_{eff} and log g. They should offer a reasonable starting point for hot stars. In the latter two sets of models, the curvature of the radiating layers was taken into account. Cassinelli computed $F_\nu(*)$ for T_{eff} = 37,496, 48,363, and 95,090°K at $\tau = 2/3$ and assumed LTE. Kunasz et al. calculated non-LTE models for stars of effective temperatures 63,000°K and 100,000°K. These models differ appreciably from those of Cassinelli. The far ultraviolet emergent fluxes depend on the choice of atmospheric abundances of He, C, N, O, and Ne. Extensive grids of pure H and pure He LTE and nonLTE model atmospheres for hot stars with high surface gravities (i.e., white dwarfs) have been calculated by Wesemael, Auer, van Horn and Savedoff (1980) and by Wesemael (1981). For the hydrogen models they chose temperatures ranging upward from 20,000°K to the Eddington limit, $T_{eff} = 6000\ \sqrt[4]{g}$, and log g from 4.0 to 9.0. For the helium models a comparable grid is calculated. These models may be useful for planetary nebular nuclei in advanced evolutionary stages. Kudritzki (1981) and his associates calculated nonLTE model atmospheres for six PNN designed to represent observed H and He line profiles. They found T_{eff} for the nuclei of NGC 4361 and 1535 to be much lower than T_Z. Shields and Searle (1978) noted that a self-consistent procedure would be to require that the illuminating stars of an H II region have the same chemical composition as the gas. These elemental abundances are then employed in model atmosphere calculations to evaluate the emergent stellar flux. Such a procedure is not necessarily appropriate for nuclei of planetary nebulae; in Abell 30 and Abell 78, the composition of the nebular envelope varies with distance from the central star and certainly differs from that of the latter. The necessary atomic parameters must be carefully chosen. These include the continuous absorption coefficients for low-lying levels, recombination coefficients (both direct and dielectronic), charge exchange cross sections, cross sections for the electronic collisional excitation of discrete levels of ions of atoms such as C, N, O, Ne, etc., and the corresponding radiative transition probabilities. These are stored in the computer program along with better known data such as spectral line wavelengths, excitation potentials, and ionization potentials. Figure 7-1 illustrates schematically some of the factors involved.

7B. Computational Procedures

Let us first consider the radiation field. It consists of two parts: the emergent radiation from the central star whose specific intensity is attenuated by absorption in the gas (mostly photoionizations of hydrogen and helium) and the diffuse radiation produced as electrons recombine on the ground levels of hydrogen or helium, and by decays from excited levels of helium. Also, one should note that the degradation of He II Lyman-alpha radiation by the Bowen fluorescent mechanism produces quanta capable of ionizing hydrogen, with an efficiency of the order of 50%. Various procedures have been proposed for handling this problem. One is by the direct integration of the equations of transfer, treating the attenuated starlight and the diffuse radiation separately. Early formulations of the problem by Milne, by Ambarzumian, and by Chandrasekhar treated the continuum as a single level. Later, Menzel et al. (1939) solved the problem, taking into account the variation of recombination rate as a function of electron velocity. Their results were quantitatively inadequate because they treated an H plasma without collisional effects or cooling ions and thus obtained too high an electron temperature. With modern computing machines, the transfer problem can be handled properly using exact equations.

At the opposite extreme is the on-the-spot (OTS) approximation in which it is supposed that each diffuse ultraviolet quantum is absorbed locally. In the same immediate neighborhood it is then converted into a quantum of Lyα at λ1216 plus quanta of the Balmer series that escape forthwith from the nebula. In high excitation objects where helium is doubly ionized, recombinations in the ground level can result in production of He II Lyα quanta at λ304. The OTS approximation neglects ionization of heavy elements by the diffuse radiation. Furthermore, it can lead to large errors in calculations where the luminosity of the central star is time-dependent.

In the "outward only" approximation, it is assumed that all diffuse photons also travel radially outwards. Thus, the diffuse component of the radiation at the ith zone is the sum of contributions from all zones with j interior to i. Tylenda (1979) modified the method to include contributions from the ith zone also and shows by a recursion relationship that one has to solve for the final physical state of, and the radiation field in the ith shell by means of an iterative procedure. For static nebulae, Tylenda's modification is not necessary but it is still worthwhile to seek an improvement on the OTS approximation. Hummer and Rybicki (1982) have developed a second-order theory for calculating the escape probability of quanta which should be quite accurate for continuum problems. The OTS approximation can be improved to include dust absorption (Petrosian and Dana 1980).

At the outset, the relevant atomic parameters and equations for physical processes, radiative transfer, etc., are stored in the program. Once the stellar radius and flux distribution, the nebular geometry, and chemical composition are given, the program integrates radially outward from the stellar surface. It calculates the absorptivity of the nebular gases. Although H and He dominate, Rubin points out that contributions of other elements such as C, N, and Ne should be considered, particularly in low-excitation objects and near the edges of He+ and H+ zones. In the first approximation, only H and He are considered, and the element of optical depth is:

$$d\tau_\nu = (1 - x)\, a_H(\nu)\, N_H\, dr + \frac{N(\text{He I})}{N(\text{He})}\, a_\nu(\text{He I})\, N(\text{He})\, dr$$

$$+ \frac{N(\text{He II})}{N(\text{He})}\, a_\nu(\text{He II})\, N(\text{He})\, dr\,. \qquad (1)$$

At any point the radiation field will be attenuated by a factor $W_g \exp(-\tau_\nu)$ where $W_g = (R[*]/2r)^2$. Next, one must calculate the ionization stage of each constituent, $N(X_i)/N(\text{element})$. The problem is simplified, of course, by the fact that the attentuation of the

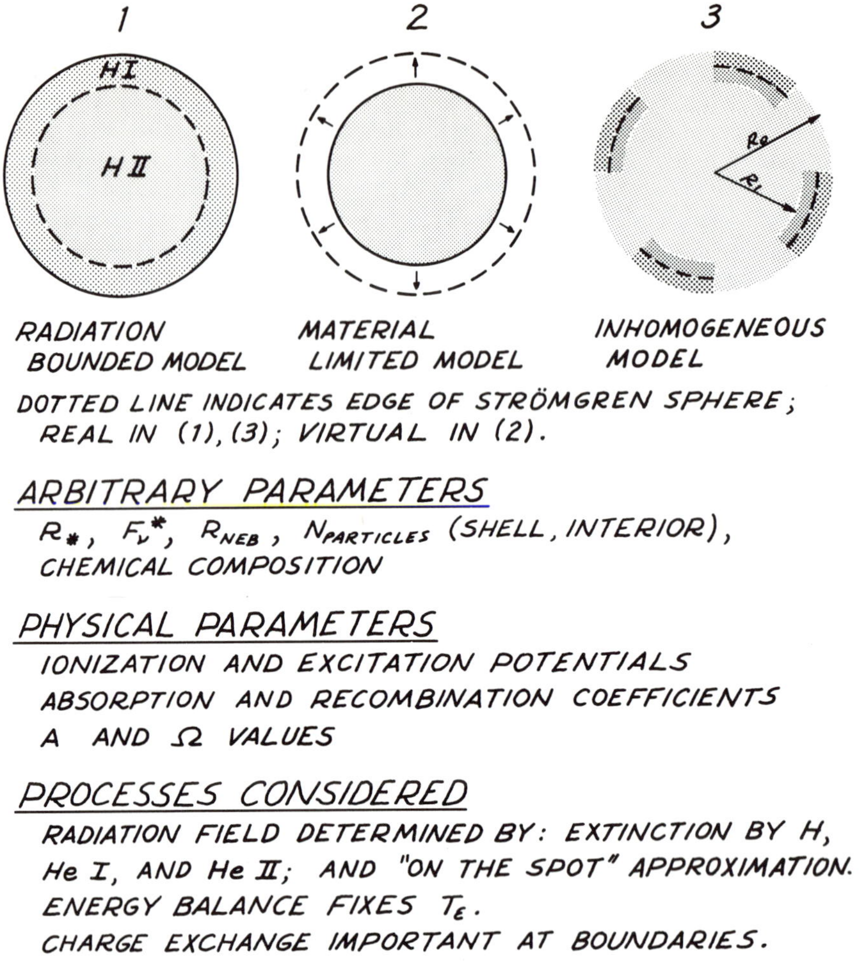
1
2
3
HI
HII
Re
Ri
RADIATION BOUNDED MODEL
MATERIAL LIMITED MODEL
INHOMOGENEOUS MODEL
DOTTED LINE INDICATES EDGE OF STRÖMGREN SPHERE; REAL IN (1), (3); VIRTUAL IN (2).
ARBITRARY PARAMETERS
R_*, F_ν^*, R_{NEB}, $N_{PARTICLES}$ (SHELL, INTERIOR),
CHEMICAL COMPOSITION
PHYSICAL PARAMETERS
IONIZATION AND EXCITATION POTENTIALS
ABSORPTION AND RECOMBINATION COEFFICIENTS
A AND Ω VALUES
PROCESSES CONSIDERED
RADIATION FIELD DETERMINED BY: EXTINCTION BY H, He I, AND He II; AND "ON THE SPOT" APPROXIMATION.
ENERGY BALANCE FIXES T_ε.
CHARGE EXCHANGE IMPORTANT AT BOUNDARIES.

Figure 7-1

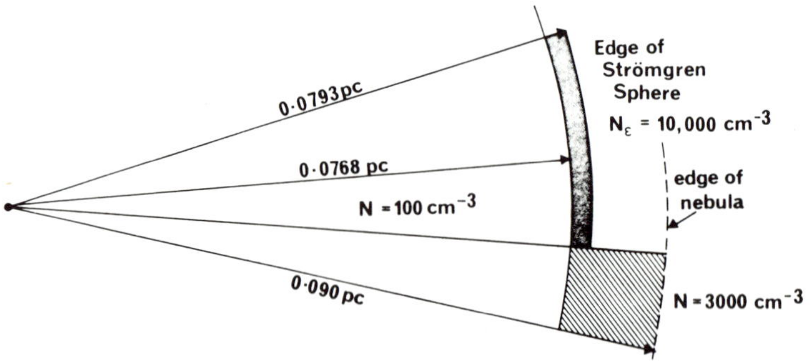
Edge of Strömgren Sphere
N_ε = 10,000 cm^{-3}
0·0793 pc
0·0768 pc
N = 100 cm^{-3}
edge of nebula
0·090 pc
N = 3000 cm^{-3}

Figure 7-2

radiation is fixed primarily by the ionization state of H and He and by geometrical factors, but we must know the electron temperature which in turn has to be found by the energy balance condition (Chapter 6, §A). The electron temperature is found at each point by an iterative procedure in which the heating and cooling rates are calculated and T_ε adjusted until equality is achieved. Heating is provided by the energy input from the radiation field, which (as noted above) consists of geometrically diluted direct starlight plus diffuse radiation). The cooling processes include free-free emission, particularly involving ions of hydrogen and helium, recombinations to discrete levels in H and He, and collisional excitation of both normal and metastable levels of abundant elements such as C, N, O, Ne, Mg, Si, S, Cℓ, and Ar. In principle, within the $H^o - H^+$ transition zone, collisional excitation of n = 2 or n = 3 levels of H^o could be significant if $T_\varepsilon \gtrsim 20{,}000°K$.

In the same package one can calculate a complete set of emission line intensities for both permitted and forbidden transitions at each step of the integration. The program not only can be devised to solve explicitly the equations of statistical equilibrium for the five levels of any p^n configuration, but also for the order of 15 levels including many of those responsible for normal and intercombination lines in the near ultraviolet.

As in all such problems in numerical integration, our formal differential equations must be replaced by difference equations. Thus, care must be exercised in the choice of integration steps, particularly in composite models or models with abrupt changes in density in order to avoid convergence difficulties. Integrations over specified frequency intervals may be replaced by Gaussian sums. For example, in integrating a radiation field over frequency, we notice that J_ν falls into three well-defined domains: $\nu_1 - 1.808\ \nu_1$, $1.808\ \nu_1 - 4\nu_1$, and $\nu > 4\nu_1$. Here ν_1 is the ionization frequency of H, $1.808\ \nu_1$ that of He, and $4\nu_1$ that of He^+. In high-excitation nebulae one must also include the Lyman α contribution of ionized helium. These quanta can ionize H, He, and ions of many elements, as can also He II Balmer continua and two-photon-emission quanta. The detailed interlocking of H and He radiation fields is often omitted in nebular models.

The integration can be terminated when the ionization of hydrogen drops below some specified value, e.g., 50%. Alternately, we can stop the integration at some intermediate point. Such models may be called material-bounded or truncated models. Since the outer shell which contains the bulk of neutral and singly-ionized atoms is stripped off, the average level of excitation will be higher than that found in radiation-bounded models (complete Strömgren spheres), see, e.g., Webster (1976). Figures 1, 2, and 3 of Chapter 6 show the ionization pattern and temperature distribution in a model intended to reproduce the spectrum of IC 4846.

Finally, the program is designed to integrate the emissivity for each line over the entire volume to obtain the total nebular flux for each transition of interest, including the ultraviolet and infrared. The total radio emission and the continuum flux at selected frequencies can also be computed. One can calculate $\langle T_\varepsilon \rangle$ for the entire nebula or for individual observed ions. Also, we can predict the intensity distribution across the image of a symmetrical nebula.

7C. Dependence of Models on Input Parameters

The objective of theoretical model calculations is to reproduce the observed spectral line and continuum intensities, and desired quantities such as $\langle T_\varepsilon \rangle$, and ideally also to represent the distribution of emissivity of various ions with distance from the exciting star. When one attempts to carry out this program using plausible input parameters, assuming geometrical symmetry and a density distribution that is dependent only on r or is constant, one often finds that although the theoretical nebular spectrum qualitatively resembles a real one, it persistently refuses to fit the actual data, especially for nebulae of moderate-to-high excitation.

Let us examine the influence of various factors, beginning with input atomic parameters. Errors in collision cross sections and A-values will enter directly into the predicted intensities. The inclusion (or omission) of charge exchange can affect markedly the predicted spectrum. Suppose one calculates a model for a moderate excitation nebula, neglecting the $O^{++} + H^\circ \rightarrow O^+ + H^+$ reaction, and tries to represent the observed spectrum as closely as possible. After a number of trials, a specific nebular chemical composition and central star energy flux is determined. Then, if this charge exchange reaction is included and the nebular line intensities are again reproduced, one will find a differing set of elemental abundances, a "bluer" central star energy distribution and perhaps a material-limited model. The charge exchange reaction shifts the ionization equilibrium, $O^{++} + \varepsilon \rightleftarrows O^+$, to favor O^+. Thus, to maintain the same I(5007)/I(3727) or $N(O^{++})/N(O^+)$ ratio as before, a "bluer" stellar flux distribution is required and it may be necessary to omit the outermost nebular strata.

Generally, it has been found that low-to-medium excitation objects, such as IC 418, are reasonably well represented by theoretical models (e.g., Flower 1969; Buerger 1973). Improvements are secured by including charge exchange. Successful models have also been obtained for a few high-excitation objects, notably NGC 7662 (Harrington 1968, 1969, 1972; Kirkpatrick 1970, 1972 and especially Harrington, Seaton, Adams, and Lutz 1982 -- see Chapter 11).

The model of a radiation-bounded nebula is not sensitive to the radius of the central star, other parameters being held fixed. The mean density $\langle N_H \rangle$ and the density distribution affect the ionization structure and the emergent spectrum in a well-defined and qualitatively easily understood way. Truncation of the emitting shell lowers the integrated brightness and -- as mentioned above -- raises excitation by weakening or suppressing lines of [N II], [O II] and [S II]. It also destroys any simple relationship between the level of nebular excitation and central star temperature T(*) unless one is prepared to assume that <u>all</u> nebular shells are optically thick.

Adjustments in the stellar far ultraviolet flux distribution, F_ν, can have a profound effect on the emergent spectrum as has been pointed out by many workers. First, we are not even sure of the proper chemical compositions to use. It is not always safe to assume that the abundances are the same as in the surrounding nebula (consider e.g., A30). The He/H ratio may also differ between the stellar atmosphere and the nebula. Some planetary nebulae nuclei may be composed mostly of C, N, or O. Theoretical stellar atmospheres, even when calculated under specifically non-LTE postulates, often assumed hydrostatic equilibrium. Yet atmospheres of hot, early-type stars are known to be sources of brisk stellar winds but as yet we have no sound theoretical bases for secure calculation of theoretical fluxes. Shock phenomena somewhere between the stellar surface and the inner edge of the surrounding nebular shell may produce high frequency line and continuum quanta that could influence the nebular ionization structure (see Sec. G). An ad hoc empirical procedure is sometimes used to modify the ultraviolet $F_\nu(x)$ distribution so as to reproduce certain excitation-sensitive ratios, e.g., He II λ4686/He I λ5876, [O III] λ5007/[O II] λ3727 and [Ne V] λ3426/[Ne III] λ3868 (see Chapter 11). Note, as mentioned in Chapter 3 that the finally-adopted stellar flux distribution may be used to obtain an effective temperature for the star. The thus-determined temperatures have the advantage of being consistent with the emitted nebular spectrum, but there appears to be a range of solution, particularly if we invoke nonhomogeneous dusty nebular models with condensations.

Establishment of the chemical composition of a nebula is often an important objective of a model calculation (see Chapter 11). Since the ratio N(X)/N(H) enters in the energy balance, there is often no simple linear relation between the intensity of a collisionally-excited line and the abundance of the corresponding element. Increasing the abundance of an element that is an effective coolant may lower T_ε effectively and not permit a realistic model to be obtained. Thus, within the gamut of input parameters it is often possible to find a model that will represent most of the observed lines reasonably well. Does this mean that we have the unique and correct model? Not necessarily!

7D. Some Complicating and Troublesome Factors for Nebular Model Constructions

By necessity, theoretical nebula models are geometrically (usually spherically) symmetrical. Observed nebulae certainly are not. Even when an overall symmetrical structure exists as in NGC 7009, the nebula is not spherically symmetrical. An almost ubiquitous filamentary pattern, and density condensations, adds further complications.

It is possible to calculate inhomogeneous models, e.g., a thin shell surrounding a denser one works well for objects such as NGC 7662. Furthermore, one may introduce filamentary structures into concentric shells. Composite models (cf. Fig. 7-2) are sometimes useful. When abrupt density variations are introduced, care is necessary in the application of integration schemes. Another problem with these more intricate structures is the matter of uniqueness. As has been demonstrated, particularly for NGC 7662, refinements in construction of geometrical models permit fairly good representation in the line intensities. On the other hand, with some objects such as M1-1, no suitable theoretical models have been obtained.

Much greater complexities in geometrical patterns and filamentary structures are encountered in H II regions than in planetaries. In the former, the illuminating stars are rarely located in a symmetrical position within the nebula. On one side, the nebula may be material-bounded; another denser side may be radiation-bounded. Planetary nebulae originate from and are excited by a well-defined parent star so that the distribution of radiating gases is usually roughly bilaterally or spherically symmetrical. An H II region may be excited by an association of stars, although in some such nebulae, a single star may dominate. If we use Kurucz model atmospheres, for example, the most successful models of H II regions in M101 usually require more than one illuminating star.

The problem of dust can be extremely troublesome for model calculations of H II regions. We examine the role of dust in more detail in Section F and Chapter 8; here we remark only on special aspects. Dust can scatter nebular light as well as starlight so an observer can detect lines of [O III], for example, from locations in an H II region where this radiation is not actually produced. Dust grains absorb both starlight and nebular light. For example, in theoretical calculations in pure gaseous envelopes, it is often found that the energy density in Lyα becomes considerably enhanced as these quanta are scattered back and forth before they can escape in the wings of the line profile. The presence of dust modifies the picture as Lyα quanta are absorbed and heat the dust.

In principle, one can handle dust absorption in a model nebula in a straightforward way (see, e.g., Balick 1975) but obvious complications are caused in H II regions where dust and gas appear to be enhanced together in distinct condensations and clouds. In nebulae without distinct blobs and dense regions, dust and gas may not be evenly mixed either, but may tend to be enhanced in outer regions. Such a situation appears to occur in NGC 7027 where Osterbrock (1974) found the red- and blue-shifted components of Balmer lines to show the same intensity. Thus, the red-shifted, receding, region was not measurably dimmed as compared with the blue-shifted, nearer, approaching region.

Another type of model that has been considered by van Blerkom and Arny (1972) is one in which denser blobs of material shield regions of the nebular gas from direct starlight; the sole source of energy in such regions is diffuse radiation scattered perhaps partially by dust from zones that are directly illuminated. Such models were explored in an effort to explain how lines of [N II], [O I], [O II], [S II] and [S III] could be stronger in certain nebulae than predicted by models, perhaps even when charge exchange was considered. The most definitive results appear to be those of Mathis (1976) who concluded that although radiations from these ions would be favored in the shadows, the level of brightness would be so low that they would never produce large enough fluxes of low-excitation radiation to explain the observed intensities.

Emissions of neutral atoms, e.g., [O I], [N I], [C I], and Mg I have not been fully explained quantitatively in any satisfactory manner. Since the ionization potential of nitrogen, 14.54 eV exceeds that of N, these lines could originate in regions where H is still substantially ionized. In those objects where N_ε has been found from the λ5198/λ5200 ratio, there is a tendency for the derived N_ε to be less than that found from other ions, suggesting that [N I] emission may arise at the edge of a transition region. Strong [O I] lines are found in many planetaries. Since O and H have the same ionization potential, charge exchange at the edge of H II zones may play an important role here. The Mg I λ4571 line has been found in many planetaries ranging from low-excitation objects such as NGC 40 up to high-excitation objects such as NGC 7027. Since the ionization potential of Mg is 7.61 eV, this radiation must originate in zones well protected from quanta shortward of λ1650 Å. The Mg I emission varies strongly from point to point, suggesting that the shielded volumes must be very small.

Carbon has an ionization potential of 11.27 eV, the [C I] λ9823, 9850 lines are observed in the Orion nebula and planetaries such as NGC 6720, NGC 6210 and 6543. The mechanism for the production of these lines is still in dispute. Possibly different excitation schemes hold for H II regions and planetaries. Among the processes suggested are collisional excitation in shadowed, partially ionized zones (Jewitt et al. 1983), charge exchange rather than radiative processes (Osterbrock AGN, Hippelein and Munch 1978), recombination in a thin interface between an H II region and a cool H I zone (such a zone is suggested by C II radio-frequency recombination lines), and shock excitation.

Strata of neutral gas frequently lie outside ionized H II regions. In objects such as Orion (see Chapter 10), this cool zone involves a dense molecular cloud material that is not yet condensed into stars. In planetaries this cool zone of ejected stellar material contains hydrogen, presumably H_2, molecules such as CO and dust. Curiously, atomic hydrogen has not been detected, even in NGC 7027 where its mass must be less than 0.001 m(sun)! The Q and S branches of molecular hydrogen have been observed. In particular, the vibration $v = 1 \rightarrow 0$ S(1) transition at 2.122 μm has been measured not only in dense objects such as Hb 12, BD + 303639, NGC 6210, 6572, and 7027, but also in the outer regions of NGC 6720. It may be produced by shock excitation and appears to be concentrated just outside the H II zone. Pure rotational quadrupole lines may be excited at lower temperatures ($50 < T < 200$) but their measurement would require observations from above the Earth's atmosphere. Eventually, theoretical models must be able to handle these neutral regions, particularly when they are generalized to treat such topics as the advance of an ionization front into the neutral zone.

7E. Time-Dependent Models

Perhaps one of the most serious defects in theoretical nebular models, at least for H II regions, is the assumption of a static situation. When a massive star is formed in the interstellar medium, it quickly moves to the zero-age main sequence. As it approaches steady-state hydrogen burning, the surface temperature rises rapidly. Thus, we can regard the star as being turned on suddenly. A Strömgren sphere is formed; its edge advances at first rapidly into the surrounding interstellar medium and later slows down (see Problem No. 2). A steep pressure difference between the ionized region and the as-yet-unperturbed gas is established and a shock wave may advance through the cool region (see Chapter 9). Actually, although time-dependent effects are important dynamically in H II regions, they probably do not introduce nonequilibrium effects that are important for interpretations of spectra (Harrington 1977).

Planetary nebulae expand with velocities typically of the order of 5 - 30 km/sec and are treated generally as static or at least steady-state configurations, on the assumption that the brightness or temperature of the illuminating star changes very slowly. If the character of the stellar radiative field changes abruptly as appears to have occurred with FG Sagittae, or if the central star suffers periodic outbursts as suggested by some theoretical workers, time-dependent effects in the surrounding nebulae shell can become important.

In a nebula of moderate excitation, the critical factor is the recombination time scale for hydrogen:

$$t_R = \frac{1}{N_\varepsilon \alpha} \sim \begin{matrix} 76{,}000/N_\varepsilon \text{ yrs (Case A)} & T_\varepsilon = 10{,}000°\text{K} \\ 120{,}000/N_\varepsilon \text{ yrs (Case B)} & \end{matrix} \qquad (2)$$

Thus, we see that for $N_\varepsilon \sim 1000/\text{cm}^3$, the recombination time scale is of the order of 100 years. As the gas cools it recombines more rapidly, but the time scale of cooling is more difficult to evaluate because it depends on the character of the ions that are present.

The equation governing the temperature and density of the recombining gas can be obtained as follows: The energy contained in a volume element is simply the kinetic energy $3/2(N_{ion} + N^o + N_\varepsilon)$ kT, where N^o is volume density of neutral particles. The rate of change in energy consists of the net loss carried away by recombining electrons, $3/2\ kT(dN_\varepsilon/dt)$, and the net gain or loss by radiative and collisional processes which we may write as G-L. We generalize Eq. (6-13) to include the contribution of helium which can supply electrons, and take care to include states of all ions and atoms that can dissipate energy. Thus, G is simply the left-hand side of generalized Eq. (6-13), and L the right-hand side thereof. Then:

$$(N_{ion} + N^o + N_\varepsilon)\ k\ \frac{dT_\varepsilon}{dt} = \frac{2}{3}(G - L) - kT_\varepsilon \frac{dN_\varepsilon}{dt} \qquad (3)$$

Since electrons of lower energy are preferentially removed by recombination, we might expect T_ε to rise, but losses by collisional level excitation tend to lower T_ε. The G-term involves the changing radiation field of the illuminating star and the secondary line and continuum radiation produced in the recombining gas. Generally, this term is nearly cancelled by the L term in the cooling plasma.

The planetary nebula PK 60° - 7°1 associated with the remarkable star FG Sge offers an example in point. The star rose in brightness by about four magnitudes from 1895 → 1970. Since 1955 its bolometric magnitude seems to have remained constant, although its effective temperature fell as the spectral class changed from B4I to F6Ip in 1973. Subsequently, lines of s-process elements appeared in the spectrum of the star. A star of spectral class B4I is too cool to produce the type of nebular spectrum observed in PK 60° - 7°1. Harrington and Marionni (1976) suggested that we are observing a "fossil" nebula, one excited by a much hotter star but now slowly recombining. Introducing the present-day properties of the nebula into a time-dependent theory, they tried to reconstruct a scenario of what has happened. With an adopted distance of 2.4 kpc, the 38 km/sec expansion velocity of the nebula combined with an angular radius of 14" implies an expansion age of about 5,000 years.

Their proposed scenario is that the nebula evolved in a steady-state fashion in thermal and ionization equilibrium for about this length of time. Then, rather abruptly, sometime near 1894, the temperature of the central star, which had been between 50,000°K and 75,000°K, began to fall linearly with time. The luminosity, about 4,000 times that of the sun, may have remained the same. Since the ultraviolet radiation falls off so rapidly, recombination phenomena dominate, and the ionization and excitation state of the gas will depart substantially from equilibrium. The behavior of line intensities with time will not reflect any equilibrium situation! For example, He II λ4868 is produced by recombination and its rate of decline indicates the rate at which He^{+} is formed from He^{++}. The more rapid decline of [Ne III] and [O III] after about 1944 (when the star presumably ceased to produce ionization radiation) is caused not by recombination of Ne^{++} and O^{++} ions but by a decline in T_{ε}. They calculated several models and showed that if T_{*} exceeded 75,000°K, the 4686 line would have been stronger than observed, while the time rate of change of [O III]/Hβ is a sensitive measure of N_H and indicates it must have been about $160/cm^3$. They concluded that the N/O ratio is very high and that this enrichment must have occurred more than 5,000 years ago when the nebula was formed and long before the recent surge of s-process elements to the surface. A more recent calculation by Tylenda (1980) who took charge exchange into account leads, however, to a normal N/O ratio.

An hypothesis adopted by Tylenda (1979) is that planetary nebulae nuclei (PNN) have repeated helium shell flashes with a duration in the range of 10 - 1,000 years, which embraces the recombination times of a typical planetary nebula. Hence, if the energy output of ionizing quanta is considerably enhanced in a flash cycle, the ionization and thermal structure can be far from equilibrium. It is assumed that the timescale of a flash is of the order of 10 to 100 times shorter than the interflash interval. We would expect most nebulae to be in steady states for which an equilibrium theory is valid.

Several types of models were calculated including some in which the normal temperature of the star is taken too low to produce a nebular-type spectrum in the surrounding gas. At the time of a thermal pulse, T(*) rises to 70,000°K or 80,000°K, thereby exciting a typical nebular spectrum. As the star cools, recombination sets in; regions of different density recombine on a different time scale. One interesting point is that if you play the game of trying to represent line intensities in time-dependent models by stationary models, the latter would tend to predict lines of [N II], [O I], and [O II] that were far too faint. Thus, the 3727/4959 ratio is 2.5 → 4.0 times smaller in stationary than in time-dependent ones. Also, the equilibrium models predict too low an [Ne V]/[Ne III] ratio. Tylenda suggests that IC 2120 may be a good candidate for the scenario observed in FG Sagittae. The central star is of G type and has both low-excitation and density (Lutz 1977). Although Tylenda's models specify a rather special character of stellar variation, for which there is no direct observational confirmation, the calculations are useful in demonstrating the manner in which fairly rapid variations in ionizing flux can affect the spectrum of a gaseous nebula.

7F. Resonance Transitions and Transfer Problems

Quanta from recombination and forbidden transitions observed in the optical region of the spectrum overwhelmingly escape the nebula without reabsorption. Some attention was paid to transfer problems in Ly α of H I and He II and to the Bowen fluorescent mechanism, but checks on the calculations were necessarily indirect. Finally, when resonance lines were observed in the UV, adequate treatments of the transfer equation became necessary.

The radiative transfer problem for a gaseous nebula differs markedly from that encountered in a stellar atmosphere. Since scattering only is involved, it is more akin to that found in a planetary atmosphere problem. The local gas kinetic temperature, T_i, enters only through its effect on the broadening of line profiles. Usually, one also assumes noncoherent scattering or complete redistribution in frequency. A quantum that is absorbed in one part of a line profile may be reemitted in another. The probability of reemission in a particular interval $d\nu$ at $(\nu - \nu_o)$ is governed by the shape of the line profile. The atomic absorption coefficient is given by:*

$$a_\nu = \alpha_o \frac{a}{\pi} \int_{-\infty}^{+\infty} \frac{e^{-y^2}}{a^2 + (u - y)^2}\, dy = a_o\, H(\underline{a},u); \tag{4}$$

where

$$a_o = \frac{\omega_j}{\omega_{j'}} A_{jj'} \frac{\lambda^2}{8\pi^{3/2}} \frac{1}{\Delta\nu_o}, \quad \Delta\nu_o = \frac{\nu_o}{c}\sqrt{\frac{2kT}{M}}, \quad \underline{a} = \frac{\Gamma}{4\pi\Delta\nu_o}, \quad u = \frac{\nu-\nu_o}{\Delta\nu_o} \tag{5}$$

Here $A_{jj'}$ is the Einstein A-coefficient for a transition $j - j'$, $T = T_{ion}$ may be taken equal to T_ε, M is the mass of the ion, and Γ is the radiative damping constant for level j.

The assumption of a complete redistribution in ν is a reasonable hypothesis for resonance lines of carbon ions, for example. It may not be valid for Ly α. (Recall the discussion in Chap. 4, F.) We are concerned here with resonance lines of carbon, nitrogen, and oxygen ions for which the optical depths at the line centers are typically 10^3 rather than 10^6 as for Ly α He II. For these heavier, less abundant ions, we need consider only Doppler broadening.

---

*See, e.g., Aller, L.H. Atmospheres of the Sun and Stars, 1963, p. 322, New York, Ronald Press Co. or Mihalas, D., 1978, Stellar Atmospheres, p. 279, San Francisco, Freeman Co.

Then:

$$a_\nu = \frac{\pi\varepsilon^2}{mc} f \frac{1}{\Delta\nu} \frac{1}{\sqrt{\pi}} \exp\{ - [\frac{\nu - \nu_o}{\Delta\nu_o}]^2 \} \qquad (6)$$

Lines of C, N, O, Si, and Ne ions for example are produced by collisional excitation to the upper level followed by radiative escape to the ground level. If the transition probability is high, as for λ1548, 1550 of C IV, the quanta are scattered from atom to atom a number of times before they can escape from the nebula. Several effects are important:

a) Formal solutions for the scattering of the radiation have been given only for geometrically simple configurations, such as flat slabs or spherically symmetrical shells. A real nebula is irregular and quanta are going to escape through holes.
b) Extinction of radiation by solid grains can play an important role.
c) In most instances the nebular shell is expanding. If the velocity of expansion greatly exceeds the thermal velocity, the quanta simply escape with little hindrance. The limiting case of $V_{exp} = 0$ can be handled by the Sobolev approximation.

It is anticipated that effects of geometrical irregularites can be handled, if ever, only in a statistical fashion with the aid of advanced computer programs. If the optical depth, τ, in individual clouds or filaments remains large, perhaps the effect can be assessed roughly by adopting some mean value $\langle\tau\rangle$.

Absorption by solid grains can have marked and unexpected effects. The efficiency of extinction by dust depends on the total path length traveled by the photon before it escapes from the nebula. When the optical depth in the line is large, this path length is much greater than the physical diameter of the nebula. Hummer (1982) pointed out that the fraction of line photons that escape depends on the extinction cross section of the grains only through the product of the cross section and the mean path length $\langle\ell\rangle_0$ for no absorption, that is, on a parameter he defined as:

$$\alpha_d = \frac{n_d \sigma_d \langle\ell_o\rangle}{k(\text{line})}$$

where k(line) is the line absorption coefficient. The larger the value of α_d, the smaller the fraction of photons that escape.

Note that a photon escapes most easily in the wings of a line, and the chance of it being shifted to a line wing is the greater the number of scatterings that it suffers. Now, the longer the physical distance a photon has to travel between scatterings, the greater will be the chance it gets absorbed by dust. On the other hand, the number of scatterings needed for a photon to escape depends weakly on the line strength. The quantitative effect is that the stronger component of a doublet such as C IV λ1948, 1950 may suffer less extinction than the weaker one. Harrington et al. (1982) deduced $\tau(\text{dust}) \sim 0.1$ in NGC 7662 from the C IV doublet. When effects of nebular expansion are also considered, some unexpected results emerge. One might suppose that photons near the line center would escape more readily as a consequence of Doppler shifts. Bonilha et al. (1979) find, however, that dust tends to destroy these photons effectively.

The general problem of scattering with subsequent escape of photons is difficult. One approach is to impose the escape-probability approximation so we can estimate the likelihood that a photon would escape from a given point in the nebula (Grachev 1978; Hummer and Rybicki 1982). For lines where complete redistribution occurs, Hummer and Rybicki obtained a new expression for the source function in the medium, with the aid of which the escape probability can be estimated.

7G. Non-Radiative Excitation

Our models have assumed that all the energy radiated in gaseous nebulae has been derived ultimately from UV quanta emitted by illuminating stars. Long ago, Minkowski noted that the wide emission lines in the Wolf-Rayet spectra of central stars of certain planetaries implied ejection velocities of the order of 1000 or 2000 km/sec, whereas the emission lines in the surrounding nebulae indicated expansion velocities of only 15 to 40 km/sec. Where and how is the wind slowed down? More recently IUE observations have shown P Cygni-type profiles in the spectra of several planetary nebulae nuclei and radio frequency data have given clues to mass loss rates. Terminal velocities can be estimated from line profiles. Thus, if we can also get mass loss rates, dM/dt, we can find the momentum and energy supplied to the surrounding nebula. The best procedure appears to be to estimate mass loss rates from line profiles. Thus, Castor, Lutz, and Seaton (1981) found the limiting ejection velocity $V(\infty) = 2150 \pm 100$ km/sec, and $dM/dt \sim 10^{-7}$ m(sun)/yr for the nucleus of NGC 6543. The profiles of many other PNN nuclei also indicate mass loss. Perinotto (1982) finds a very large spread in these objects, $10^{-11} < dM/dt < 10^{-7}$ m(sun)/yr. We might anticipate that the winds would be more pronounced in Wolf-Rayet stars than in those of O or Of spectral classes. Not surprisingly, in the WC8 central star of NGC 40, Benvenuti, Perinotto, and Willis (1982) find $v(\infty) = 2370$ km/sec and estimate $dM/dt = 6 \times 10^{-7}$ m(sun)/yr.

The Kwok-Purton-Fitzgerald model for the development of a planetary nebula (cf. Chapter 12) envisages a gentle wind of a few kilometers per second blowing outward from a red giant in an advanced evolutionary stage. Then the outer envelope is detached and suddenly the material is exposed to a hot core that is settling down to become a white dwarf. The stellar flux, rich in high-frequency quanta, ionizes and excites the nebular shell, which is buffeted now also by a stellar wind that is many times larger than the solar wind. We seek answers to the following questions:

1) At what point in the nebula is the wind slowed down and momentum transferred to the nebular shell?
2) Where is the kinetic energy dissipated?
3) How does the energy release affect the observable spectrum?

In most instances, models based on theoretical stellar fluxes and conventional theories of radiative nebular excitation account quantitatively for the observed spectrum. For some nebulae small but definite modifications of theoretical fluxes are required. Most often we need an excess of emission shortward of λ228 Å. This excess is often interpreted as radiation produced by a hot corona or a high temperature in the wind plasma. Köppen (1982) argues that λ4686 is excited by a direct stellar wind in NGC 1535, although it appears possible to represent the spectrum with a model with a reasonable stellar flux. What might we anticipate in the way of effects?

Theoretical studies of high-velocity winds impinging on nebular shells or the local interstellar medium (see Kahn 1982; also Chapters 9 and 10) indicate that gases in the inner regions will be heated by shocks to temperatures in excess of 10^6°K. In such a medium we might expect to find radiations of [Fe X] and [Fe XIV] as well as ultraviolet signatures of a hot gas. These radiations are indeed found in the Network Nebula and other supernova remnants, but we are concerned here with nebulae excited by stellar fluxes, not SNR's. The highest excitation lines found in planetary nebulae are those of [Fe VII] and [Ne V]; these emissions can be explained by radiative models. Possibly the densities in the presumably scoured-out cores of PN are too low to permit appreciable emissivities in these "coronal" stages of iron. Failure to detect [Fe X] or [Fe XIV] cannot be taken as conclusive evidence for the absence of a hot, shocked zone.

If stellar winds play important roles in the excitation of nebular spectra, the most promising place to look for evidence might be planetaries with nuclei of the Wolf-Rayet type since these objects yield a nonradiative energy flux considerably in excess of that emitted by an O or Of star. In particular, we might expect the WC 8 nuclei of NGC 40 or BD + 30 3639 to show He II λ4686, yet this line is missing and [O III] is much weaker than [O II]. A stellar radiation field appropriate to a temperature of about 30,000°K would appear capable of reproducing the observed spectrum.

Yet, we cannot write off nonradiative energy sources as unimportant. The spectra of some planetaries such as NGC 6886 and of (86 - 8°1) cannot be represented by straightforward theoretical radiative models. An even more striking example is the kinematically active object NGC 6302. R. Minkowski and Hugh Johnson (1967) were the first to note the high-velocity fields in this object. Meaburn and Walsh (1980) noted an 800 km/sec-wide component of [Ne V] 3426, which is emitted near the core. They interpret this profile as evidence for a radiatively ionized stellar wind. In the outer regions of NGC 6302, the profiles of Hα, [N II], [S II], and [O II] give separate velocity components with heliocentric velocities ranging from - 186 to + 140 km/sec. Mass loss rates as high as ~ 10^{-5} m(sun) have been suggested, but no direct check is possible since the central star is not visible, presumably because of dust extinction. IUE observations (Chapter 2, Fig. 8) show no trace of the central star. Attempts at theoretical models show no homogeneous structure with any reasonable energy flux will reproduce the nebular spectrum. Nor does it seem likely that inhomogeneous models will work either, unless perhaps we go to very elaborate structures. Can direct excitation by the wind play any role here? The high electron temperature, in the neighborhood of 20,000°K, noticeably exceeds that of most planetaries and may be at least partly maintained by the dissipation of mechanical energy. The chemical composition of NGC 6302 is unusual (Aller et al. 1981) and does not fit our usual scenarios of stellar evolution.

Thus, the dynamical evolution of gaseous nebulae must be considered in any fully developed theoretical approach. Lack of space precludes a proper treatment of this important subject, so vital to an understanding of classical H II regions, planetary nebulae and supernova remnants. Phenomena of advancing ionization fronts, pressure waves, and the establishment of instabilities must all be considered. Some observational aspects are presented in Chapter 10.

As one example, we consider Capriotti's (1973) work on the evolution of an outer H I shell in an idealized spherical planetary nebula. The inner region is ionized so that the shell is accelerated outwards by thermal and dynamical pressure on its inner boundary. He found that this ionization front becomes unstable when the nebular gas has become so rarefied that recombination and ionization processes are no longer able to damp out either outward or inward displacements of the ionization front. A Rayleigh-Taylor instability develops akin to that observed when a denser incompressible fluid overlies one of less density. As the ionization front deforms, plumes of dense, cool, neutral gas penetrate into the underlying H II region. Velocity differences between the expanding ionized gas and the plumes produce shearing forces upon the latter, which cause them to break off from the H I region and form dense globules. Subsequently, the globules may erode due to ionizing radiation from the central star. Thus, Capriotti was able to explain the fine condensations and associated radial filaments detected in Baade's photographs of NGC 7293, the Helix Nebula. It is not clear whether larger structures can be explained in this way. Many of them must originate from irregularities in the initially expelled shell. Vladimirov and Khromov (1978) considered the filaments and condensations in NGC 7293 as relic formations from epoch of ionization of the nebulae.

Problems

1. Suppose that a model nebula is calculated by a program such that $N(X^i)/N(X)$ for ions of elements including H, He, C, N, O, Ne, Mg, Si, S, Cℓ, and Ar are available at each point. T_ε and N_ε are also given. Derive an expression for calculating a mean value of T_ε, appropriate to getting $\langle N(X^i)\rangle/\langle N(H+)\rangle$ from $I_\lambda(X^i)/I(H\beta)$ where $I_\lambda(X^i)$ and $I(H\beta)$ refer refer to the total intensities of a collisionally-excited line of ion X^i and of $H\beta$, respectively.

Problems (Continued)

2. Consider the growth of a Strömgren sphere about a hot star which is suddenly "turned on." If Q_S is the number of stellar quanta emitted per second with $\nu > \nu_1$ ($h\nu_1$ = ionization energy of hydrogen) then in equilibrium:

$$Q_s = \frac{4}{3}\pi r_S^3 N_\varepsilon N(H^+)\alpha_2(T_\varepsilon)$$

gives the radius of the Strömgren sphere. Show that the rate of growth of the H II zone in a time dt is given by:

$$Q_s\, dt = 4\pi r^2 N_o dr + \frac{4}{3}\pi r^3 N_\varepsilon N(H^+)\alpha_2(T_\varepsilon)\, dt$$

If you assume that $N_\varepsilon = N(H^+) = N(H^o)$ where $N(H^o)$ is the density in the neutral gas, show that the solution is:

$$(r/r_s)^3 = [1 - \exp(- t/t_R)]$$

where $t_R = [N_\varepsilon\ \alpha_2(T_\varepsilon)]^{-1}$ is the recombination time scale for a single hydrogen ion. Let:

$$Q_s = 2 \times 10^{49}\ s^{-1},\ \alpha_2(T_\varepsilon) = 2.6 \times 10^{-13}\ cm^3\ s^{-1},\ N_o = 1000/cm^3$$

Calculate the times required for r/r_S = 0.90, 0.96, and 0.99. Assume T_ε = 10,000°K. How does the rate of growth of r compare with the sound speed as r/r_S increases from 0.3 to 0.995?

References

Models of H II regions have been calculated for example by:

Hjellming, R.M. 1966, Ap. J., 143, 420.
Balick, B. 1975, Ap. J., 201, 705.
Stasińska, B. 1978, Astron. Astrophys. Suppl., 32, 429.
Shields, G. and Searle, L. 1978, Ap. J., 222, 331 (M101).
Kwitter, K.B. and Aller, L.H. 1981, M.N.R.A.S., 195, 939 (M33).
Rubin, R.H., Hollenbach, D.J., and Erickson, E.F. 1983, Ap. J., 265, 239.

For gas excited by synchrotron power law radiation, see:

Williams, R.E. 1967, Ap. J., 147, 556.
MacAlpine, G.M. 1972, Ap. J., 175, 11.

Numerous models have been calculated for planetary nebulae. Some of the "classical" papers are:

Hummer, D.G., and Seaton, M.J. 1963, M.N.R.A.S., 125, 437; 1964, 127, 217.
Goodson, W.L. 1967, Zeits. f. Astrofis., 66, 118.

References (Continued)

Harrington, J.P. 1968, Ap. J., 152, 943; 1969, 156, 903; 1972, 176, 172.
Hummer, D.G., and van Blerkom, D. 1967, M.N.R.A.S., 137, 353.
Flower, D.R. 1969, M.N.R.A.S., 146, 171, 243.
Kirkpatrick, R.C. 1970, Ap. J., 162, 33; 1972, 176, 381.

Further models were calculated by:

Buerger, E.G. 1973, Ap. J., 180, 817.
Webster, L. 1976, M.N.R.A.S., 174, 157.
Shields, G. 1978, Ap. J., 219, 565.
Keyes, C.D. and Aller, L.H. 1978, Astrophys. Space Sci., 59, 91; 1980, 72, 203.
Koppen, J. 1979, Astron. Astrophys., 80, 42; Astron. Astrophys. Suppl., 35, 111; 1982, I.A.U. Symposium No. 103.
Harrington, J.P., Seaton, M.J., Adams, S., and Lutz, J.H. 1982, M.N.R.A.S., 199, 517. (NGC 7662)

Stellar fluxes suitable for nebular calculations have been calculated by: Hummer, D., and Mihalas, D. 1970, M.N.R.A.S., 147, 339 (see also "Surface Fluxes for Model Atmospheres for the Central Stars of Planetary Nebulae," JILA Report No. 101, 1970).

Cassinelli, J.P. 1971, Ap. J., 165, 265.
Kunasz, P.B., Hummer, D.G. and Mihalas, D. 1975, Ap. J., 203, 92.
Wesemael, F., Auer, L.H., van Horn, H.M., and Savedoff, M.P. 1980, Ap. J. Suppl., 43, 159.
Wesemael, F. 1981, Ap. J. Suppl., 45, 177.
Méndez, R.H., Kudritzki, R.P., Gruschinske, J., and Simon, K.P. 1981, Astron. Astrophys., 101, 323, and references therein cited.

Effects of dust on models of H II regions and planetaries:

Sarazin, C. 1976, Ap. J., 204, 68; 208, 323.
Petrosian, V. and Dana, R.A. 1980, Ap. J., 241, 1094.
Ferch, R.L. and Salpeter, E.E. 1975, Ap. J., 202, 195.

For a discussion of the dust distribution in NGC 7027, see:

Osterbrock, D.E. 1974, Publ. Astr. Soc. Pac., 86, 609.

Effects of shadows have been considered by:

van Blerkom, D. and Arny, T. 1972, M.N.R.A.S., 156, 91.
Hummer, D.G. and Seaton, M.J. 1973, Mem. Soc. Roy. Sci. Liège, 6th series, 5, 225.
Mathis, J. 1976, Ap. J., 207, 442.

Time-dependent effects:

Harrington, J.P. 1977, M.N.R.A.S., 179, 63.

References (Continued)

Harrington, J.P. and Marionni, P.A. 1976, Ap. J., 206, 458.
Tylenda, R. 1979, Acta Astronomica, 29, 355.
Capriotti, E. 1973, Ap. J., 179, 495.
Vladimirov, S.B. and Khromov, G.S. 1978, Astrofizika, 14, 307.

Resonance Transitions and Transfer Problems:

The OTS approximation was introduced by Zanstra (1951, B.A.N., 11, 341) and improved by Hummer and Seaton (1963) and by van Blerkom and Hummer (1977). Scattering of line radiation is a time-honored problem in radiative transfer that has been treated in many texts. See, e.g., Mihalas, D., Stellar Atmospheres, 1978, San Francisco, Freeman and Co.

Among recent papers are:

Hummer, D.G. 1982, I.A.U. Symposium No. 103, Planetary Nebulae.
Gracheve, S.I. 1978, Astrophysics, 14, 63.
Bonhila, J.R.M., Ferch, R., Salpeter, E.E., and Slater, G. 1979, Ap. J., 233, 649.
Hummer, D.G., and Rybicki, G.B. 1982, Ap. J., 254, 767.

Stellar winds are discussed in Chapters 10 and 12. In the present context, the review by M. Perinotto (1982), I.A.U. Symposium No. 103 and references therein are useful. Two examples of estimates of stellar winds in planetary nebular nuclei are:

Castor, J.I., Lutz, J., and Seaton, M.J. 1981, M.N.R.A.S., 194, 547.
Benvenuti, P., Perinotto, M., and Willis, A.J. 1982, I.A.U. Symposium No. 99, Wolf Rayet Stars, C.W.H. de Loore and A.J. Willis, eds. Dordrecht, D. Reidel Publishing Co., 1982.

For an account of the theory of the interaction of fast winds with a planetary nebular shell, see:

Kahn, F.D. 1982, I.A.U. Symposium No. 103, Planetary Nebulae.

NGC 6302 is an example of a planetary nebula where radiative models do not appear to work and where strong stellar winds may contribute to the excitation of the nebular shell.

Minkowski, R., and Johnson, H.M. 1967, Ap. J., 148, 659.
Meaburn, J., and Walsh, J.P. 1980, M.N.R.A.S., 191, 5P; 193, 631.
Aller, L.H., Keyes, C.D., O'Mara, B.J., and Ross, J.E. 1981, M.N.R.A.S., 197, 95.

List of Illustrations

Fig. 7-2. A Typical Composite Model. A hypothetical nebula is presumed to consist of an attenuated inner zone of radius 0.0768 pc. The outer shell has a radius of 0.090 pc but has two types of segments, one with a density of 3000 cm^{-3}, the other with a density of 10,000 cm^{-3}. The central star T_{eff} = 48,000°K, R = 0.8 R(sun) ionizes the lower density segment to the boundary of the nebula, but in the dense shell the ionized zone extends only to 0.0793 pc.

Chapter 8

DUST AND RELATED PHENOMENA IN GASEOUS NEBULAE

The importance of dust in the interstellar medium has long been recognized. The Coal Sack and the Great Rift in the Milky Way are two manifestations most obvious to the unaided eye. Barnard's classical photographs, for example, and much later surveys such as the Stromlo Atlas of the Southern Milky Way demonstrate the omnipresence of dust and the extremely irregular way in which it is distributed. The Bok globules and the dark lanes in the Trifid Nebula attest to association of dust with HII regions. The dust/gas ratio appears to be of the order of 0.01 by mass and reasonably constant from one part of the galaxy to another. Radio, infrared, and UV observations have disclosed dust in many planetary nebulae. In some of these nebulae, a large fraction of the less volatile metals may be tied up in these grains which play important roles in the interstellar medium. Since they lock away atoms of many heavy elements such as Fe and Ti, and also lighter ones such as C, O, and Si, many atoms which could play roles as coolants for the gas are not there. Accordingly, the energy balance of the interstellar medium is affected. Grains also serve as catalysts for molecule formation. The formation of H_2 and other molecules on grain surfaces may be an important process.

How do we observe and measure it? How much is there, where does it come from, of what does it consist and what effect does it have on the physics of nebulae and the interpretation of their spectra? We are concerned here primarily with dust in H II regions and gaseous nebulae, rather than with the broad problem of dust in the general interstellar medium. Lack of space precludes discussion of some important topics: theories of grain formation and destruction, their size distribution functions, and the role of grains as catalysts in interstellar chemistry.

8A. Extinction Effects Produced by Dust

One way to study interstellar dust is by an examination of its effects on the radiation impinging upon it. Dust particles absorb and scatter light; they are also heated by radiation and re-emit it in the far infrared. Optical effects of dust were first studied in the context of the extinction and reddening of light from distant stars. Internal dust in planetary nebulae (which often evades optical detection) sometimes is revealed through its infrared radiation.

First consider how dust grains falsify measurements of optical and ultraviolet lines and continua. We can write the extinction law in the form:

$$\frac{I(\lambda)}{I_o(H\beta)} = \frac{F(\lambda)}{F(H\beta)}\, 10^{Cf(\lambda)} \tag{1}$$

Here $F(\lambda)$ and $F(H\beta)$ are the measured fluxes at λ and $H\beta$, respectively, as corrected for extinction in the earth's atmospheres. $I(\lambda)$ is the intensity of a spectral line (or unit interval of the continuum) referred to that of Hβ. C is the logarithmic extinction at Hβ and f(λ) involves the wavelength dependent factor. Since the grains extinguish selectively, roughly as 1/λ in the visible region of the spectrum, $f(\lambda) > 0$ for $\lambda < \lambda 4861$ and $f(\lambda) < 0$ for $\lambda > \lambda 4861$. We measure the extinction law by comparing flux distributions from obscured and unobscured stars. We must use care in extending such measurements to the infrared because results could be falsified by thermal emission from hot dust clouds around stars.

The extinction law seems to be fairly uniform at optical wavelengths in many different directions in the galactic plane. There are exceptions: in Orion, towards ρ Oph and HD147889, the extinction is abnormal. Although a standard curve is useful in a first reconnaissance and may even turn out to be justified, one should always check available UV extinction data to see if peculiarities appear in the region of the sky involved (Meyer and Savage 1981; Bohlin and Savage 1981). For example, the extinction curve for σ Sco falls far below the standard curve for $1/\lambda > 3.0$.

If A_λ is the extinction in magnitudes at wavelength λ, the color excess at any wavelength λ is $E(\lambda - V) = A_\lambda - A_V$ and in particular,

$$E_{B-V} = A_B - A_V \tag{2}$$

where $1/\lambda_B = 2.31$ and $1/\lambda_V = 1.83$, the wavelength, λ, being given in μm. We can express interstellar extinction as $E(\lambda-V)/E(B-V)$ versus 1/λ. Thus: $A_\lambda/E(B-V) = E(\lambda-V)/E(B-V) + R$ where $R = A_V/E(B-V)$ is the ratio of extinction-to-color excess. Numerically, $R \sim 3.1$ (Savage and Mathis 1979). Seaton (1979) adopted $R = 3.2$ and wrote:

$$X(x) = \frac{A_\lambda}{E(B-V)} \quad \text{where} \quad x = 1/\lambda(\mu m) \tag{3}$$

The extinction at any wavelength is $10^{-C[1+f(\lambda)]}$ where:

$$f(\lambda) = \frac{X(x)}{X(H\beta)} - 1 \;,\; X(H\beta) = X(1/0.4861) = 3.68 \tag{4}$$

Thus, $f(\lambda_{H\beta}) = 0$, and $C = 0.4X(H\beta)E_{B-V} = 1.47\ E_{B-V}$ (5)

Figure (8.1) illustrates the behavior of X(x) in the near ultraviolet. Seaton has given an empirical formula for X(x) based on discussions by Code (1976) and Nandy _et al._ (1975, 1976). See also Savage and Mathis (1979), Table 2. If the functional form of f(λ) is accepted, a number of methods are available for getting C.

Table 1 summarizes some of the line ratios that may be employed to determine extinction. In nebulae of high excitation, the 1640/4686 ratio is particularly useful. Others may be of greater interest in estimating dust extinction in quasars; Draine and Bahcall (1982-see table 1) give a more complete list.

Table 8-1.

Some Spectroscopic Line Ratios Useful for Determining Extinction Factor C

(I) Methods Involving Line Ratios in Hydrogen and Helium

					notes
Balmer decrement		Chap. 4, Table 1			(1,2)
Paschen/Balmer ratio		Chap. 4, Table 2			(1)
$\frac{I(1640)}{I(4686)}$ predicted ratio	HeII	6.25	N = 0,	T = 10,000	(3)
		6.60	$N = 10^4$	T = 10,000	
		7.14	N = 0.0	T = 20,000	
		7.45	$N = 10^4$	T = 20,000	

(II) Methods Involving Ratios of Collisionally Excited Lines

Line ratio	ion	ratio	type of transition	
2321/4363	[OIII]	0.24	TA/aur	(4)
$\frac{I(4076)+I(4068)}{I(10.3\ \mu m)}$	[SII]		TA/aur	(5)
I(3722)/I(6312)	[SIII]	0.609	TA/aur	(6)
I(2855)/I(7171)	[ArIV]	6.72	TA/aur	(7)(4)
I(1575)/I(2972)	[NeV]	2.79	TA/aur	(8)(4)

(1) These ratios are relatively insensitive to temperature and density, at least for the Balmer lines; an interative procedure can be employed — Brocklehurst, M. (1971) M.N.R.A.S. 153, 471; (2) Clarke, W. (1965); (3) Seaton, M.J. (1978), M.N.R.A.S., 185, 5P; (4) Draine, B.T. and Bahcall, J.N. (1981) Ap. J., 250, 579; (5) Miller, J.S. (1968) Ap. J., 154, L57; the ratio is $1.186\ (1 + 5.08b^*)/(1 + 2.88b^*)$ where $b^* = b(^2P_{3/2})/b(^2P_{1/2})$. The b-factor measures the population deviation from thermodynamic equilibrium (see Eq. 5-36). With lower dispersions, λ4068 and 4076 can be blended with C II and O II. Here 10.3 μm refers to the sum of the four [S II] infrared auroral-type transitions, λ10321, 10287, 10373, and 10339A; (6) Since λ3722 is usually badly blended with a Balmer line, this ratio is difficult to use; (7) The 2855 line is usually weak and difficult to observe; (8) In most sources the lines are weak.

Note that the broad interstellar absorption structure centered near λ2160 can be used in objects where we anticipate the continuous spectrum to be well-defined. Feibelman (1982) employed it in planetary nebulae where the central star had a strong, uncomplicated continuum. This λ2160 feature is generally attributed to extinction by small graphite grains (see, e.g., Mathis et al. 1977).

Extinction is also often measured by comparing radio frequency and Hβ fluxes. For example, we may compare the ratio of the 5 GHz flux density to Hβ fluxes. The free-free radio emission in ν to ν+ dν over all directions from a unit volume radiating at a temperature T_ε is (Oster 1961):

$$E_\nu d\nu = 4\pi j(\nu)d\nu = \frac{N_\varepsilon N_i 32\sqrt{2\pi}\, Z^2\varepsilon^2}{3m^2c^3}\left(\frac{m}{kT_\varepsilon}\right)^{1/2} \ln\left[\left(\frac{2kT_\varepsilon}{\gamma m}\right)^{3/2} \frac{m}{\pi\gamma Z\varepsilon^2\nu}\right]d\nu \tag{6}$$

Here N_ε and N_i are the electronic and ionic densities, respectively; here γ denotes Euler's constant, 1.78. Z is the ionic charge, while k, ε, C, and m have their usual meanings. Consider a plasma that consists of ions of H^+, He^+, and He^{++}. Then:

$$4\pi j(\nu)d\nu = 3.75 \times 10^{-40}\, t^{-1/2}\, N_\varepsilon \left\{\ell n\ 2.48 \times 10^{13} \frac{t^{3/2}}{\nu}\, 4N(He^{++})\right.$$

$$\left. + \ell n \left(4.96 \times 10^{13} \frac{t^{3/2}}{\nu}\right) \left[N(H^+) + N(He^+)\right]\right\}d\nu \text{ erg cm}^{-3} \text{ sec}^{-1}$$

$$t = \frac{T_\varepsilon}{10{,}000}\ . \tag{7}$$

If we approximate E(Hβ) as

$$E(H\beta) = 1.23\, t^{-0.9}\, N(H^+)\, N_\varepsilon \times 10^{-25} \text{ erg cm}^{-3} \text{ sec}^{-1}\ ,$$

then the ratio of the fluxes will be:

$$\frac{S(\nu)\Delta\nu}{F(H\beta)} = 3.05 \times 10^{-15}\, \Delta\nu\, t^{0.4} \left\{\ell n\left(2.48 \times 10^{13} \frac{t^{3/2}}{\nu}\right) \frac{4N(He^{++})}{N(H^+)}\right.$$

$$\left. + \ell n \left[4.96 \times 10^{13} \frac{t^{3/2}}{\nu}\right] \left[1 + \frac{N(He^+)}{N(H^+)}\right]\right\} \tag{8}$$

In particular, we may compare the ratio of the 5 Ghz flux density to the Hβ flux density (Milne and Aller 1975):

$$\frac{S(5\text{ GHz})\ [Wm^{-2}\ Hz^{-1}]}{F(H\beta)\ [\text{erg cm}^{-1}\ s^{-1}]} = 3.05 \times 10^{-18}\, t^{0.4}\, \ell n(9900\, t^{3/2})\, K_{He}, \tag{9}$$

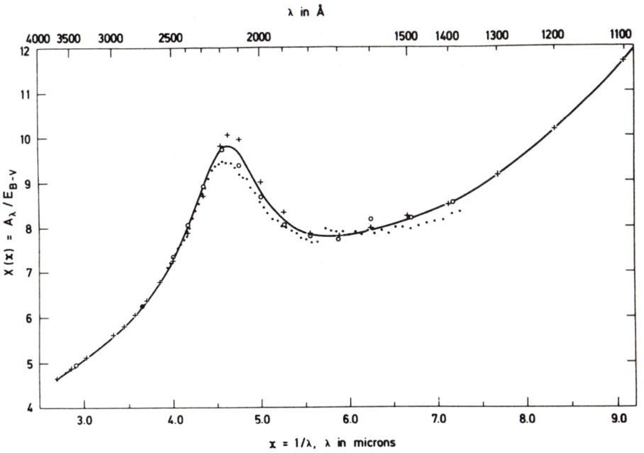

Figure 8-1

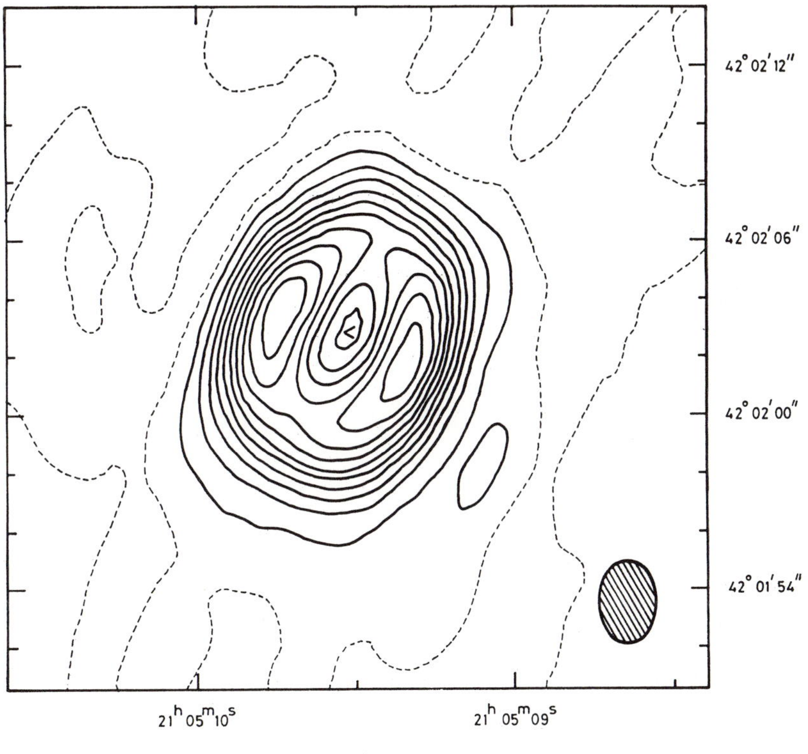

Figure 8.2.a

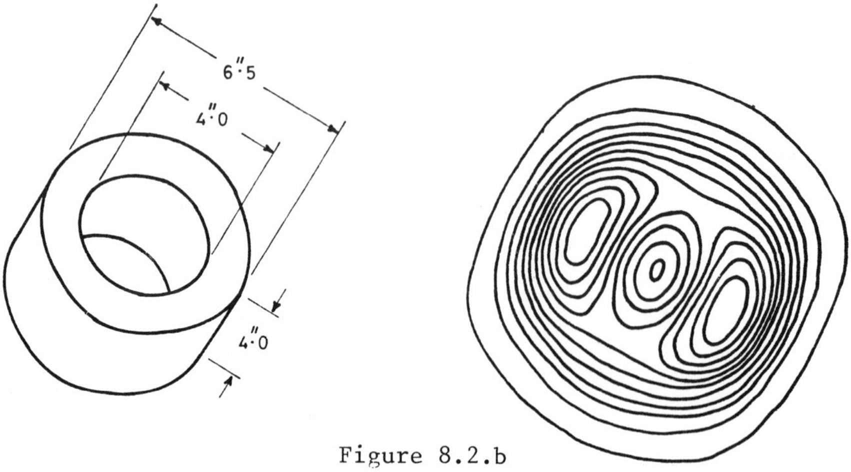

Figure 8.2.b

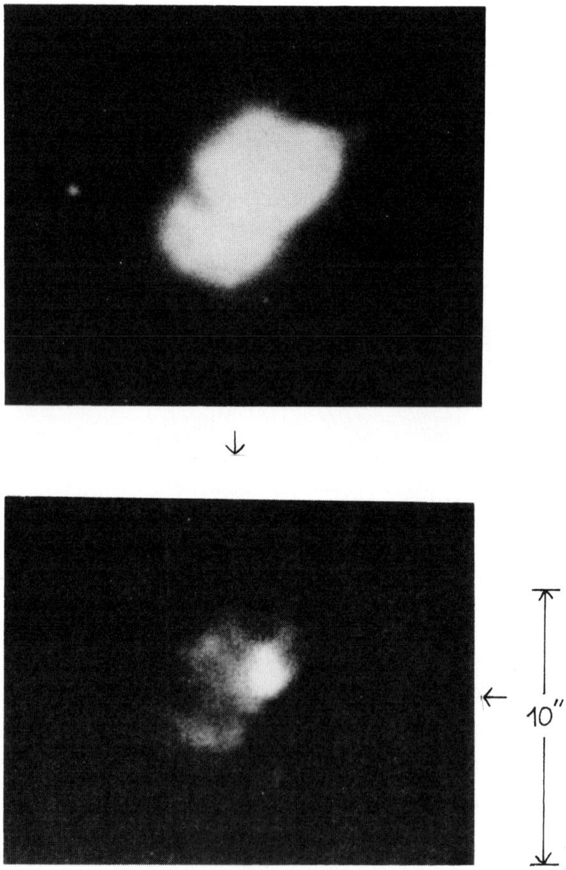

Figure 8.2.c

where:

$$K_{He} = \left[1 + (1 - x'')y + 3.7\ x''y\right] \quad (10)$$

Here hydrogen is assumed to be fully ionized, y is the helium/hydrogen abundance ratio N(He)/N(H), and $x'' = N(He^{++})/N(He)$. Total extinctions derived by this procedure have been given for a large number of planetary nebulae.

It is often found that the value of C found by comparing radio frequency and Hβ fluxes differs from values obtained, for example, by fitting the Balmer decrement in the sense that the former is numerically larger. Consider an H II region behind an obscuring cloud that cuts out all radiation from half the object and only, say, 10% of the light from the other half. The Balmer decrement gives a low value of the extinction since it is observed only in the relatively unobscured region but the ratio of Hβ to radio frequency flux gives some kind of an average over the entire nebula. On the other hand, if the radio frequency flux is measured at a low frequency in an optically thick nebula such as Orion, it will show effects of self-absorption. Consequently, C will be underestimated. For a weak, confused source, one might have a tendency to overestimate the flux.

A good illustration of how dramatic a role may be played by dust is provided by NGC 7027, which Jones _et al._ (1980) JMSW have described as revealing more intricately interesting detail to infrared observers than any other object in astronomy. Compare Minkowski's direct photograph (Fig. 8-2) with radio contours at 50 GHz measured by Scott (1973). Because of surrounding dust the optical picture bears little resemblance to the actual shape of the H II emitting region. On the other hand, measurements by Becklin _et al._ (1973) at 10 μm show contours revealing that emissions from hot dust are spatially distributed in a manner similar to that of the ionized gas. The observed optical extinction, however, is produced partly by the interstellar medium and partly by a circumstellar cloud. This envelops may be identical with a CO cloud found by Mufson _et al._ (1975) which extends over an area 30" × 60" and also appears to contain H_2. Long Hα exposures (Atherton _et al._ 1979 [AHRRW]) showed an extended region of emission with a major axis of about 50", presumably coincident with the CO cloud. Evidently, this emission is Hα radiation scattered in an outer dust envelope.

8B. Dust and Infrared Spectra of Gaseous Nebulae

Some of the most significant clues pertaining to the dust are offered by the infrared spectrum which includes the following contributions: 1) fine structure lines of atomic transitions in H (Paschen, Brackett and Pfund series), He II, [N I] 1.042 μ, [Mg V] 5.61 μm, [Ar III] 24.28 μm, [O IV] 25.87 μm, [O I] 63.2 μm and molecular transitions: N_2 (Q and S branches) and CO; 2) an atomic continuum; 3) a continuum believed to arise from thermal emission of dust grains with temperatures typically of the order of 100°K; 4) diffuse lines at 3.07 μm and 9.07 μm, attributed to ice and silicates, respectively, which have been observed in both absorption and emission; 5) broad unidentified purely emission features at 3.27 μm, 3.4 μm, 6.2 μm, 7.7 μm, 8.6 μm, and 11.3 μm.

Processes leading to the emission of atomic lines and continua seem to be well understood. Molecules appear to be formed in cooler, neutral regions. Although the infrared continuum in the neighborhood of 1.5 to 2 μm arises mostly from atomic line and continuum photon emission), this source fails by factors of ~ $10^2 \rightarrow 10^3$ to account for radiation in the medium- and long-range infrared 10 → 100 μm, where thermal emission from dust grains must dominate.

What might we anticipate about the nature of these grains? When dust was first discovered in the interstellar medium, attempts were made to explain how solid grains could be formed there in situ. No really satisfactory mechanism was ever proposed. It is now generally believed that solid grains are formed in atmospheres of cool, highly evolved giant and supergiant stars when the temperture falls below a critical value. Eventually, this outer stellar envelope, which may often evolve into a planetary nebula, is ejected into the interstellar medium. M stars tend to supply silicate grains, while carbon stars furnish grains of soot, possibly other forms of solid carbon, and Si C. A knowledge of grain characteristics in planetaries could be valuable for many problems. Do they play decisive roles in nebular evolution, especially in the early stages? The material in the interstellar medium, on the other hand, has been accumulated from many sources at many different epochs.

Direct identification of grain ingredients from the spectra in the middle infrared 2 to 13 μm range (accessible to ground-based observation) is difficult. Definite identifications have been made in dusty regions in the interstellar medium for features at 3.07 and 9.7 μm as ice and silicates, respectively. Since silicates are relatively refractory and can persist to temperatures of the order of 800°K, the broad 9.7 μ line has been seen in both absorption and emission. Interstellar ice, which must be very impure, is amorphous rather than crystalline, exists only at low temperatures, is seen in absorption, and appears to be associated with inner regions of molecular clouds. It is never seen in planetary nebulae.

The 10 μ silicate feature which is seen, for example, in the Orion Trapezium region is found in a number of very low-excitation (VLE) planetaries such as MI-26, SW St-1, Hb 12 and objects suspected of being in the pre-planetary nebulae stage, such as HM Sge and V 1016 Cyg. In these objectsd, [O II] is prominent and Hβ is stronger than [O III]. Cohen and Barlow (1980) note that the counterparts of these VLE objects in the Magellanic Clouds have $M_V \sim -4$ to −5, characteristic of Population I objects. They suggest VLE objects may represent a different type of object than more ordinary planetary nebulae. Presumably, these nebulae are oxygen-rich. Carbon-rich objects such as NGC 7027, BD + 30°3639, NGC 6790, NGC 6572, and IC 418 show a feature in the 10 μm to 14 μm region attributed to Si C.

The Orion nebula presumably has the same bulk chemical composition from one point to another. The Trapezium region shows the silicate emission feature characteristic of conventional O-rich interstellar medium material. Yet, the ionization front region of the

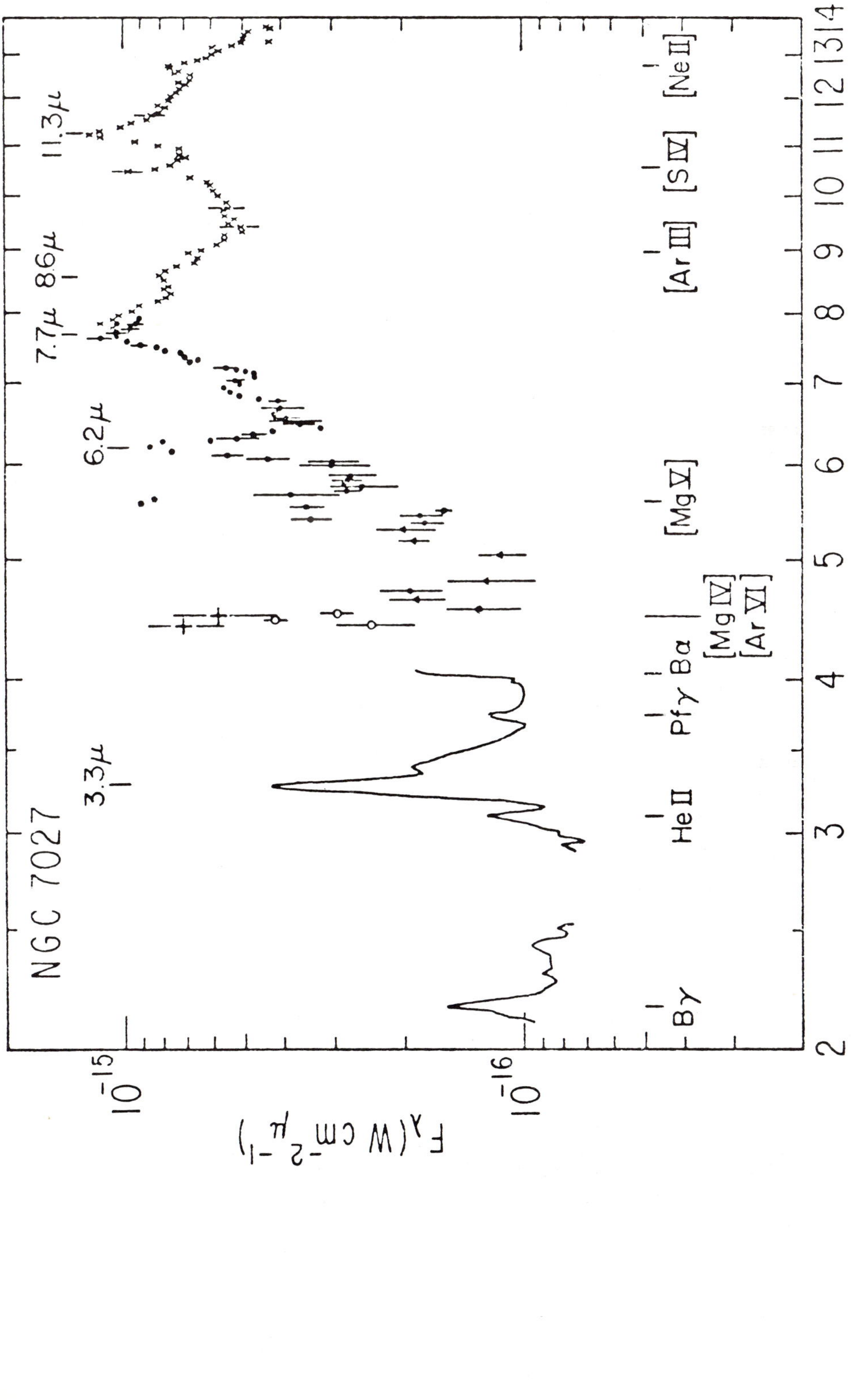

NGC 7027
$F_\lambda (W\ cm^{-2} \mu^{-1})$
10^{-15}
10^{-16}
$\lambda (\mu)$
3.3μ
6.2μ
7.7μ
8.6μ
11.3μ
Bγ
He II
Pfγ
Bα
[Mg IV]
[Ar VI]
[Mg V]
[Ar III]
[S IV]
[Ne II]
2
3
4
5
6
7
8
9
10
11
12
13
14

Figure 8-3

Orion nebula has an infrared spectrum very similar to that of NMGC 7027 (ARSJ and JMSW), which is regarded as a carbon-rich object.

The broad unidentified IR emission features (cf. Fig. 8-3) have the following characteristics:

i) Their line breadths, 01 μm - 1 μm, are intrinsic; that is, they are not closely packed lines of free or loosely-hold molecules on grain surfaces;

ii) They are absent in absorption, even in compact IR sources seen through large column densities of cold material;

iii) They are observed in regions illuminated by hot stars or other sources rich in UV radiation, e.g., nuclei of active galaxies;

iv) Their relative intensities are not fixed, but show spatial variations, differing from the underlying thermal continuum. In Orion where these bands are prominent, the 3.3 μm and 11.3 μm radiation comes from the interface of H I and H II regions, but is not found in the H II zone itself. Planetaries seem to show a similar type of behavior.

How are we to interpret these observational data? Items ii) and iii) would seem to be in harmony of fluorescence from molecular impurities on interstellar grains, but Dwek <u>et al</u>. (1980) argue that these properties can be explained by emission from a mantle of small ($\lesssim$ 0.01 μm) efficiently emitting grains at T ~ 300°K. These grains would constitute only a small percentage of the total present.

Many suggestions have been made for these spectral features: "oily plastic" mantles (Salpeter), silicon carbide, hydrocarbon fragment impurities on grain surfaces (3.3 μm feature is consistent with the C- stretching mode and the 7.7 μm feature is consistent with an N-C-N bending mode). Water of hydration in crystals has been invoked to explain the 6.2 μm line. These diffuse bands appear to come from different molecular structures.

Aitken <u>et al</u>. (1979 [ARSJ]) suggested that the observed infrared spectra of planetaries in the 8 to 13 μm region could be reproduced by combining three continuous sources (each derived from non-PN sources). These include: 1) silicate features seen in the Trapezium; 2) the Si C feature seen in the 10 to 140 μm region; 3) a continuum with an energy distribution $f(\lambda) \sim \lambda^{-n}$ which seems to be produced by thermal emission from hot grains, plus 4) the diffuse 8.7 μm and 11.3 μm features that are particularly prominent in NGC 7027 and BD + 30°3639. <u>Ad hoc</u> combinations of these empirical sources can give satisfactory representations of observed energy distributions. At the present time, no identifications can be given for these diffuse emission features. The long-studied diffuse absorption features such as the λ4430 structure in the optical region so far likewise have defied identifications.

8C. <u>Temperature and Energy Balance of the Dust</u>

Flux measurements over a wide range of the infrared are needed to establish dust cloud temperatures. Of particular interest are the measurements by Moseley (1980) covering the range 37 μm to 108 μm and data by Cohen and Barlow (1977 to 1980) in the range 10 μm to 20 μm.

We do not expect energy distributions to fit black body curves because: a) the dust cloud can be optically thick for $\lambda > 35$ μm so that the curve is narrower than a Planckian one; b) clouds can differ in temperature because of a difference in grain size and changes in the radiation field with distance from the star; the dust does not radiate as a gray body. For $\lambda > 35$ μm, there is some evidence that the emissivity goes as λ^{-2}. Naturally, we would expect the dust nearer the illuminating star to be hotter than that further away. But grains of different temperature may exist side by side since absorption and emission efficiencies depend on size. Large, massive grains may absorb a smaller fraction of the available radiation and be relatively cold. Chemical composition also plays a role. For example, carbon grains may have a different temperature from silicate grains.

Although much of the emission from NGC 7027 corresponds to $T \sim 90$°K, there is some evidence for a hotter component with $T \sim 140$°K and also a cooler dust cloud in the outer CO envelope with $T < 35$°K. In other planetaries such as NGC 2392 and NGC 7662, the hot dust component seems to be much smaller. Characteristic dust temperatures in planetary nebulae seem to be near 100°K (Moseley 1980).

The heated grains that appear to produce the observable infrared spectrum in planetaries appear to be co-existent with the H II regions. Moseley finds little evidence that much of the flux originates outside the region of the optical nebula, although in IC 418, the dust seems more heavily concentrated in the outer [N II] shell than in the inner [O III] regions. In IC 418 the spectral energy curve falls off steeply beyond 35 μm possibly because the dust has a different composition than in other planetaries.

The mass of the dust producing the far infrared emission is extremely difficult to determine. Estimates are very sensitive to the assumed dust temperature and depend also on optical properties of the grains, on their size distributions and on their densities. Moseley (1980) concluded that since we did not know the necessary parameters, about the best we can say is that the dust/gas ratio $\simeq 0.005$ (or about the same as in the interstellar medium). Although absolute masses are difficult to obtain, except for grain-deficient IC 418, most of the planetaries observed by Moseley seemed to have similar masses of dust.

A popular hypothesis invoked to assess the energy balance of the dust is that the heating is supplied by Lyα quanta produced in the H II region of the nebula and scattered repeatedly until absorbed by grains. We define the infrared excess IRE by the equation:

$$\mathrm{IRE} = \frac{L_{IR}}{h\nu_\alpha \ \alpha_B^*(T_\varepsilon)\int N(H^+)N_\varepsilon dV} \tag{11}$$

Here L_{IR} is the total infrared flux from the solid grains, and $h\nu_\alpha$ is the energy in a Lyα photon. The total number of Lyman α photons will be the total number of recombinations in the second and higher levels, $\alpha_B(T_\varepsilon)\int N(H^+)N_\varepsilon dV$ multiplied by the fraction which gives rise to Lyα. Thus, under ordinary circumstances (cf. Sec. 4E and Eq. 4-72):

$$\alpha_B^*(T_\varepsilon) = \alpha_B(T_\varepsilon)[1 - X_o\beta_{2q}] \tag{12}$$

If the nebula is optically thick, Eq. 3-56 holds and we can relate the UV flux directly to the Hβ flux (Eq. 3-99), provided we know the interstellar extinction. Alternately, we may use radio-frequency flux data which avoids the problem of extinction in the optical region.

We can write $S_\nu(r,f)\ \Delta\nu \sim \int f_n(T_\varepsilon,\nu)N_\varepsilon\ N(H^+)\ dV$ where $f_n(T_\varepsilon,\nu)$ is a known function (cf. Chapter 3, Eq. 64) and $P_{UV} \geqslant \int N_\varepsilon\ N(H^+)\alpha_B(T_\varepsilon)dV$. Rubin (1968) uses the expression:

$$S_\nu(r,f) = \frac{8.61 \times 10^{-76}}{D(pc)^2\ \nu^{0.1}} \int N_\varepsilon\ N(H^+)\ T^{-0.35}\ dV\ \text{ergs cm}^{-2}\ \text{sec}^{-1}\ \text{Hz}^{-1} \tag{13}$$

where $D(pc)^2$ is the distance of the nebula in parsecs and ν is in GHz. Using the approximation $\alpha_B = 4.10 \times 10^{-10}\ T^{-0.8}\ \text{cm}^3\ \text{sec}^{-1}$, Rubin finds:

$$P_{UV} \geqslant 4.76 \times 10^{65}\ \nu^{0.1}\ D(pc)^2\ S_\nu \frac{\int N_\varepsilon\ N(H^+)\ T_\varepsilon^{-0.8}\ dV}{\int N_\varepsilon\ N(H^+)\ T_\varepsilon^{-0.35}\ dV}\ \text{photons/sec} \tag{14}$$

If T_ε is known and assumed constant throughout the volume, L_{IR}/P_{UV} can be calculated from $L_{IR}/S_\nu(rf)$. If we assume that all quanta in the Lyman continuum are absorbed, IRE = $L_{IR}/[PUV(\alpha_B^*/\alpha_B)]$. Infrared excesses greater than 1 imply that there is some absorption of continuous stellar radiation as well as of Ly α. An IRE value exceeding 3 requires significant absorption of continuous photons; such occurs, for example, in young, dense objects such as IC 4997 or BD + 30°3639. Cohen and Barlow (1980) concluded that absorption of Ly α photons is a dominant grain heating mechanism in most planetary nebulae. On the other hand, Panagia (1974) found that in compact H II regions, dust mixed with gas absorbs about two-thirds of the Lyc radiation. He argued that Ly α absorption is not the dominant source of heating unless the total amount of heating is small. Panagia constructed a series of useful models of dusty H II regions, considering explicitly different contributions to the radiation field, the stellar Lyman continuum flux, stellar flux longward of the Lyman limit, quanta of Lyman α, Balmer continuum and collisionally-excited lines created within the gas, and also the effect of an outer blanket of scattering grains. For H II regions, the IRE is a sensitive function of the spectral class of the exciting stars, of the optical thickness in the Lyman continuum, and of that of the surrounding shell of dust. From a comparison of then-available observations with theory, Panagia concluded that interstellar grains had a high infrared absorption efficiency and sizes between 0.05 μm and 0.15 μm. In these dusty H II regions, the gas-to-dust ratio was about 100 by mass. Furthermore, the dust-to-gas ratio in H II regions seemed to exceed the value of about 0.01 appropriate to the interstellar medium. From an analysis of Moseley's (1980) data, Natta and Panagia (1981) concluded that for the most of the observations in his sample more than 80% of the photons are absorbed by the gas. Hence, dust does not have a great influence on the energy balance. In some objects, such as BD + 30°3639, though, dust may absorb more than 50% of the far UV quanta. Hence, they found that neglect of the dust could cause a substantial underestimate of the luminosity of the central star and of its effective temperature: 800 $L_\odot$ versus 570 $L_\odot$ and 34,200°K versus 31,500°K from a direct application of Zanstra's method. See also Helfer et al. (1981). Natta and Panagia also suggested that planetaries contain a lower fraction of dust than is found in the interstellar medium and will supply it with dust-depleted matter. Furthermore, they argued that the average properties of the grains changed as the nebulae evolved. Possibly the particles may fragment or suffer partial destruction.

Physical and chemical processes involving dust grains may be complicated. Particles almost certainly will become electrically charged. Photoelectric effects tend to produce a positively-charged grain, while collisions with charged particles in the ambient medium will produce a negatively-charged one. Which situation prevails depends on the radiation field and electron density. Grains may accrete ions and atoms and they may also be destroyed or modified in chemical composition by positive ion bombardment if they carry a negative charge, and lose volatile elements by photo-detachment effects or evaporation. The temperature that a grain may achieve will probably depend on its size, chemical composition, and on collisions with other particles as well as the radiation field. Silicates and carbon grains, heavily contaminated by other elements probably dominate.

Properties of grains found in planetary nebulae may be further investigated through their extinction effects on the energy distribution in the spectrum of the central star. A beautiful example is provided by Abell 30. Since the nebula lies at a galactic latitude of 33°, absorption by the foreground interstellar medium is negligible. Strong infrared emission is produced by a conversion of UV radiation by a dust cloud. There is also an internal extinction which matches the usual reddening law to 2500A, but thereafter departs from it. The extinction peaks at $\lambda 2470$ and possibly is concave upward at $\lambda < 2000$. The absorption law most clearly fits the curves for amorphous carbon smoke, i.e., soot with mean particle radii of 0.013 μm. Such particles have negligible scattering and a low albedo (Greenstein 1981).

8D. Optical and Near UV Observations of Diffuse Nebulae

Continuous spectra of the ionized nebular plasmas of planetaries are produced by processes described in Chapter 4. Even in the dustiest objects, reflection by grains is negligible. The situation is quite different in many diffuse nebulae. Consider, for example, the Orion Nebula. It consists essentially of an H II region (excited by the Trapezium stars) which is expanding into a cool dense region on the far side from the observer. The background continuum includes both the "atomic" continuum and starlight scattered by the solid grains.

As we have seen in Chapter 4, the absolute intensity of the "atomic" continuum (which includes free-free, bound-free, and two-photon emission) can be calculated if the absolute flux in Hβ or Hα is given. Thus, if we measure the continuum spectral distribution in a diffuse nebula such as M8 or M42, and the intensity of Hβ in the same units, we can calculate the atomic continuum, subtract it off and obtain the residual spectrum due to scattered starlight.

We may describe the scattering properties of those dust grains in terms of the spherical albedo, γ, and a parameter, g, which measures the asymmetry of the scattering. Many workers have used the Henyey-Greenstein phase function (1941):

$$\Phi(\alpha) = \frac{\gamma(1 - g^2)}{4\pi} \left(1 + g^2 - 2g \cos \alpha\right)^{-3/2} \tag{15}$$

where α is the angle of scattering with respect to the forward direction (thus, $\alpha = \pi$ is pure backscattering). Here g is a measure of the asymmetry of the phase function:

$$\gamma g = \int \Phi(\alpha) \cos \alpha \, d\omega \tag{16}$$

Note that $g = 0$ for isotropic scattering; $g = +1$ if all radiation is scattered in a forward direction. The Henyey-Greenstein formula is an idealization. If we consider a mixture of spherical particle scattering by the Mie theory, we would find that the form of the phase function would be different from this simple expression and would require more than two parameters.

In principle, one could determine γ and g from observations of the scattered light in a diffuse nebula provided the details of the incident radiation field are known together with the total optical depth of the dust slab and the geometry. The latter offers one of the most severe stumbling blocks. In spite of the fact that many reflection and diffuse nebulae have been studied in detail, few firmly established results have been obtained for γ and g. For example, if we assume that the scattering cloud associated with the Merope nebula is behind the star as is usually done, we get results in contradiction to those obtained if we assume it is in front of the star.

The Great Orion Nebula M42 offers a number of advantages for extinction studies. It is seen against an extensive, opaque molecular background cloud which apparently contributes but little of the nebular light. Ionizing radiation from the exciting stars of the Trapezium presumably creates a roughly spherical cavity as the Strömgren sphere grows. Most of the light near the center of the nebula is produced by forward scattering grains between the Trapezium stars and the observer. At Hβ the total extinction amounts to 2.03 magnitudes of which 1.86 magnitudes are contributed by the nebular dust. The general interstellar medium contributes only 0.17 magnitudes.

When the observed optical and ultraviolet continuum is corrected for the "atomic" contribution, the resultant data may be analyzed by multiple-scattering theoretical models (Schiffer and Mathis 1974; Mathis *et al.* 1981). Best results are found with a hollow shell model with a radius of about 3'. Since the extinction, t_λ, is well-established and the scattering cloud is known to be in front of the star, γ is fairly well-determined. In the ultraviolet $\gamma \sim 0.45$ and is constant across the extinction feature at λ2200. At λ4600, $\gamma \sim 0.6$ to 0.7. Alas, g is poorly determined since a cloud of particles near the star that scattered well at large angles (large g) would produce the same surface brightness distribution as forward-throwing particles in a hollow shell. The latter situation is believed to be more likely.

A disadvantage of the Orion nebula is that its extinction law differs appreciably from that shown by the general interstellar medium for which no reliable determinations of γ and g are available. Presumably, the size distribution and perhaps also the chemical compositions of the particles differ.

8E. Further Properties of Interstellar Dust Grains

We now describe some properties of the dust that have been derived from studies of dark clouds and the general interstellar medium, but which may be relevant to problems of H II regions and planetary nebulae. Many attempts were made in the 1930's to obtain a mean absorption A_V per kpc for the Milky Way, mostly by star count methods. The efforts were largely wasted; it is meaningless to average extinction values between obscured and transparent regions.

1) Dust/Gas Ratio. One may correlate column densities $N(H_{total})L$ in H^o and H_2 as obtained from interstellar line strengths (as observed with an instrument such as the Copernicus satellite) with the color Bohlin, Savage, and Drake (1978) found:

$$< \left[N(H^o) + 2N(H_2)\right] L/E(B-V)> = 5.8 \times 10^{21} \text{ atoms cm}^{-2} \text{ mag}^{-1} ,$$

although in some regions, e.g., ρ Oph, where large grains appear to exist, the ratio is higher. Then using a theory of grain size distribution due to Mathis, Rumpl, and Nordsieck (1977) to fit the interstellar extinction law in the range $0.11 \ \mu m < \lambda < 1 \ \mu m$ (including the λ 2160 feature), Savage and Mathis (1979) found the lower limit of the dust-to-total-mass ratio to be 0.007. Larger particles contribute negligibly to the color excess, but they might possibly contribute substantially to the mass. Such large particles are formed by coagulation and can actually decrease the A_V/N_H ratio since grains tend to hid behind one another (Jura 1980a).

2) Depletion of Refractory Elements in the Interstellar Medium. The formation of solid particles appears to remove a large fraction of heavy elements from the gaseous phase. These grains may grow by accretion under nonequilibrium conditions in diffuse clouds. Later they may be destroyed by collisions between gas clouds and by shock waves (Chapter 9). The depletion pattern is complex, varying from one part of the galaxy to another. Furthermore, seemingly similar elements elements do not go in lockstep with one another. Thus, although the depletions of Ca, Ti, and Fe are correlated, implying that processes of grain formation involving these metals are similar, absorption line profiles of Ca II and Ti I lines are sufficiently dissimilar to suggest that gas phase concentrations differ from one cloud to another and that grain compositions may also differ.

Detailed results are available for a few clouds such as those in front of ζ Oph and ζ Pup. Refractory metals such as Aℓ, Ca, Ti, Fe, and Ni are depleted by factors of 1.7 to 4.0 dex; whereas, Li, Mg, Si, P, K, and Mn are down by factors of 1 to 1.5 dex. In planetary nebulae, where the C/O ratio often exceeds 1, much of the carbon may be incorporated into grains, although the depletion seemingly does not seriously affect the C/H ratio (Chapter 11). Carbon depletion in the interstellar medium is not known; certainly, CO is one of the most abundant and stable chemical compounds there. In dense clouds some CO may solidify. Not much nitrogen appears to be lost in grains. Oxygen is not heavily depleted; perhaps it is bound up in silicates rather than as ice! Studies of abundances in diffuse interstellar clouds with $E(B-V) < 0.5$ and $T \sim 100°K$ suggest that the depletions occur mostly after the grains enter the interstellar medium (Jura 1980b). Most refractory elements are locked in grains to which they adhere for a long time. Volatiles such as N and Cℓ are depleted by factors of the order of 2, the depletion showing a scatter comparable with the variability of the dust/gas ratio. Such elements do not appear to stick to grains for long time intervals.

3) Linear polarization of light of distant stars lying behind obscuring interstellar material (Hiltner-Hall effect) shows not only that scattering particles are elongated, but they are also aligned, evidently by an interstellar magnetic field. This polarization is wavelength-dependent, reaching a maximum at some λ_{max}. Serkowski, Mathews, and Ford (1975) found an empirical relationship for the polarization, $p(\lambda)$, viz.:

$$p(\lambda)/p_{max} = \exp\left[-\ 1.15\ \ell n^2\ (\lambda/\lambda max)\right]$$

The 9.7 μm feature shows a polarization similar to the continuum in its neighborhood. It appears that polarization is produced by absorption in grains formed of disordered silicates. Since $p(\lambda)$ shows no perturbation near the λ2200 dip; the grains that produce this feature are evidently not elongated and aligned. Shafter and Jura, A. J., **85**, 1513, 1980) suggested that circular polarization may be a useful tool for studies of circumstellar envelopes.

4) Variation of the Ratio; $R = A_V/E(B-V)$. Although $R \sim 3.1$ throughout much of the galaxy, it can become as high as 5.7 in thick, dense clouds containing larger grains. Serkowski et al. found a tight linear polarization correlation with R; $R = (5.8 \pm 0.4)\ \lambda_{max}(\mu m)$. In dense clouds (e.g., Orion, Ophiuchus, and Scorpio), λ_{max} can reach 0.80 or more, while in the Cyg OB2 association, it can be as low as 0.41; the mean value is ~ 0.545.

5) Are the Interstellar Media of Other Galaxies Similar to Those of Our Own? There is some evidence that the λ2200 feature is weaker in a number of these systems, particularly the Large Magellanic Cloud. Koornneef and Code (1981) found also that for $\lambda < 2000$ the observed extinction law rises 1.8^m between λ 1850 and λ 1450, while the galactic curve is almost flat; for $\lambda < 1400$, the two types of curves are similar. They suggest that the discordances may be produced by differences in particle sizes rather than in bulk chemical composition. For example, solid carbon particles may fall in the size range from 0.02 μm to 0.4 μm rather than 0.01 μm to 0.25 μm, while silicates fall

in the range 0.01 μm to 0.25 μm. The overall abundances of C, N, and O appear to be smaller in the LMC than in our galaxy and processes of particle formation and destruction may be quantitatively different. In some galaxies, dust may be so plentiful as to conceal the stars; such objects would appear as an infrared rather than as optical sources (see, e.g., Jura 1981).

Solid grains must play significant roles in H II regions and in planetary nebulae. They often appear to be well-mixed within the ionized zones, although existing in greater proportions in surrounding H I envelopes. They can affect the energy balance and complicate chemical composition estimates by withdrawing elements from the gaseous phase (see Chapter 11).

Selected References

Interstellar Extinction

The fundamental investigations were those of R.J. Trumpler (1930), Publ. Astron. Soc. Pac., **42**, 214, 267; Lick Obs. Bull., **14**, 154, in which he showed that A_λ increased as the wavelength decreased (but not as steeply as λ^{-4} [Rayleigh law]). He also estimated a mean photographic (blue) region absorption of 0.67 mag/kpc. Subsequent investigations by Schalen, Greenstein, van de Hulst and many others attempted to define the character of the obscuring particles (see Vol. 7 of Stars and Stellar Systems, "Nebulae of Interstellar Matter," 1968, ed. B. Middlehurst and L.H. Aller, Chicago, University of Chicago Press -- especially articles by H.L. Johnson and M. Greenberg). An excellent bibliography with special emphasis on topics relevant to gaseous nebulae is given by D.E. Osterbrock in Astrophysics of Gaseous Nebulae. A comprehensive treatment of the subject may be found in:

Spitzer, L. 1978, Physical Processes in Interstellar Medium, Chapters 7, 8, and 9. New York: Wiley.

A concise summary with extensive bibliography, and giving a definitive interstellar extinction law, is to be found in:

Savage, B.D., and Mathis, J.S. 1979, Ann. Rev. Astron. Astrophys., **17**, 73.

Our discussion employed a version of the extinction law given by:

Seaton, M.J. 1979, M.N.R.A.S., **187**, 75P.

Important recent references that emphasize the variation of the ultraviolet extinction within the galaxy are:

Meyer, D.M., and Savage, B.D. 1981, Ap. J., **248**, 545.
Bohlin, R., and Savage, B.D. 1981, Ap. J., **249**, 109.

Methods for estimating the extinction in gaseous nebulae are given in Table 1 and references therein. Additional examples are given by:

Selected References (Continued)

Feibelman, W. 1982, A.J., 87, 555 (λ2200 feature).
Milne, D., and Aller, L.H. 1975, Astron. Astrophys., 38, 183, (radio frequency/Hβ fluxes).

The planetary NGC 7027 is a rich source of information for studies of dust and molecules and their effects on nebular structure. See:

Becklin, E.E., Neugebauer, A., and Wynn-Williams, C.G. 1973, Ap. J., 15, 87.
Osterbrock, D.E. 1974, P.A.S.P., 86, 609.
Mufson, S.L., Lyon, J., and Marionni, P.A. 1975, Ap. J. (Letters), 201, L85.
Treffers, R.R., Fink, V., Larson, H.P., Gautier, T.N. 1976, Ap. J., 209, 793.
Telesco, C.M., and Harper, D.A. 1977, Ap. J., 211, 475.
Atherton, P.D., Hicks, T.R., Reay, N.K., Robinson, G.J., Worswick, S.P., and Phillips, J.P. 1979, Ap. J., 232, 786 (AHRRW).
Condal, A., Fahlman, G.G., Walker, G., and Glaspey, J. 1981, P.A.S.P., 93, 191.
Melnick, G., Russell, R.W., Gull, G.E., Harwit, M. 1981, Ap. J., 243, 170.
Smith, H.A., Larson, H.P., and Fink, V. 1981, Ap. J., 244, 835.

For pioneer infrared measurements of a planetary (NGC 7027) see:

Gillett, F.C., Low, F.J., and Stein, W.A. 1967, Ap. J. (Letters), 149, L97.

A review of work up to 1977 and appropriate references are given by:

Rank, D.M. 1978, Planetary Nebulae (ed. Y. Terzian), p. 103, Reidel.

Among more recent papers involving measurements of the dust component we call attention to papers in I.A.U. Symposium No. 103 (especially the review article by M.J. Barlow) and:

ARSJ Aitken, D.K., Roche, P.F., Spenser, P.M., and Jones, B. 1979, Ap. J., 223, 925.
Cohen, M., and Barlow, M.J. 1980, Ap. J., 238, 585.
Grasdalen, G.L. 1979, Ap. J., 229, 587.
JMSW Jones, B., Merrill, K.M., Stein, W., and Willner, S.P. 1980, Ap.J., 242, 141.
MFH McCarthy, J.F., Forrest, W.J., and Houck, J.R. 1978, Ap. J., 224, 109.
RSW Russell, R.W., Soifer, B.T., Willner, S.P. 1978, Ap. J., 220, 568.
WJPRS Willner, S.P., Jones, B., Puetter, R.C., Russell, W., and Soifer, B.T. 1979, Ap. J., 234, 496.
Dwek, E., Sellgren, K., Soifer, B.T., and Werner, M.W. 1980, Ap. J., 238, 140.

Selected References (Continued)

Discussions pertaining to calculation and interpretation of infrared excess (IRE), including also the influence of dust on temperature determinations for central stars, are given by:

Rubin, R.H. 1968, Ap. J., 154, 392.
Panagia, N. 1974, Ap. J., 192, 221.
Natta, A., and Panagia, N. 1981, Ap. J., 248, 149.
Helfer, H.L., Herter, T., Lacasse, M.G., Savedoff, M.P., and van Horn, H.M. 1981, Astron. Astrophys., 94, 109.

The Henyey-Greenstein scattering formula is given by them in Ap. J., 93, 70, 1941. It has been employed by many investigators.

The anomalous extinction in Abell 30 is described by:

Greenstein, J.L. 1981, Ap. J., 245, 124.

Scattering properties of dust in the Orion Nebula were assessed by:

Schiffer, F.H., and Mathis, J.S. 1974, Ap. J., 194, 597.
Mathis, J.S., Perinotto, M., Patriarchi, P., and Schiffer, F.H. 1981, Ap. J., 249, 99.

The gas-to-dust ratio in the interstellar medium has been the subject of many researches; we quote here results from:

Bohlin, R.C., Savage, B.D., Drake, J.F. 1978, Ap. J., 224, 132.

For the particle size distribution law in the general interstellar medium, see, e.g.:

Mathis, J.S., Rump, W., and Nordsieck, K.H. 1977, Ap. J., 217, 425.

Heavy element depletion in the interstellar medium, formation, destruction, and properties of dust grains are discussed in a series of papers in Annual Reviews of Astronomy and Astrophysics. In addition to the previously cited summary by Savage and Mathis, we mention:

Aanestad, P.A., and Purcell, E.M. 1973, A.R.A.A., 11, 309.
Salpeter, E.E. 1977, A.R.A.A., 15, 267.
Spitzer, L., and Jenkins, E.B. 1975, A.R.A.A., 13, 133.
Jura, M. 1980a, Ap. J., 235, 63; 1980b, Highlights of Astronomy, 5, (ed. A. Wayman), p. 293, Reidel.

Empirical relationships involving polarization and extinction properties of grains are given by:

Serkowski, K., Mathewson, D.S., and Ford, V.L. 1975, Ap. J., 196, 261.

Selected References (Continued)

There is some evidence from the work of Code, Nandy, and their respective associates and from other investigators that the extinction law in the Magellanic clouds is different from that in the galaxy. See, e.g.:

Koornneef, J., and Code, A.D. 1981, Ap. J., 247, 860.

Dust in galaxies has been discussed by many workers, see, e.g., Jura, M., 1981, Ap. J., 243, 108.

List of Illustrations

Fig. 8-1. The Extinction in the Far Ultraviolet. $X(x) = A_\lambda/E_{B-V}$ is plotted against $x = 1/\lambda_\mu$ for the region λ3800 to λ1100. Seaton's (1979) mean curve is represented by empirical formulae:

$$2.70 \leqslant x \leqslant 3.65, \quad X(x) = 1.56 + 1.048x + 1.01 \left\{(x-4.6)^2 + 0.28\right\}^{-1}$$
$$3.65 \leqslant x \leqslant 7.14, \quad X(x) = 2.29 + 0.848x + 1.01 \left\{(x-4.6)^2 + 0.28\right\}^{-1}$$
$$7.14 \leqslant x \leqslant 10, \quad X(x) = 16.17 - 3.2\ x + 0.2975\ x^2.$$

For $x < 2.7$,

x =	1.0	1.1	1.2	1.3	1.4	1.6	1.8	2.0	2.2	2.4	2.5	2.6	2.7
X(x) =	1.36	1.44	1.84	2.04	2.24	2.66	3.14	3.56	3.96	4.26	4.4	4.52	4.64

Fig. 8-2. Effect of Extinction on the Image of NGC 7027.
a) Radio frequency contours of NGC 7027 at 5 GHz. The shaded ellipse shows the half-power beam area. The contour interval is 650°K.
b) Cylindrical model proposed for NGC 7027 (left) and brightness distribution found after convolution by telescope beam (right). Compare with a). The axis of the cylinder makes an angle of 30° with the line of sight. The labeled dimensions in (') are those that would be observed if each dimension were measured normal to the line of sight (P.F. Scott, 1973, M.N.R.A.S., 161, 3P).
c) Direct photograph of NGC 7027 secured by R. Minkowski in Hα with the Hale Telescope. The radio center of the nebula lies at the intersection of the arrows.

Fig. 8-3. The Spectrum of NGC 7027 in the Near-to-Middle Infrared. Note the 3.27 μm, 3.4 μ, 6.2 μ, 7.7 μ, 8.6 μ, and 11.3 μ features, the strong continuous infrared spectrum and both permitted and forbidden lines (courtesy Russell, Soifer, and Willner, 1977, Ap. J., 217, L149, University of Chicago Press, © 1978).

Chapter 9

SHOCK WAVES

9A. Introduction

Early successes of the theory of stellar radiative excitation of gaseous nebulae led to an expectation that the emission from all galactic gaseous nebulae could be explained in this way. One well-known object which resisted this interpretation was the Network Nebula in Cygnus, now recognized as a supernova remnant. Already by the 1930s intensive searches for an illuminating star had proven fruitless. No plausible candidate could be found and the suggestion that such an object might be hidden behind a dark cloud is not substantiated by star counts in the region.

The spectrum of the Cygnus loop, optical, X-ray and radio frequency yields firm evidence that radiative excitation cannot play a dominating role. A huge range of excitation is observed, from Ca II λ3933 (not seen in planetaries), and [Mg I] λ4562, 4571, [S II] to [Fe V] λ4227 and [Ne V] λ3426. By itself, this would not prove the network spectrum to be abnormal, since a large range in excitation is found in inhomogeneous objects for which the radiative theory seems to be valid, e.g., NGC 6302. On the other hand, such a range of excitation is not found in diffuse galactic nebulae. Miller's careful measurements of the emission line intensities show: a) forbidden lines, especially those of [O II], [O III], and [S II] are very strong relative to permitted lines (compared to the H II regions); and b) the [O III] lines indicate T_ε of the order of 40,000°K, the [N II] and [O II] temperatures fall in the neighborhood of 12,000 to 13,000°K, while T_ε([S II]) ~ 8500°K to 10,500°K in two representative filaments. Similar phenomena are displayed by other supernova remnants. Further clues are provided by the rapid motions of the filaments ~ 100 km/sec, nonthermal radio emission, [Fe XIV] coronal-type emission lines, and X-rays.

These data can be understood if we regard the Cygnus Loop as a remnant of an old supernova detonation. the delicate lace-like structure which gave this object an alias, the Veil or Network Nebula, is an important datum. (See Figure 9-1.) These intricate wisps are sometimes called filaments, but detailed studies indicate that their thickness along the line of sight is perhaps an order of magnitude greater than their projected widths on the plane of the sky, so we are dealing with sheets or slabs of material moving in a direction roughly perpendicular to their largest dimensions.

In a detonation, be it a terrestrial explosion or a supernova, the surrounding gas is hit by a compression front that moves through the medium with a speed greater than the ambient sound speed. The ratio of the compression front velocity to the sound speed is called the Mach number, M. The compression front is usually called a shock wave; we use that designation though in this instance the "wave" refers to a unique event and not to a repetitious phenomena as is usually associated with the concept of a wave. In that sense a shock wave reminds us of an earthquake-induced water wave or tsunami. The gas in the path of the onrushing wave has no inkling of impending disaster -- except in some astrophysical-type situations, where it is exposed to radiation from the hot plasmas in and behind its shock front.

Immediately following a supernova detonation, the ejected envelope expands freely without hindrance from the surrounding interstellar medium. The Crab Nebula appears to be in this phase. Eventually, the expanding shell sweeps up more material than was contained in the original ejection and begins to slow down. If radiative cooling is not yet important, the shell enters the phase of adiabatic expansion when it expands at nearly constant energy. Then as the growth of the shell continues, radiative cooling becomes significant and the shell enters a (sometimes inappropriately called isothermal) radiative stage in which the thermal energy of the heated interstellar medium is radiated away.

The adiabatic phase can generally be described analytically. The gas cools rapidly because of expansion even at this stage. When the temperature of a mass element drops faster from radiation than from expansion, cooling begins to dominate the blast wave. Then the cooling of the gas causes the pressure behind the shock to drop. The shock velocity decreases, most of the mass of the cooling remnant gets compressed into a shell as it accumulates behind the shock front. The shock front slows down; more and more interstellar material piles up on the outer edge.

The inner region comprises a heated gas at a nearly uniform pressure. It expands almost adiabatically as the shell grows in size. Although radiative losses are low, the emitted radiation can be important for the observer. Much of the shell radiation escapes as high-frequency (i.e., UV and X-ray) quanta and ionizes the gas in front of the advancing shock front. Thus, in calculations of shock phenomena in such SNR, we can treat the impinging gas as ionized.

We shall see that the Cygnus Loop and similar supernova remnants can be understood as examples of shock waves on the grand scale. The lace-like features are basically planar shock fronts, sheets of gas seen edge-on and compressed and heated to incandescence as the disturbance ploughs its way through the interstellar material. The Cygnus Loop is an especially precious object because its spectrum seems to be totally a product of shock-wave events. No significant component of its emission seems to arise from a general photoionization source. Shock wave theory gives a good explanation of the spectrum of the Cygnus Loop, not only qualitatively but also quantitatively.

9B. The Structure of Shocks

We are concerned here with structures of shocks, not with their motions, except insofar as they influence the emitted spectrum of the shock-heated gas. We consider a steady state situation in which a shock wave moves into an undisturbed gas at a supersonic velocity. The following presentation of the basics of one dimensional shock wave theory is due to Donald Cox of the University of Wisconsin and is reproduced here with his generous permission:

In a frame of reference moving with the shock front, the conditions behind the shock can be related simply to those in front.

Figure 9.1

The resultant relations are called the Rankine-Hugoniot equations. The underlying expressions represent the conservation laws for mass, momentum, and energy. The mass conservation equation is particularly straightforward. The mass flux into the transition region is $\rho_1 v_1$ (where the subscripts 1 and 2 are used used to denote pre-shock and post-shock values) and that flowing out is $\rho_2 v_2$. Mass conservation then dictates that:

$$\rho_1 \, v_1 = \rho_2 \, v_2 \ . \tag{1}$$

The momentum per gram in the flow is v, so the momentum flux is ρv^2. The momentum flux, however, is not a constant across the transition surface because the pressure jump from P_1 to P_2 provides a net upstream force $P_2 - P_1$ per unit area. Thus, $P_2 - P_1$ is the loss of momentum per unit area per second and this must equal $\rho_1 v_1^2 - \rho_2 v_2^2$. Hence

$$P_1 + \rho_1 v_1^2 = P_2 + \rho_2 v_2^2 \ , \tag{2}$$

is the momentum equation. The pressure in this equation includes both thermal pressure and magnetic pressure ($B^2/8\pi$) for a magnetic field perpendicular to v) if the latter is appreciable.

The energy in the flow is in several forms: thermal energy, ionization energy, magnetic energy, and kinetic energy. If we lump the first three terms into a quantity called ξ for the non-kinetic energy per gram, then the energy flux is $\rho v(\xi + v^2/2)$. As with momentum, the energy flux is not conserved across the transition, even in the absence of radiative losses. This is because the pressure forces do work on the flowing gas. The power output of a force is $\vec{F} \cdot \vec{v}$ so the upstream presure adds energy to the transition region at a rate, $P_1 v_1 = \rho_1 v_1 (P_1/\rho_1)$ and the downstream pressure subtracts energy (the pressure pushes back upstream on the transition region) at a rate $P_2 v_2 = \rho_2 v_2 \, (P_2/\rho_2)$. If, in addition, an amount of energy Q is radiated per unit mass during the transition, the change in energy flux is

$$\rho v(\xi_2 + \frac{1}{2} v_2^2) - \rho v(\xi_1 + \frac{1}{2} v_1^2) = \rho v(\frac{P_1}{\rho_1} - \frac{P_2}{\rho_2} - Q)$$

where we have made explicit use of Eq. (1). Gathering terms we have

$$\xi_1 + \frac{P_1}{\rho_1} + \frac{1}{2} v_1^2 - Q = \xi_2 + \frac{P_2}{\rho_2} + \frac{1}{2} v_1^2 \ . \tag{3}$$

Once again, a magnetic field perpendicular to $\vec{v}$ will add $B^2/8\pi\rho$ to each ξ and P/ρ. Together, Eq. (1), (2), and (3) constitute the basic relations for one dimensional flow-through shock fronts. The solution of these equations is particularly instructive for the special case of no magnetic field, constant ionization energy, and a perfect gas for which the thermal energy per gram is:

$$\xi = \frac{1}{\gamma-1}\frac{P}{\rho}, \quad \text{so} \quad \xi + \frac{P}{\rho} = [\gamma/(\gamma-1)]\frac{P}{\rho}. \tag{4}$$

The solution to the equations can be conveniently reformulated in terms of a compression factor, $y = \rho_2/\rho_1$. With this factor, Eqs. (1) and (2) become:

$$\rho_2 = y\rho_1, \quad v_2 = \frac{v_1}{y}, \quad \rho_2 v_2^2 = \frac{\rho_1 v_1^2}{y},$$

$$P_2 = P_1 + \rho_1 v_1^2 \left(1 - \frac{1}{y}\right) = P_1 \left[1 + \gamma M_1^2 \left(\frac{y-1}{y}\right)\right], \tag{5}$$

where we have introduced the Mach number of the shock front $M_1 = v_1/c_1$, $c_1^2 = \gamma P_1/\rho_1$. The energy equation is now:

$$\frac{\gamma}{\gamma-1}\frac{P_1}{\rho_1} + \frac{v_1^2}{2} - Q = \frac{\gamma}{\gamma-1}\frac{P_1}{\rho_1}\frac{1}{y}\left[1 + \gamma M_1^2 \left(\frac{y-1}{y}\right)\right] + \frac{v_1^2}{2}\frac{1}{y^2}, \tag{6}$$

where P_2 ρ_2, and v_2 have been replaced on the right hand side using relations (5). Now multiplying through by $(\gamma-1/\gamma)y^2\rho_1/P_1$ and once again replacing $\rho_1 v_1^2/P_1$ by γM_1^2 we finally have as the energy equation:

$$\left[1 + \frac{\gamma - 1}{2} M_1^2 - \frac{(\gamma - 1)\rho_1 Q}{\gamma P_1}\right] y^2 = [\gamma M_1^2 + 1]y - \frac{\gamma + 1}{2} M_1^2. \tag{7}$$

Since Eq. (7) is quadratic, there are two solutions. For $Q = 0$, one solution is simply $y = 1$; the outgoing and incoming parameters are the same; there was no shock in the region we are considering! Of course, this is a perfectly valid possibility. As Q is increased from zero, the solution corresponding to $y = 1$ at $Q = 0$ is altered and correctly follows the changing conditions in a radiating flow with no shock front.

For $M_1^2 > 1$, the second solution is a shock front. For $M_1^2 < 1$, the second solution is nonsense. It corresponds to a time-reversed shock front and is forbidden because entropy always increases across a shock front, turning highly ordered flow velocities into random thermal velocities. Let us define a quantity

$$\delta = \frac{(\gamma-1)\rho_1 Q}{\gamma P_1 \left[1 + \frac{\gamma-1}{2} M_1^2\right]} = \frac{Q}{\frac{\gamma}{\gamma-1}\frac{P_1}{\rho_1} + \frac{v_1^2}{2}}. \tag{8}$$

Physically, δ is the ratio of radiated energy to total energy, internal plus kinetic. Note that $\delta = 1$ would correspond to radiation carrying off all the available energy. The shock solution to Eq. (7) can be written:

$$y = \frac{(\gamma M_1^2 + 1) + \sqrt{(\gamma M_1^2 + 1)^2 - (\gamma + 1)\, M_1^2\, [2 + (\gamma - 1)\, M_1^2]\,(1 - \delta)}}{[2 + (\gamma - 1)\, M_1^2](1 - \delta)} \,. \qquad (9)$$

This equation contains a wealth of information as it is applicable to a wide variety of situations. As a first example, consider the limit as M_1^2 increases without bound as $\delta \to 0$. This situation corresponds to an extremely fast shock front which is adiabatic in the sense that it does not radiate. It is not, however isentropic and $(P/\rho)^{5/3}$ is not a constant across the front because of the irreversible generation of entropy. In this situation $y = (\gamma + 1)/(\gamma - 1) = 4$ for $\gamma = 5/3$. All of the other parameters follow from Eq. (5). Thus, for a strong adiabatic shock in a gas with $\gamma = 5/3$,

$$\rho_2 = 4\rho_1, \quad v_2 = \frac{1}{4} v_1, \quad P_2 = \frac{3}{4} \rho_1 v_1^2 \,. \qquad (10)$$

Taking $P = NkT$ and $\rho = mN$,

$$T_2 = \frac{P_2}{N_2\, k} = \frac{m}{k} \frac{P_2}{\rho_2} = \frac{3}{16} \frac{m}{k} v_1^2 \,. \qquad (11)$$

If the shock is in an astrophysical plasma with a He/H ratio of 0.1 by number, and if H & He are everywhere singly ionized, then

$$N = 2\, N_H + 2\, N_{He} = 2.2\, N_H \,,$$

$$\rho = m_H N_H + 4 m_H N_{He} = 1.4 m_H N_H, \quad m = \frac{\rho}{N} = \left(\frac{1.4}{2.2}\right) m_H \,,$$

so that

$$T_2 = \frac{3}{16} \left(\frac{1.4}{2.2}\right) \frac{m_H}{k} v_1^2 = 1.44 \times 10^5 \left(\frac{v_1}{100 \text{ kms}^{-1}}\right)^2 \,{}^\circ\text{K} \qquad (12)$$

The precise post-shock temperature thus depends on the degree of ionization since this affects the number of particles sharing the energy. If the ionization differs across the shock, the higher ionization behind the shock represents a loss of thermal energy (to ionization energy) and can be found by including the ionization energy as Q in Eq. (9). This has the effect of lowering the post-shock temperature and raising the compression factor.

For a shock front which is followed by a zone of radiative cooling, Eq. (9) provides the compression factor variation within that zone as a function of the radiated energy per gram, Q. If the volume emissivity is $E_r N^2$, then the emissivity per gram is $E_r N/m = \dot{Q}$. Let the location of the discontinuous adiabatic shock be at $x = 0$, then

$$Q(_{rad}) = \int_0^t \dot{Q} dt = \int_0^x \frac{\dot{Q}}{\dot{x}} dx = \int_0^x \frac{E_r N}{mv} dx = \frac{1}{\rho v} \int_0^x E_r N^2\, dx \,. \qquad (13)$$

This equation illustrates that for a steady flow, the time integral for a parcel going through the structure is equivalent to a spatial integral over all parcels at one time. It is clear from comparing Eq. (13) with (9) that even if E_r, the emissivity coefficient, were known explicitly, knowledge of y between between 0 and x is required to find Q which in turn is needed to find y(x). We have a differential equation that must be solved numerically in any practical situation. On the other hand, a host of differential equations for the non-equilibrium ionization state must also be solved numerically to find $E_r(x)$ so one more equation to find y(x) adds little to the burden.

One result is immediately apparent from Eq. (9), namely that as $\delta \to 1$, y increases without bound. This corresponds to the radiation carrying off all the available energy. The density then increases without bound, the velocity goes to zero, the pressure to $\rho_1 v_1^2$ and the temperature goes to zero. This means physically that the gas simply cools down, piles up, and comes to rest relative to the shock front.

Considered in the frame of reference of the pre-shock gas, the shock moves into the gas at velocity v_1, heating and compressing the material to T_2, N_2 and accelerating it to a velocity $v_1 - v_2 = v_1(y-1)/y$ or three-fourths of the shock velocity for an adiabatic shock preionized material (actually constant ionization) with $\gamma = 5/3$. If the post-shock material radiates significantly in a time scale of interest (that is, if the cooling time scale is shorter than the time the shock has been present), then as the post-shock gas cools its temperature falls, its pressure rises slightly towards the value $P_1 + \rho_1 v_1^2$, the density rises dramatically (until the compression is stopped by magnetic pressure or net cooling ceases) and the material is further accelerated by the slight pressure gradient to a velocity close to that of the shock.

9C. <u>Intricacies and Complexities of Some Astrophysical Shock Phenomena</u>

Let us now examine implications of some of the above in the context of applications. Laboratory investigations of shock phenomena usually involve controlled conditions. For example, shock-heated gases whose temperatures, densities and ionization stages are accurately known, since they achieve momentary thermodynamic equilibrium, frequently have been used for measurements of atomic transition probabilities. The situation is quite otherwise in the interstellar medium where such equilibrium is virtually impossible to attain.

Interstellar shocks can be produced by collisions of gas clouds or perhaps even by density waves in spiral galaxies. We shall be concerned here, however, only with shock waves produced by stars and by supernova remnants. Thus, when a luminous early-type star is born it quickly ionizes the gas in its neighborhood and raises its temperature. For example, if the initial neutral medium was in pressure equilibrium at a kinetic temperature of 90°K, and the ionized gas is raised to $T_\varepsilon \sim 9000$°K, there is a pressure differential of a factor of 20 between the two zones. This may produce a shock wave that moves into the surrounding cool medium at a speed that is slow compared to SNR shocks. A stellar wind from the newly formed star may strengthen this shock.

In 7G we have already discussed planetary nebulae nuclei with strong broad Wolf-Rayet lines, or with P Cygni profiles, all indicating stellar winds with velocities of the order of 1000 km/sec. Yet, nebular expansion velocities are at least an order of magnitude smaller.

Shocks may occur in QSO's and other objects with pre-shock densities of $\sim 10^6$ atoms/cm^2 or in the interstellar medium with a density of ~ 0.10 atoms/cm^3. Shock velocities may range from a few km/sec (as in the expanding H II region mentioned above) to 10^4 km/sec. Mach numbers can range up to 10^3 and immediate post-shock temperatures can achieve 10^8°K.

What happens when material goes through a collisional shock? Donald Cox has given a concise description: The particle velocites are largely randomized by collisions and this appears as the large temperature increase. In calculating the temperature in Eq. (12), it was assumed that ions and electrons have equal temperatures. The first randomizing collisions, however, will leave them instead with roughly equal velocities, so that the proton energies and temperatures exceed those of the electrons by a factor of 1836 (for strong shocks). Immediately after the shock the effective proton temperatures is about a factor of 2 higher than that given by Eq. (12) while the electron temperature is much lower. Subsequently, coulomb collisions between protons and electrons equilibrate their temperatures to the value given by Eq. (12). For very young SNR's the time to achieve this equilibration by coulomb collisions can be much longer than the remnant age, but for older remnants with radiating shock waves the equilibration is rapid and can be considered to be part of the shock discontinuity. This picture, however, ignores the complications of plasma interactions in the shock fronts. In fact, the electrons may equilibrate much more rapidly with the protons so that their temperatures are nearly equal even in young remnants.

9D. Post-Shock Cooling Processes

Collisional excitation of discrete atomic levels is the primary cooling mechanism in the post-shock gas. In the high-temperature zone behind the shock front, collisional ionizations and excitation of H, He, He^+, and ions of C, N, and O are very significant. Collisional deexcitation is insignificant and the cooling rate goes as N^2. When the temperature drops below about 5,000°K, excitation of fine structure levels of elements of the carbon and silicon rows C II, O I, and Si II, etc., becomes important. Molecules such as H_2, CH^+, OH, H_2O, SO, H_2S, and SiO may also form in the cooling gas behind the shock front. These molecules can play important roles in cooling processes. Intricate chemical effects and condensation of nonvolatile elements such as C, Si, Ca, etc., into grains can also complicate the cooling scenario.

When cooling becomes important, ions and electrons do not recombine sufficiently rapidly to maintain an equilibrium appropriate to the ambient temperature, T_ε. One must also consider effects of radiative processes which interact with collisional events in a non-trivial complicated way. Detailed calculations have been carried out with specific models.

9E. Radiative Transfer Effects

Recombination of electrons with protons and He ions, for example, produces continuous radiation in the visible, ultraviolet and infrared. Collisional excitation of discrete levels is followed by line emission which may produce photoionization of atoms and ions. Several important aspects are to be noted.

Radiation from the shock zone may heat and ionize the pre-shocked gas as well as the "downstream" post-shock gas. In the latter domain it tends to maintain the ionization level above that appropriate to collisional processes at the local temperature. For most elements, resonant scattering of the line radiation does not affect the total number of photons and therefore is not important. For H and He I the resonance lines are scattered many times. The λ304A He II resonance line can attain great strength for shocks with velocities of the order of 100 km/sec. It can play an important role in cooling, keep oxygen doubly ionized down to $T \sim 30,000°K$, and cause a weakening of the O II lines. Another point is that the radiative field is highly anisotropic. The shock front has a dimension L perpendicular to its direction of motion much larger than the thickness of the relaxation layer D. The optical thickness of the slab is an important parameter. An interesting illustration of some of the complexities that might occur is offered by the Balmer decrement in the Network Nebula. We might expect the atomic levels to be excited by collisions, thus tending to a collisional decrement. Actually, a radiative decrement is observed. Why? Immediately behind the shock front the temperature is very high. This region produces many high-frequency quanta that can ionize H in the cooler regions downstream where many protons have recombined with electrons. The Balmer lines are excited by recombination and cascade.

9F. Collisionless and Nonradiative Shocks

In certain low-density plasmas, the shock front can be thinner than a mean free path of a charged particle. Energy is dissipated by hydromagnetic turbulent phenomena, rather than by collisions; hence, the designation collisionless shocks. When the solar wind hits the earth's magnetosphere it produces what is probably the best-known astrophysical or perhaps geophysical example of collisionless shock. Other examples may exist in some young supernova remnants.

Shocks are called nonradiative if the emission of the radiation does not affect the dynamics of the gas downstream from the shock front. They occur in young supernova remnants with shock velocities in excess of 1000 km/sec. X-ray spectra are observed. These regions also emit coronal lines: [Fe X] and [Fe XIV], which have been observed in the Cygnus Loop, Puppis A, Vela X and in the Large Magellanic Clouds.

9G. Theoretical Models for Shock-Excited Nebulae

Is it possible to construct plausible theoretical models that will represent the observed spectra of objects such as the Network Nebula? Conversely, from the observed spectrum of such a supernova

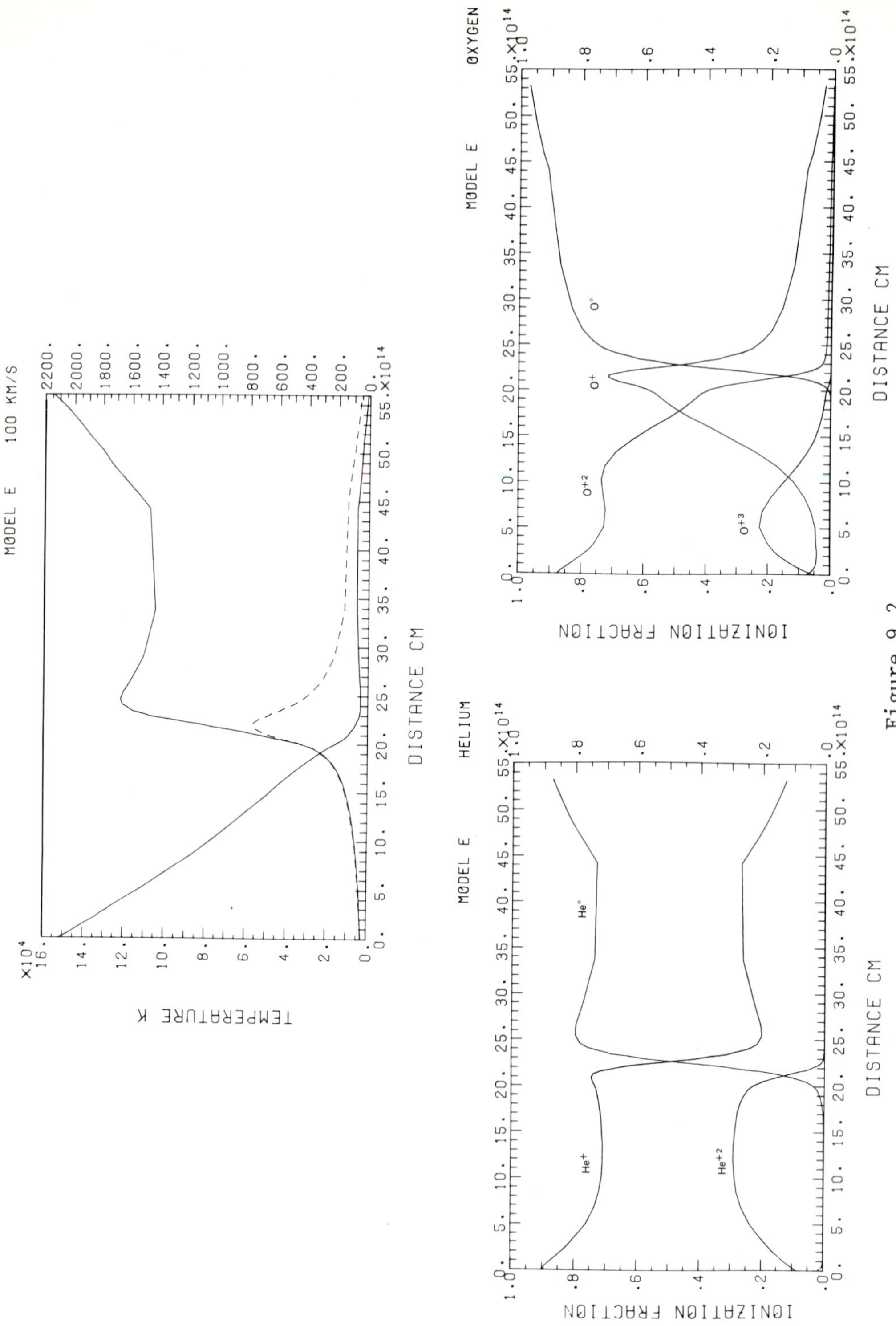
MODEL E 100 KM/S
TEMPERATURE K
DISTANCE CM
MODEL E HELIUM
He°
He+
He+2
IONIZATION FRACTION
DISTANCE CM
MODEL E OXYGEN
O°
O+
O+2
O+3
IONIZATION FRACTION
DISTANCE CM

Figure 9.2

remnant can we reconstruct the shock conditions and such parameters as the chemical composition of the emitting gas? To the first question we can give a rather affirmative answer; in reply to the second question it would appear that there are too many initial uncertainties such as the magnetic field, the shock-front area, inhomogeneities in the interstellar medium, and the condition of the gas before it encounters the shock. Furthermore, a single observed region may consist of contributions from parcels of different shock velocities and pre- and post-shock densities. Various investigators, e.g., Cox and Raymond have constructed models for shocks in old SNR's.

As an example, we consider Raymond's Case E (see Figs. 9-2a, b, c). The adopted pre-shock gas density and shock velocity are $N_1 = 10^{-3}$ and U = 100 km/sec, respectively. The width of the shock front is taken as five times its thickness. Also, $B_1 = 1\mu$ gauss. The pre-shocked gas is taken as primarily singly ionized. At the shock front the gas heats up to 150,000°K. About 17 years after passing through the shock front, about 29% of the helium is doubly ionized, while carbon reaches a maximum ionization of 0.65 C III and 0.31 C IV. These concentrations are much higher than would be appropriate to equilibrium at the corresponding temperature. In fact, the presence of C III, C IV, and He II in the heated post-shock gas increases the radiative cooling rate by an order of magnitude. There is also an enhanced photoionization rate of He I and other elements by the He II λ304 A photons. In passing down through the post-shock gas, an average He atoms recombines 2.7 times compared to 2.0 times for an H atom. By the time the gas has cooled to 30,000°K, the principal cooling mechanism has become the collisional excitation by Ly α, near ultraviolet intercombination lines and forbidden lines. As the temperature falls yet further, recombination and infrared forbidden lines become important. Then the number of neutral H atoms increases, photoionization by quanta arising from the hotter regions becomes significant, energy is thus pumped into the gas, and the average thermal energy per particle actually rises for a while, until the high-frequency photons are cut off and thermal energy is gradually drained by emission in far infrared fine structure lines of ions of C II, Si II, Fe II, etc. The ionization pattern can become quite complex as T_ε sinks to 1000°K.

This particular shock model gives an Hβ emission of 5.34×10^{-6} erg cm^{-2} s^{-1} str^{-1}, of which 97% arises from recombination versus 3% from collisional excitation. The [O III] I(4959 + 5007)/I(4363) ratio corresponds to $T_\varepsilon = 20,000$ while the [N II] nebular/auroral line ratio indicates 15,000°K. We can now understand some of the spectral differences between a gaseous nebula excited by shock waves and one excited by radiation. Strong lines of [O I], [N I], [O II], [N II], [S II] (as compared with Hβ) appear along with highly excited ions such as [Ne V]. In H II regions and planetary nebulae, Hβ is produced throughout the entire ionized zone, while [O II], etc., emission occurs over only a relatively limited region. In the shock situation, both Hβ and the low-excitation lines are produced in the recombination zone. High-excitation lines are produced in the small, hot, well-ionized zones. Further, the range in T_ε is very large compared with that existing in an H II region or planetary nebula, thus accounting for the large differences in T_ε between the [O III] and [N II] zones.

How does the predicted spectrum change as the input parameters are varied? If the gas is pre-ionized and other parameters are held fixed, the total emission in Hβ is roughly proportional to the square of the shock velocity. I(4686)/I(Hβ) increases with v_1 until v_1 = 140 km/sec, while I(5876)/I(Hβ) reaches a maximum for $80 < v_1 < 120$ km/sec where I(λ304) attains its greatest strength. The [I(4959) + I(5007)/I(4363) ratio decreases monotonically as v_1 rises from 50 km/sec to 200 km/sec. Unfortunately, it also depends on other model parameters as thus is not a good velocity indicator. If the pre-shock gas is largely neutral, much of the kinetic energy of the flow will be required to ionize the gas. Consequently, the post-shock temperature will be lower. Secondary effects can become complicated.

We can scale the flow structure in cooling time and distance over which cooling occurs by $1/N_1$, where N_1 is the pre-shock density. Since collisional ionization recombination rates and dT/dt are proportional to N, the ionization state at any given temperature is independent of N_1. At high densities dT/dt is no longer proportional to N because collisional deexcitation of metastable levels becomes important. Because radiative transfer depends on the size and shape of the emitting volume, the area of the shock front can be important, and also its orientation with respect to the line of sight can affect intensities of UV resonance lines (e.g., C IV 1548, 50) of finite optical thickness. A serious limitation is imposed by inhomogeneities in the medium into which the shock wave penetrates. Such irregularities cause serious instabilities. Finally, since the lines of force of a magnetic field are frozen into an ionized plasma, any such field would reduce compression, hinder recombination, and thus shift the ionization equilibrium.

9H. Interpretation of Optical Spectra

Detailed theoretical models such as those by Raymond (1979) indicate that the optical spectrum depends in such a complex way on shock parameters that it will not be easy to determine both shock conditions and chemical abundances from measured line intensities. Certain optical lines yield important information, e.g., $I(H\beta) \propto N_1 v_1^2$. Helium lines arise mostly from recombination, photoionization by He II λ304 is important for $80 < v_1 < 120$ km/sec. Then I(5876)/I(Hβ) > 0.13, if n(He)/N(H) = 0.085. Also, I(4686) indicates the maximum degree of double ionization of He. Forbidden lines of oxygen may supply useful shock diagnostics: [O II]/[O III] and [O III]/Hβ give estimates of v_1. If $v_1 < 70$ km/sec, oxygen cannot be doubly ionized, while if $v_1 > 140$, I(Hβ) becomes very large and I([O III])/I(Hβ) is small. If the pre-shocked gas contains some O^{++}, however, [O III] can become strong even if v_1 is small. [Ne III] λ3868 behaves somewhat like [O III]; Raymond finds that except in slow shocks N(Ne)/N(O) = 1.35 ± 0.20 × I(3868)/I(5007). The [S II] lines yield information on a region in the shock structure that is cooler and denser than the domain where [O II] is produced. If calcium is not held in grains, lines of Ca I, Ca II can appear; lines of [Fe II], [Fe III], and [Fe V] are found in SNR's.

Dopita (1977) tried to develop a systematic procedure for determining chemical abundances in SNR. He sought to identify line ratios that depended on: 1) shock conditions (e.g., pre-shock density $N_1 = N_2/4$ for strong shocks, and T_2, the immediate post-shock temperature, and 2) abundances of major cooling agents, C, O, and S. For example, $F(H\beta)/N_1$ is a unique function of T_2. The familiar [S II] and [O II] ratios give N_ε in the range $10^2 - 10^4$ in the post-shock gas; hence, with the aid of a shock model we can reconstruct N_1. The 5007/3727 ratio can be used as a temperature indicator, but the total heavy element abundance affects this and all other temperature-sensitive ratios by changing the ionic balance. From line ratios involving [O I], [O II], [N I], and [N II] at moderate densities, the N/O abundance ratio can be estimated. Dopita used only the singly-ionized species, O^+, N^+, S^+ to find the corresponding elemental abundances. He and his associates have tried to construct shock models and obtain chemical abundances for SNR's including some in the Magellanic Clouds and M33. The agreement between O/H abundances found in the M33 SNR's and those found from H II regions appears satisfactory (Chapter 11).

The difficulties in the analysis are formidable. First we have uncertainties arising from our ignorance of the physical state of the not-yet-shocked gas, the presence or absence of magnetic fields, and the lateral extent of the shock front. There are chemical and physical effects, e.g., the depletion of C and Si in grains, that can retard cooling. There remains the possibility (particularly for SNR's observed in external galaxies) that in any one structure we may be observing the combined contributions from several shock fronts, with a spread in v_1 values. Nevertheless, applications of diagnostic diagrams, such as those proposed by Dopita, may be worthwhile for a first reconnaissance, provided we are ever mindful of possible pitfalls. A summary of the difficulties encountered in deriving shock conditions and elemental abundances in SNR's are documented by Raymond and in the review by McKee and Hollenbeck (1980).

The optical spectrum of the Network Nebula may be compared with theoretical predictions by Cox and Raymond, for example, on the assumption that shock waves associated with an old SNR are involved. The hypothesis is that this nebula has already passed far beyond the initial stage of rapid expansion when it had yet swept up little of the interstellar gas, then the stage called adiabatic expansion when the remnant had nearly constant energy (since radiation losses are unimportant), and has entered the isothermal stage when it is radiating copiously. Various features of the Network Nebula spectrum suggest it is a nonhomogeneous structure. The Cygnus Loop emits X-rays and [Fe XIV] λ5303 appropriate to a gas at 2×10^6°K, and presumably produced by shock waves moving at 300 to 400 km/sec in a medium with a density near 1 cm^{-3}. Such a gas would require 10^5 years to cool enough to produce the radiation observed in the Network "filaments." Since the Cygnus Loop is much younger than this, one may conclude that the veil-like features are caused by a supernova shock running into inhomogeneous denser material. Raymond's analysis of Miller's data suggested a density ~ 10 to 20/cm^3 and a shock with v_1 ~ 70 km/sec with some depletion of C and Si. For the same positions as observed by Miller, Benvenuti et al. (1980) found from UV lines a range in shock velocities between 90 km/sec and 130 km/sec. Strong [Ne III] and [S IV] lines suggest that neon and sulfur may be overabundant in this SNR.

Problem

Show that the equation for the conservation of energy may be written in the form:

$$\frac{dh}{dt} = - LN_{\varepsilon} \frac{N_H}{N}$$

here $h = \varepsilon_1 + \frac{B^2}{4\pi N} + \frac{mv^2}{2} + \frac{5}{2}\left(1 + \frac{N_{\varepsilon}}{N}\right) kT$

is the enthalpy per particle. Normally, enthalpy is written as $H = E_I + PV$ where E_I is the internal energy, P is pressure and V is the volume. Here ε_I is the mean ionization energy per particle, B is the component of the magnetic field perpendicular to V, and L is the net cooling coefficient in ergs cm^{-3} s^{-1}. L involves terms for both heating and cooling. Since the magnetic field is frozen into the gas, $B_2 = N_2 B_1/N_1$.

It is assumed that the plane parallel approximation can be applied and that the flow parameters are constant while the gas cools from its post-shock temperature to one low enough for optical emission.

References

The classical reference on shock waves is Supersonic Flow and Shock Waves, Courant, R., and Friedrichs, K.O., 1948, New York, Interscience.

Our discussion is based largely on basic papers by:

Cox, D.P. 1972, Ap. J., 178, 143, 169.
Raymond, J.C. 1979, Ap. J. Suppl., 39, 1.

See also Spitzer, L. 1968, Diffuse Matter in Space, New York: Interscience.

That the excitation of the Network Nebula was probably caused by a shock wave of velocity ~ 100 km/sec passing through the interstellar medium was suggested by J. Oort (1946, M.N.R.A.S., 106, 159). Extensive optical observational studies by E. Hubble (1937), J.W. Chamberlain (1953, Ap. J., 117, 399), R. Minkowski (1958, Rev. Mod. Phys., 30, 1048; 1968, Stars and Stellar Systems, 9), D.E. Harris (1962, Ap. J., 135, 661), S.B. Pikel'ner (1954, Crimea Obs. Publ., 12, 93), R.A. Parker (1967, Ap. J., 139, 493; 149, 363; 1969, 155, 359), A. Poveda (1965, Bol. Tonanzintla y Tacubaya, 27, 49), and J. Miller (1974, Ap. J., 189,

References (Continued)

239) have been extended to X-ray, radio and ultraviolet regions; see, e.g., P. Benvenuti, S. D'Odorico and M.A. Dopita (1979, Nature, 277, 99), S. Rappaport et al. (1974, Ap. J., 194, 329), B.E. Woodgate et al. (1974, Ap. J., 188, L79). For further SNRs see, e.g., D. Osterbrock and R. Dufour (1973, Ap. J., 185, 441), N49 in LMC, D.E. Osterbrock and R. Costero (1973, Ap. J., 184, L71). Other references may be found in papers cited below.

For a review on interstellar shock waves, see:

Shull, J.M., and McKee, C.F. 1979, Ap. J., 227, 131.
McKee, C.F., and Hollenback, D.J. 1980, Ann. Rev. Astr. Astrophys., 18, 219, and references therein.

Attempts to derive chemical element abundances in SNRs have been made, for example, by:

Dopita, M.A. 1976, Ap. J., 209, 395; 1977, Ap. J. Suppl., 33, 437.

Supernova remnants have been discussed by many writers. Some useful references are:

Chevalier, R.A. 1977, Ann. Rev. Astr. Astrophys., 15, 175 (Interaction of SNR and interstellar medium).
Chevalier, R.A., Kirshner, R.P., and Raymond, J.C. 1980, Ap. J., 235, 186 (Optical Emission of a Fast Shock Wave).

List of Illustrations

Fig. 9-1. The Network or Veil Nebula in Cygnus. (Lick Observatory Photograph)

Fig. 9-2. Properties of a Representative Shock Wave. The shock parameters are v_1 = 100 km/sec. Density of pre-shocked gas equals 10 atoms/cm^3, interstellar magnetic field equals 1 microgauss. T_s = 152,000°K. Hydrogen is ionized in the pre-shocked gas. Helium is 2% neutral and 89% singly ionized. Each hydrogen atom recombines an average of 2.02 times in the post-shock gas, each helium atom 2.76 times.

a) The left-hand curve gives the temperature behind the shock front. The solid right-hand curve gives the density and the dashed curve gives the electron density.

b) Fractional ionization of helium in the post-shock gas.

c) Fractional ionization of oxygen in the post-shock gas. In each instance, distance is reckoned from the shock front. J.C. Raymond, 1971, Ap. J. Suppl., 39, 1. Courtesy, University of Chicago Press.

Chapter 10

DIFFUSE NEBULAE, WIND-BLOWN SHELLS, AND PLANETARIES

10A. Introduction:

In preceding chapters we have reviewed physical processes associated with thermal excitation of spectra of various types of gaseous nebulae, H II regions, planetaries and supernova remnants. We must now examine more carefully some of the distinctive characteristics of these objects. We consider first certain of the classical diffuse galactic nebulae, properties of shells blown by stellar winds, and finally basic statistics, structures, and internal kinematics of planetary nebulae. In Chapter 12 we return to these latter objects in the context of stellar evolution; here we shall be concerned with some fundamental empirical data.

The term diffuse nebula, or diffuse galactic nebula, was used traditionally to describe objects such as the Orion, Trifid, Lagoon and η Carina nebulae. Many of these objects give the impression of dense clouds of initially cold neutral gas suddenly subjected to a rich ultraviolet photon flux from newly ignited stars, often near their edges. Sometimes, compact dark globules (as they were called by Bok) are seen in projection against the bright nebulosity. Conventional theories of nebular excitation seem adequate to explain the optical radiations of the H II region which is radiation bounded on the side towards the dense cloud where masses of neutral H and molecules exist. Various 21-cm observations indicate that dark clouds often contain little atomic hydrogen, H I. Hydrogen atoms appear to have reacted on surfaces of dust grains to form H_2 molecules. Other types of molecules are revealed by their microwave spectra, the most prevalent of which is carbon monoxide. Inside the molecular clouds, young stars in a very luminous phase of their birth process are observed. Some of these may evolve to power and ionize their own nebulae within 10^{5-6} years.

Care must be taken to distinguish between galactic nebulae that are associated with star formation (e.g., Orion, M8, Trifid nebulae) and those like the Gum Nebula or supernova remnants (SNR) that are associated with the demise of a star. NGC 6888, an intricate filamentary shell surrounding a Wolf-Rayet star, is another representative of a type of nebula associated with a highly evolved star. In that sense it resembles a planetary nebula, but the masses involved are much greater and interaction with the interstellar medium is pronounced (Kwitter 1981).

Dynamical effects are present, as in the Orion nebula, for example, but their manifestations do not seem to be so spectacular as in some other classes of objects. Aside from bona fide supernova remnants, there occur some aesthetically beautiful nebulae with graceful arches and complex patterns where dynamical effects have obviously played an important role.

As examples we mention the Rosette Nebula in Monoceros, NGC 604 in M33, and 30 Doradus in the Large Magellanic Cloud. In all of these the character of the optical spectra is consistent with radiative excitation by stars, but the topology of the structures indicates that the shells and arches were put in place by mechanical forces far beyond the capabilities of gas pressure effects in expanding H II regions. Stellar winds can be important.

Discrimination between SNRs on the one hand and H II regions (with perhaps associated neutral hydrogen and associated molecular clouds) on the other is sometimes straightforward as for the Cygnus Loop of "Network" Nebula. Here the radio frequency spectrum is non-thermal (spectral index $\alpha < 0.3$), the optical spectrum may show a huge range in excitation with a large [S II]/Hα ratio, and a gas kinetic temperature of 30,000°K. Furthermore, the gas may be moving with a high velocity. There may be soft X-rays and "coronal" lines of highly ionized atoms (e.g., [Fe XIV]). In contrast, a radiatively excited region on the edge of a dense, extensive neutral cloud (e.g., Orion) has a gas kinetic temperature of about 10,000°K, the exact value depending on the energy balance, a thermal radio-frequency spectrum, identifiable exciting O and B stars, and usually thermal and microwave radiation emerging from an associated dust and molecular cloud. Internal motions are typically of the order of a few km/sec as compared with expansion velocities of often hundreds of km/sec in SNR's. It is worth nothing, however that if the velocity of the gas exceeds the sound speed (~ 10 km/sec), either there will be energy dissipation raising the electron temperature above the value corresponding to energy balance (about 10,000°K) or there will be a nonthermal process, i.e., a wind. Stellar winds, originating in hot, luminous stars are often capable of producing shock waves, clouds of outflowing material moving at high velocities, and regions of high temperature. Optical emissions from shells excited purely by winds are likely to be small.

The M42 or Orion Nebula is often cited as a classical H II region, but Balick has pointed out that it has some puzzling features: Orion has a striking network of shells 30' SW of the Trapezium which bears some resemblance to structures usually attributable to SNR's. It is hard to fit the Barnard Loop in the outer region of Orion into a classical H II region scenario. Could be it the remnant of a supernova of long ago?

The Orion complex and structures such as 30 Doradus remind us that in the interstellar medium there can exist regions in which all three types of phenomena occur simultaneously: thermal emission from H II regions excited by embedded stars, stellar winds, and SNR's. Sometimes it is difficult to tell whether arcs or filaments, for example, arise from brisk winds from Wolf-Rayet stars or whether they are to be traced to supernova events. We first discuss situations in which complications produced by supernova detonations apparently are not involved, at least recently, e.g., Orion, Sharpless 155 and the associated Cepheus OB-3 association, i.e., "classical" H II regions plus associated dust and molecular clouds. Then we examine some objects whose topologies suggest violent events.

When spatial resolution is missing, as in observations of small gaseous nebulae in nearby galaxies or the central bulge of our galaxy, it is difficult to distinguish low-excitation planetaries from H II regions around hot, massive stars, unless we can observe the continuum fluxes of the latter.

10B. Some Brief Comments on Molecular Clouds

Great quantities of cool gas and dust are often associated with H II regions as in Orion. Molecules are important in the cool, nonionized gas. CO, which is more stable against photodissociation than most compounts, is found over a larger volume of space than any molecule except H_2. It is easier to observe than H_2. Carbon monoxide and nitrogen freeze out only on grains cooler than about 17°K. Complex molecules (with which we shall not be concerned here) are found only in massive clouds at very large optical depths where they are protected from photodissociation.

Molecular clouds are not insignificant. Those that contain H_2 and CO but not necessarily complex molecules comprise about half the mass of the interstellar medium. We must distinguish between individual, isolated clouds and large complexes such as occur in Orion and throughout the constellation of Taurus. One of the former has a diameter ranging from 1 to 5 pcs. It can be seen in emission if the particle density is greater than about 1000 per cm^3. If no internal heat source exists, the temperature tends to lie between 10°K and 20°K. The temperature does not sink lower because grains are heated by heavy (large Z) particles in cosmic rays, which can easily penetrate to centers of dense clouds. Observations of HCO+ and γ-ray emission from molecular clouds show that cosmic ray heating can be important. Frequently, the gas in a molecular cloud is heated by a nearby star or it may contain an embedded one.

A cloud of 1 parsec diameter, $N_H = 1000/cm^3$, and T = 10°K can contract under its own gravitational field, unless inhibited by a magnetic field. Large complexes can have extents of 10 - 20 parsecs, masses of hundreds or even thousands of suns, and large optical depths. They can evolve into whole stellar associations.

It is generally believed that the dust/gas ratio in molecular clouds is about the same as in the general interstellar medium, i.e., about 0.01, but no direct measurements are available. In dark clouds it is possible to measure ratios such as ^{12}CO/dust, but not H_2 easily.

10C. Relations Between H II Regions and Exciting Stars

10C-1. Some Basic Considerations

In Chapter 3 we briefly described the theory of the Strömgren sphere and derived theoretical relations between P_{UV}, the number of ultraviolet photons, N_ε, N(H+) and s_0, the radius of the Strömgren sphere or the excitation index, U (Eq. 3-58). Equations 3-56 or 3-57 are valid if certain conditions are fulfilled:
1) exciting stars are located within H II regions that are ionization bounded, 2) the region is dust-free, and 3) the ionization and heating are provided only by photons. Then all the Lyman continua quanta are absorbed and:

$$P_{uv} = \alpha_B(T_\varepsilon) \int N(H^+)N_\varepsilon dV = N_{rec}^{neb} \qquad (1)$$

where N_{rec}^{neb} is the number of recombinations on the second and higher levels. This is proportional to U^3 where the excitation parameter, U, is defined by:

$$U = s_o N_\varepsilon^{2/3} \quad (2)$$

In practice, U is often derived from radio-frequency measurements. Schraml and Mezger (1969) wrote:

$$\left[\frac{U}{\text{pc cm}^{-2}}\right] = 4.5526 \left\{a(\nu, T_\varepsilon)^{-1} \left[\frac{\nu}{\text{Ghz}}\right]^{0.1} \left[\frac{T_\varepsilon}{^\circ\text{K}}\right]^{0.35} \left[\frac{S\nu}{\text{f.u.}}\right] \left[\frac{D}{\text{kpc}}\right]^{2}\right\}^{1/3} \quad (3)$$

where a is a factor of the order of 1 which is tabulated by Mezger and Henderson (1967). S_ν is the radio-frequency flux measured at the frequency ν. Clearly, we can apply this equation only if another further condition is fulfilled, viz.: 3) the optical thickness of the nebula in free-free radiation $\tau_{ff} << 1$. For nebulae in equilibrium, the difference $P_{UV} - N_{rec}^{neb}$ should be small compared to P_{UV} since most photons will be used up to keep the gas ionized. In many instances the exciting star may be placed asymmetrically in the nebula allowing UV photons to freely escape over a large solid angle, then $N_{rec}^{neb} < P_{UV}$.

The conditions given by Eq. 3-56 offer an interesting opportunity to probe relationships between H II regions and their exciting stars: Georgelin et al. (1975) selected a group of H II regions presumably satisfying conditions 1), 2), 3) and 4) and for which the photoelectric, spectroscopic, and luminosity characteristics of the exciting stars were known. They used model atmosphere theory to predict the ratio of the Lyman continuum photon flux to the energy emitted in the photometric V-band, $L(uv)/\pi_{FV}$. For stars of a given chemical composition, theory gives the emergent flux, $\pi F(\nu)d\nu$ as a function of T_{eff} and g. One must adopt a calibration of the MK spectrum-luminosity classification of stars as a function of T_{eff}. Then, for a given star of magnitude V (corrected for interstellar extinction), one can predict from the spectral classification and photometric data values of P_{uv} and of the excitation parameter, U. Furthermore, if the nebula is optically thin in the radio-frequency range, its T_ε and its distance are known, the radio flux alone will yield a measure of P_{uv}. Thus, if the conditions required by Eq. (1) are fulfilled, we can determine Zanstra-type temperatures for the exciting stars. Table 10-1 lists values of P_{uv} and U for luminosity class I and V stars, as predicted by Georgelin et al. More recent work by Pottasch et al. (1979) in which satellite UV data are used, led to a greatly improved scale of T_{eff}. They carefully selected H II regions for which good radio flux data were available. Model atmospheres may not give good UV stellar predictions for low surface gravities. Columns 2 and 3 give the effective temperature scales by Georgelin et al. and by Pottasch et al., respectively. The values for U, P_{UV}, etc., given by Georgelin et al. should be revised, but Table 10-1 can still give a good idea of the photon flux, provided we use the later effective temperature calibration.

We have here a powerful tool for studies of H II regions when we know the characteristics of the exciting stars. As we saw in Chapter 8, dust and ionized gas can occur together and in dense H II regions absorption by dust can become troublesome. Furthermore,

Table 10-1. Visual Absolute Luminosities Excitation Parameters and Numbers of Ionizing Photons for Main Sequence and Supergiant Stars (Georgelin et al. 1975; see also Pottasch et al. 1979)

	Luminosity Class V					Luminosity Class I		
Spectral Class	$T_{eff}/10^3$ G.	P.	M_V	U pc cm^{-2}	Log P_{UV} (Photons s^{-1})	M_V	U pc cm^{-2}	Log P_{UV} (Photons s^{-1})
05	(43.4)	41.5	-5.5	86	49.93	-6.8	128	49.95
06	39.4	38.0	-5.2	66	49.09	-6.8	108	49.73
07	36.8	34.7	-5.1	55	48.85	-6.7	90	49.49
08	35.0	31.5	-4.9	45	48.59	-6.6	76	49.27
09	33.0	30.0	-4.4	33	48.19	-6.5	62	49.00
095	32.1	29.5	-4.0	26	47.87	-6.3	51	48.75
BO	30.9	28.5	-3.7	20	47.53	-6.2	43	48.53

free-free self-absorption effects can become important when the condition $\tau_{ff} \ll 1$ is no longer fulfilled. More commonly, difficulties are encountered because of the irregularities of structures of H II regions. For example, the exciting star may lie on one side of a dust and molecular cloud as in Orion or Sharpless 155. In the cloud direction, the H II region is ionization bounded. The ionizing radiation penetrates only partly into the cloud. In the opposite direction, where the density may fall off rapidly with distance from the star, the cloud is material-bounded. Frequently the true exciting stars are obscured by the dust clouds in close proximity to the nebula, e.g. in the case of M17, or by foreground dust, e.g. W51.

Another situation occurs when dense structures are located in a region of very much lower density, e.g., 2 to 5 cm^{-3}. Powerful, high-resolution radio telescopes such as aperture synthesis instruments detect the smaller, denser clouds, but may miss an underlying structure of lower density. The condition of ionization balance, however, requires that we take into account regions of both high and low density. Careful observations with single dishes can, however, detect the total radio emission.

The most tenuous H II regions that can still be investigated as discrete objects are possibly structures such as the Monoceros loop (Kirshner et al. 1978) and the λ Orionis cloud which has a mean density of 2 cm^{-3} and an extension of 440 pcs (Reich 1978). Another example is the Gum nebula ($N_\varepsilon \sim 1$ to 5 cm^{-3}), which is one of the nearest of these structures (see § D2). At even lower densities and on a much larger scale is the diffuse ionized interstellar medium itself. Evidence for this attenuated thermal ionized gas in the galactic plane is supplied, for example, by: 1) pulsar dispersion effects; 2) galactic and background thermal emission; 3) He and H recombination lines found at

Figure 10.1

galactic longitudes and latitudes where no H II regions are seen; 4) emission from non-thermal galactic sources that sometimes suffer absorption at low frequencies because of a diffuse ionized gas.

Furthermore, much of the highly attenuated gas is found in "bubbles" and hot tunnels that lie in wake of supernova ejecta and stellar winds. Such material often has $T_\varepsilon \sim 1 \times 10^6$°K, and densities $N_i \cong 0.01$ cm^{-3}. Ultraviolet lines of OVI are prominent signatures of this hot gas that is so rarefied that it cannot cool or recombine for a time scale of the order of a galactic rotation! Some low-density regions may be in much the state anticipated for a plasma ionized by O and B stars outside of diffuse H II regions.

10C-2. The Orion Nebula

The most intensively studied diffuse nebula in the entire sky is certainly the Great Orion Nebula M42. (See Fig. 10-1.) It has been investigated over a spectral range from meter radio frequency waves to the ultraviolet and X-rays. It is the brightest and presumably the most active part of a complex of reflection nebulae, dark clouds, such as the Horsehead Nebula, fainter H II regions (such as NGC 2024 and the Barnard loop) and stellar associations, the whole conglomerate extending over a region of about 10° × 10° (see Goudis 1982).

The luminous M42 nebula with its H II mass of about 20 $M_\odot$ is a midget compared with the associated vast cool cloud whose total mass has been estimated to exceed 100,000 $M_\odot$. It is generally believed that the Orion Nebula has little effect on the dynamics of this cloud, although some workers believe that the nebula has sent a shock wave through the molecular cloud that may induce star formation. Fortunately, the Trapezium stars and their associated nebulosity lie in front of this cloud. At the core of this dark cloud is an infrared sourced called the Kleinman-Low Object. This feature includes H_2O and OH "point" sources with radii $\sim 10^{14}$ to 10^{15} cm. The radius of the KL object does not exceed 10^{17} cm, its density $N(H) > 10^7$ cm^{-3}. Surrounding this object ($10^{17} < r < 10^{18}$ cm) is a dense molecular cloud ($N(H_2) \sim 2 \times 10^5$ cm^{-3}), while outside the dense cloud lies the zone of cold neutral gas with a density of about 2000 atoms/cm^3. The KL object seems to be contracting gravitationally. It is much smaller and denser than the Jeans limit for stability against gravitational forces. The radial velocity of the KL object with respect to the local standard of rest, V(LSR), is 8 km/sec. The sheet of C^+ at the interface between the neutral gas zone and the H II region shows V(LSR) = + 9.7 km/sec, while for the Trapezium stars, V(LSR) = 11 km/sec. Thus, the material in the neutral cloud and the Trapezium stars themselves appears to be falling toward the molecular cloud that surrounds the KL object! On the other hand, the bulk of the ionized gas in Orion exhibits v(LSR) ~ -2 kms^{-1} this gas is flowing outwards from the cloud and past the Trapezium at the local sound speed. It is probably being driven into a region of very low density by its internal pressure (see Fig. 10-2).

Radio continuum isophotes of M42 at 2 cm (Schraml and Mezger 1969) show an orderly Gaussian pattern, on a scale of approximately 0.5 pc, plus a smooth plateau. Superposed on this regular distribution is a fine structure that is revealed by optical and by aperture

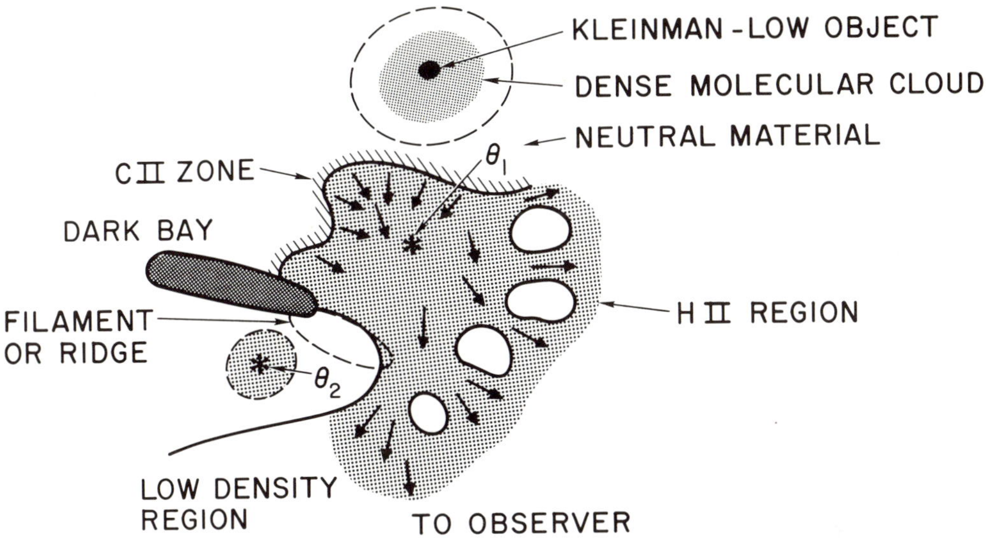

Figure 10.2

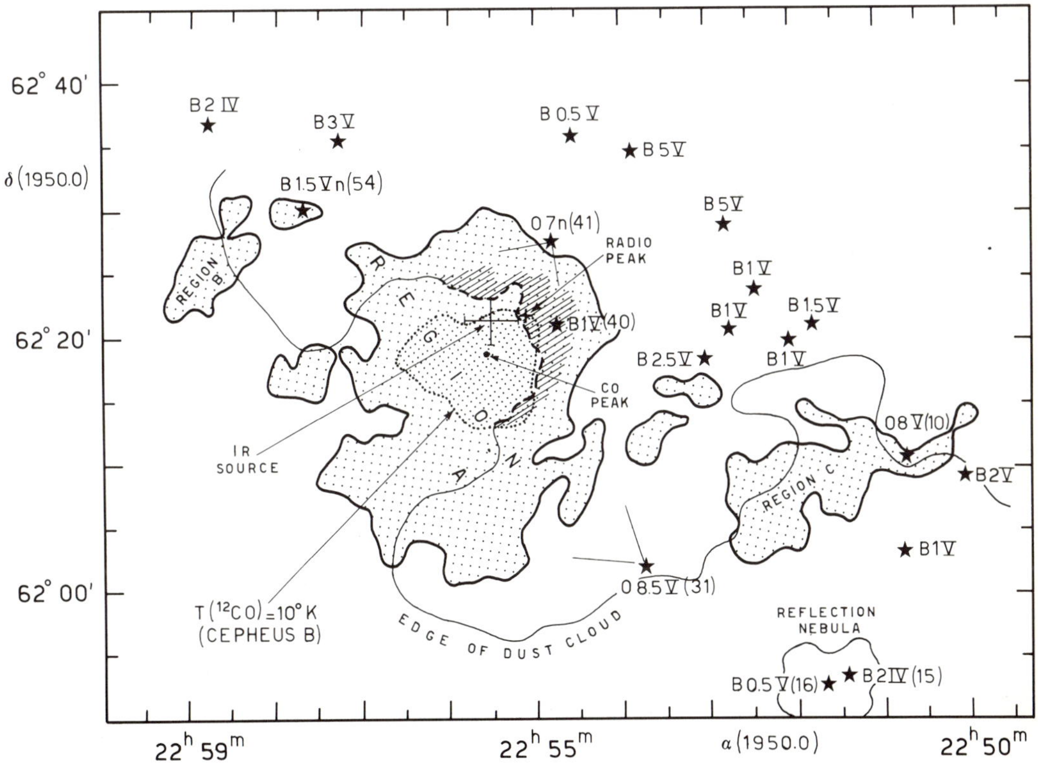

Figure 10.3

synthesis observations. Optical extinction does not seem to play an important role in affecting the observed nebular shape in the Trapezium area. VLA maps show the inner few arc minutes of Orion to have virtually the same shape as optical photographs.

Near the core, $N_\varepsilon \sim 2 \times 10^4$ cm^{-3} as indicated by both radio-frequency line data and by optical data. There are fluctuations: N_ε values of the order of 5×10^4 to 10^5 cm^{-3} have been reported from [O II] and [S II] lines in small regions. The density, as seen in projection, seems to fall off exponentially from the center, declining by a factor of 10 in about 0.3 pc (2'). Further evidence is supplied by VLA maps. The electron temperature, T_ε, has been evaluated by both optical and radio-frequency methods as 7000°K to 10,000°K. There is no firm evidence for temperature gradients. See Chapter 4, D, Fig. 4.

As previously remarked in Chapter 8, the observed nebular spectrum is strongly affected by scattering from dust. Much of the optical and ultraviolet continuum is simply scattered starlight, mostly from the Trapezium. In the outer region, the lines as well as the continuum show strong polarization, indicative of scattered light. Although the emission line spectrum in the inner region appears to be but little affected by dust, one cannot determine the ionization structure of the nebula, nor interpret spatial variations of line intensities without taking the role of dust into account.

The velocity structure of the Orion nebula is one of its most challenging features. In particular, the core region shows a considerable range in velocities, and even a spread in intrinsic line width. Here we are looking through the most highly excited and ionized region right to the surface of the neutral cloud. Transitions of ions of low ionization potential, e.g., Fe^+ and also the radio frequency C^+ lines show V(LSR) $\leq$ +9 to +9.7 km/sec while highly ionized ions such as [O III] or [Ne III] have V(LSR) $\sim$ +2 km/sec. Here V(SLR) denotes velocity with respect to the local standard of rest. Also, the C^+ lines are narrow, suggesting that the C^+ zone is cold ($\sim$ 50°K) and quiescent. Optical observations in Hα, [O III] or [N II], for example, show that velocities change rapidly between regions separated by only a few seconds of arc as well as from one level of excitation to another.

It would appear that the basic pattern of the motions can be interpreted in terms of a description given by Balick et al. (1980). (See their Fig. 3.) The inner portion of the H II region seems to be relatively free from dust (cf. 8-D). Hence, the UV radiation from the Trapezium stars falls with virtually undiminished intensity on the dense molecular neutral cloud. This neutral material, upon being ionized and heated rapidly, moves outward, i.e., away from the cold cloud and in this instance towards the observer. The velocity of this ionized flow depends on the strength of the radiation field at the edge of the neutral slab and the density of the foreground ionized gas into which it flows. The flow velocity lies between c(H II) and 2c(H II) where c(H II) is the velocity of sound in the H II region.

Presumably, the bulk of this main flow comes from a small dense region in the neutral complex. It flows into a region where N_ε lies between 1000 cm^{-3} and 30,000 cm^{-3}. In front of the Trapezium stars the density drops sharply so the gas flows out towards the observer. Balick et al. estimated the mass loss from the neutral surface on their hypothesis. Assume that $N_H vA$ = constant at the ionization front. If $N_H = 30{,}000\ cm^{-3}$, $v = 10\ km\ sec^{-1}$ and $A = 10^{35}\ cm^2 = 0.013\ pc^2$, $dM/dt = 10^{-4}\ M_\odot$/year. If the density distribution in the core of the Orion nebula were a static configuration, free expansion would level it out in about 60,000 years, a figure that has sometimes been interpreted as the age of the Orion nebula. It is more likely that something resembling a steady-state situation exists. Since the mass of the neutral cloud vastly exceeds that of M42, the nebula could last a great deal longer than 60,000 years, perhaps as long as θ_1 Orionis itself.

There is a cluster estimated to contain nearly 1900 stars brighter than V = 18 associated with the Orion nebula. Its probable mass is about 2600 $M_\odot$. The stars appear to be distributed in the same region as the CO cloud, where $N(H_2) \gtrsim 10^3\ cm^{-3}$. This seems to be sort of a critical density below which stars cannot form.

The free-fall time of the extended Orion Cloud ($r \sim 10^{19}$ cm; $n(H_2) \sim 10^3$) is about 10^7 years, which is near the age of the Orion cluster. Star formation and cloud contraction may have started at about the same time, millions of years before the present Orion nebula was turned on. An even more interesting future may lie ahead. The Kleinman low object and surronding molecular cloud may blaze forth as a new stellar association producing a new Orion Nebula that may outshine the old, provided it can break up the opaque neutral cloud that envelopes it at present.

10C-3. Sharpless 155 and a Nearby Cepheus Association

As a second example of a complex region with an H I envelope, dense molecular cloud and exciting stars, we discuss Sharpless 155 and the nearby Cepheus Association which were analyzed in detail by Felli et al. (1978). The existence of a large body of optical, infrared, mm wave (molecular), and both thermal and H I radio data make it possible to carry out a reasonably satisfactory analysis of this region. One may study interaction of bright, hot, young stars with a highly inhomogeneous complex low density plasma, with dense molecular clouds, with hot dust, individual dense clouds, and a vast H I envelope (see Fig. 10-3).

In many ways, the morphology of S 155 resembles that of other H II regions, notably Orion, where the exciting stars are located near the edge of a cool, massive cloud; we see the effects of an ionization front eating into a neutral dusty gas. There are important differences, however. In Orion, the energy of excitation of the molecules in the dust cloud arises within the cool complex itself, while Felli et al. find that the molecular cloud in S 155 is heated by the nearby stars:

Direct photographs in Hα such as from the Palomar-Schmidt survey reveal an extended nebula with a bright rim on the side facing the exciting stars of the Cepheus OB3 association and bounded on the south by an absorbing cloud. Long-exposure photographs in Hα show an extent of about 60'. The stellar association (with a spread over about 30 × 50 parsecs) and at a distance of about 800 parsecs is one of the youngest groups within 1 kpc of the sun. Blaauw divided it into two subgroups, the one denser and closer to S 155 being younger with an age of 4,000,000 years. There is a strong infrared source at the core of the complex of dust clouds and bright nebulosity.

The molecular cloud has also been detected in the radiation of ^{12}CO, ^{13}CO and H_2CO; it is associated with the dense cloud of dust and gas. ^{12}CO radiation has been measured over an area of 20 × 60 pcs!

There is a large expanding H I shell, about 8° in diameter, involving a total H mass of about 50,000 suns, that surrounds the entire Cepheus OB3 association and H II nebulosity. It appears to be dynamically coupled to the H II region.

Radio observations at 0.610 GHz (λ = 49 cm) with a spatial resolution of 1' (or about 0.2 pc) show S 155 to be a strong thermal source, extending over an area coincident with the faint Hα emission.

A logical first step in the analysis is to make a quantitative comparison of the ionization balance of the nebula, by comparing the Lyman continuum photon flux from the stars with the number of such UV quanta implied by the hot ionized plasma, N_{UV}. Felli et al. find:

$$\sum P_{UV} = 1.32 \times 10^{49} \text{ Photon/sec}, \quad N_{UV}^{neb} = 1.6 \times 10^{48} \qquad (4)$$

Since $N_{UV}^{neb} \ll P_{UV}$ we would conclude that most of the region has such a low density that it escaped detection, or that most of the UV quanta escaped. In particular, S 155 appears to be simply the ionized surface of the dense molecular cloud. Most of the ionizing energy comes from a single O7 star, HD 217086, which appears to lie on the far side of the nebula since we see the cloud of dust and gas in its crescent phase. Not all of the UV flux that impinges on the cloud from the bright star ionizes the gas; some of it is absorbed by the dust and is re-radiated as thermal energy. Felli et al. find that the total number of photons absorbed by dust and gas is about 5% of the total stellar flux, 7.9×10^{48} photons s^{-1}, from HD 217086 which suggests that the cloud intercepts a solid angle of about 25°. In this way they inferred a cloud star distance d = 3.5 pc with a cloud radius of 1.5 pc.

Starlight cannot penetrate to the inner regions of the dusty cloud. If we take the ratio of visual extinction to hydrogen column density, $A_{vis}/[N(H) + 2N(H_2)]L = 3.7 \times 10^{-22}$ (Jenkins and Savage 1974), with $N(H_2)L = 6.2 \times 10^{22}$ cm^{-2}, we find A_{vis} = 46 magnitudes!

From an analysis of ^{13}CO and ^{12}CO lines, Felli et al. constructed a model for the molecular cloud for which they find a total mass of about 2700 suns. The cloud should be unstable against gravitational collapse. Certainly, it was a recent region of star formation, and may be the site of further such events. The Orion nebula and Sharpless 155 are two examples of diffuse nebular complexes where existing stars are found near the edge of dark obscurring clouds. A dominant factor in the observed features of these objects is the radiation field of the stars.

Thin optical arcs and ridges of thermal radio emission seen in many objects such as M17 have been interpreted by Meaburn (1978) as radiation-caused ionization fronts at the edge of the HI-molecular cold shell. For a number of diffuse nebulae, including IC 1318, M17, λ Orionis, and objects in the Magellanic Clouds, he has compared predictions of stellar wind theory and has concluded that a satisfactory agreement is obtained.

10D. Wind-Driven Shells and the Interstellar Medium

10D-1. Some Theoretical Considerations

Many structures in the interstellar medium cannot be understood without invoking winds. Even in regions of star formation and dense molecular clouds, infrared observations of broad CO lines, maser lines of OH, H_2O, and SiO or emission lines of H_2 reveal velocities that may range between 10 and 250 km/sec, involving mass losses from 10^{-7} to 10^{-3} $M_\odot$/yr. Easier to study and more tractable to theory are the winds associated with: a) bubbles around single OB stars (with $R_{neb} > 10$ pcs) or "superbubbles" with $R > 100$ pcs; b) Ring Nebulae around Wolf-Rayet stars; or c) nuclei of planetary nebulae.

Early-type stars have winds with velocities ranging from 3000 km/sec for O4I stars to 1000 km/sec for O9V stars. These winds are quite capable of scouring out cavities around the stars. With mass loss rates between 10^{-4} and 10^{-8} $M_\odot$/yr, they can supply the interstellar medium with mechanical energy between 10^{48} and 3×10^{51} ergs in 300,000 to ten million years -- a total amount of energy comparable with that supplied by supernovae.

Earlier theoretical work predicted that the strong stellar wind would bulldoze the interstellar material to form a thin relatively dense shell, evidence for which has been found as sheets of absorbing material in front of early-type stars. Typical values of thickness and density of such a sheet are 0.1 pc and 500/cm^3, respectively.

Detailed theoretical studies of wind-driven circumstellar shells have been given by Castor, McCray, and Weaver (1975), and by Weaver *et al.* (1977). See especially McCray (1983), whose discussion we now follow.

The development of a wind-driven circumstellar shell follows a scenario somewhat similar to that of a supernova shell. Initially, there is a very brief interval of free expansion at the terminal wind velocity. Then there is a period of adiabatic expansion during which radiative losses by the gas are not important.

Later there is a phase during which radiative losses do become significant. The theory of McCray *et al.* assumes an idealized interstellar medium of uniform density $\rho_o = \mu M_H N_o$, where μ is the mean molecular weight. The star is assumed to be at rest with respect to the interstellar medium. Further, the wind is isotropic with a constant energy flux $L_W = 1/2\ \dot{M}_W V_W^2$ ergs/sec.

This wind produces a cavity that bulldozes the surrounding interstellar medium into a thin shell of mass $M_s(t) = (4\pi/3)R_s^3\rho_o$. Note that $M_s(t) \gg \dot{M}_V t$, the mass of material in the wind. From Newton's second law of motion:

$$\frac{d}{dt}\left[M_s(t)\ V_s(t)\right] = 4\pi\ R_s^2\ P_{int} \qquad (5)$$

where P_{int} is the internal pressure. The assumption we make about P_{int} will determine the solution to the problem. For example, suppose we assume (a) conservation of thermal energy of shocked stellar wind:

$$P_{int} \sim \frac{\text{energy}}{\text{volume}} \sim \frac{1}{3} \frac{L_W}{4\pi} \frac{t}{R^3/3} \ . \tag{6}$$

The radius, R (t) [in parsecs], of the outer shell is given approximately by:

$$R_s(t) = 27\ N_o^{-1/5} \left(\frac{L_W}{L_o}\right)^{1/5} \left(\frac{t}{t_o}\right)^{3/5} \text{ parsecs,} \tag{7}$$

where $L_o = 10^{36}$ erg/sec, $t_o = 10^6$ years, and L_W is the energy in the stellar wind, typically 10^{36} erg/sec. The age of the structure is $t \sim 0.6\ R_s/V_s$. Figure 10-4, due to McCray, explains the notation. The stellar wind, W, encounters a shock at radius R_1. The shocked gas in the region denoted as C is heated to coronal temperatures, i.e., in excess of 10^6°K. This heated gas exerts a pressure on the thin, relatively dense shell of bulldozed interstellar gas whose temperature is typically about 10,000°K. A sharp temperature gradient exists across the interface between the coronal gas and the compressed interstellar material. Since electron conduction is very efficient, heat flows from the hot coronal gas into the cooler outer shell. Gas evaporates from the shell and flows inward, now lowering region C's temperature, increasing its density and enhancing the radiative losses.

The radiative loss from the coronal gas L_{rad} increases with time until it becomes comparable with the energy loss in the wind. Now the coronal gas region, C in Fig. 10-4, begins to collapse, permitting R_1 to increase. McCray calls the epoch at which this occurs t_{rad}, and the corresponding shell radius R_{rad}. From this epoch onwards a better description of the dynamics is given by a model (b) in which we assume that internal thermal energy is radiated away; the momentum of the wind is conserved as it impinges on the shell. Thus:

$$P_{int} = \rho_W V_W^2 = \frac{M_W V_W}{4\pi\ R^2} \tag{8}$$

and

$$R(t) = 16\left(\frac{L}{L_o}\right)^{1/4} \left(\frac{V_o}{N_o V}\right)^{1/4} \left(\frac{t}{t_o}\right)^{1/2} \text{ parsecs} \tag{9}$$

where V_o = 1000 km/sec. McCray finds:

$$t_{rad} = 3 \times 10^6 \text{ yr} \left(\frac{L}{L_o}\right)^{0.4} N_o^{-0.7} \qquad R_{rad} = 50 \left(\frac{L}{L_o}\right)^{0.4} N_o^{-0.6} \tag{10}$$

Application of models (a) energy conservation and (b) momentum conservation leads to substantially different conclusions when we try to use observations of the expanding shell to deduce properties of the wind. Shells following models (b) have less energy, smaller radii, and smaller velocities than those following assumption (a).

Are such structures likely to be recognized? Bubbles may be difficult to distinguish from defunct supernova remnants except that they would contain a central star. We might expect to see i) a ring-shaped H II shell in Hα and [O III], and ii) absorption lines such as O VI λ1035 which is produced mostly in the conduction front between the hot internal region and the expanding shell. Weaver et al. (1977) predicted a path length N(O VI)L = 5 × 10^{13} cm^2, which is almost independent of the density of the interstellar medium, the energy flux in the wind and the age of the system. The predicted value is in harmony with the observed value (Jenkins 1978) but it cannot be taken as definite confirmation of the model. Although the hot inner region of the bubble produces soft X-rays, their fluxes are too small to be detected.

Superbubbles with radii from 50 pc to 1 kpc and involving energy inputs 10^{52} ergs appear to be associated with clusters of O stars such as those found around Orion and the Carina OB associations. Combined actions of stellar winds and supernova explosions in OB associations can give 10^{52} to 10^{53} ergs over an interval of ten to a hundred mission years. When an OB association is formed, stellar winds may produce bubbles or superbubbles. Later, supernova detonations may dominate the ionization of the interstellar medium. After they die away, the remaining stars may still supply sufficient ultraviolet quanta to keep the shells ionized (see Sections F, G).

Expanding rings and shells around Wolf-Rayet stars provide yet another example of stellar winds impinging on a surrounding tenuous medium. These have been studied recently by a number of observers (e.g., Kwitter 1981; Chu 1982; Johnson 1982; Heckathorn 1982). Typically, the radii of the shells lie between 3 and 10 pcs, the expansion velocities between 20 and 80 km/sec, and the lifetimes between 20,000 and 200,000 yrs. The masses of the shells seem to lie between 5 and 20 $M_\odot$, although Chu finds suggestions of much larger masses for the ring nebulae RCW 104 and NGC 3199.

Extended Wolf-Rayet shells appear to be in the momentum-conserving phase, which means that the inner regions must contain several solar masses of gas in order to radiate away the thermal energy of the shocked wind. Theoretical considerations and the observed filamentary structure suggest that the shells suffered from some instability that produced radially oriented wisps.

Why are such shells associated with Wolf-Rayet stars and not with ordinary O stars? The former are generally believed to be massive highly evolved objects that have passed through the M supergiant phase and have lost their outer envelopes. McCray suggests that their vigorous winds $V_W \sim 2500$ km/sec, $L_W \sim 5 \times 10^{37}$ ergs/sec, $\dot{M}_W \sim 3 \times 10^{-5}$ $M_\odot$/yr impinge not upon the local ISM but rather upon debris ejected during the previous red supergiant phase of the star itself. The model thus resembles that proposed by Kwok for the

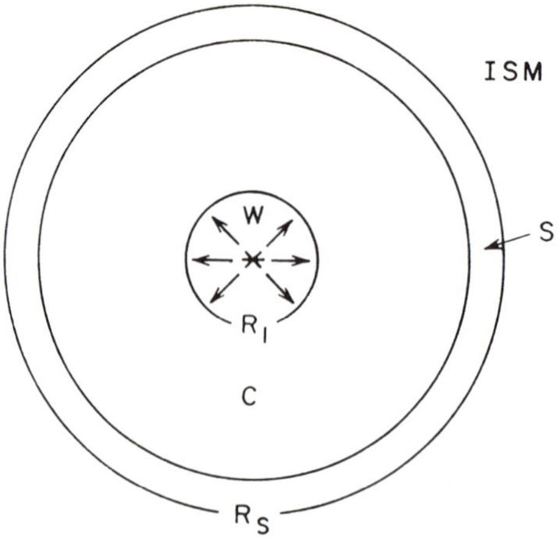

Figure 10.4

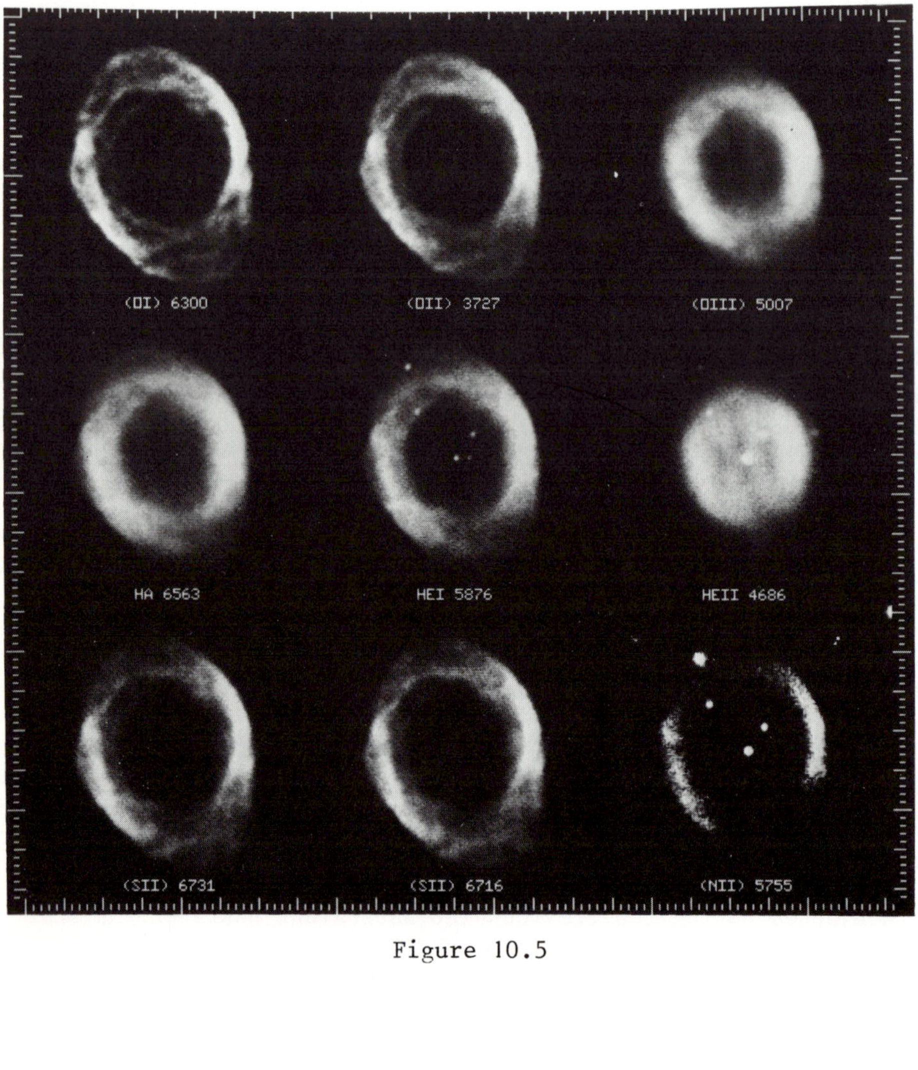

Figure 10.5

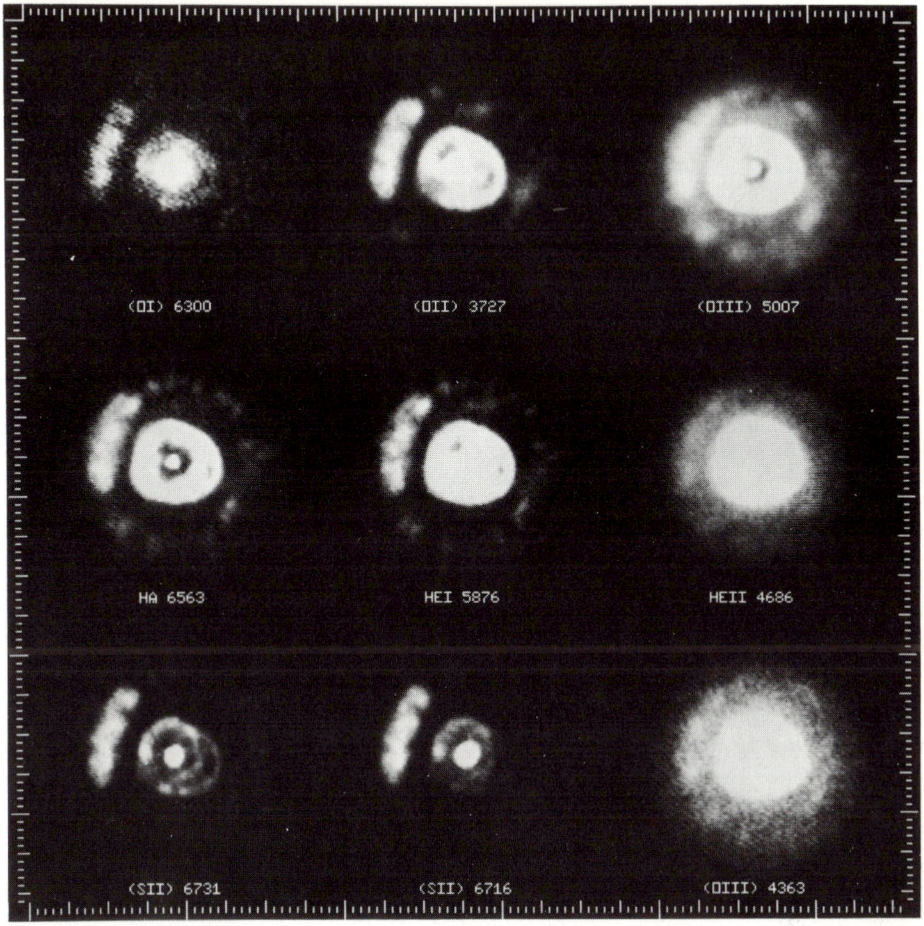

Figure 10.6

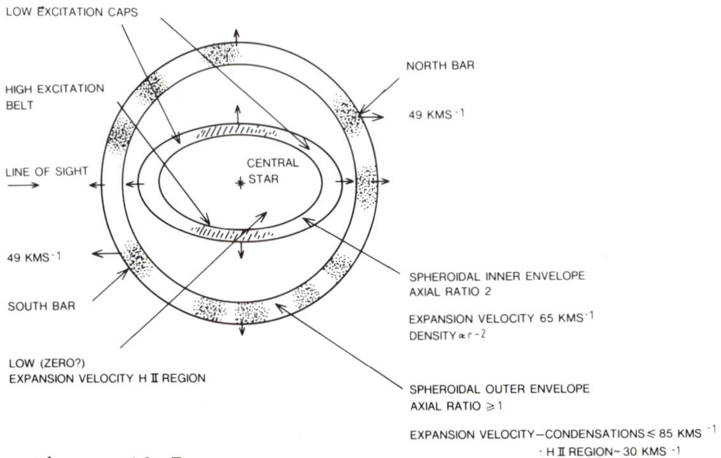

Figure 10.7

planetaries (Chapter 12). Corroborative evidence is supplied by the association of WR stars with very massive OB supergiants (Garmany 1982) and enhanced abundances of products of nuclear burning, e.g., N and He in the shell of NGC 6888 (Kwitter 1981). Let us consider some specific examples of large-scale objects where stellar winds have played important roles.

10D-2. The Gum Nebula

In apparent size this is the largest galactic H II region with an angular diameter of 36° ($240 < \ell < 276°$; $-19° < b < +17°$). It is a roughly circular object of such low surface brightness that it was detected only on Hα photographs secured many years ago by Colin Gum. If the distance to the center is 400 ± 60 pc, the diameter of the Gum Nebula is 250 pc.

The general appearance of this object, delicate filamentary structure and a hollow core suggests a supernova remnant (SNR)*. Reynolds (1976) compared the observed features of the Gum nebula with an SNR model due to Chevalier (1974). Taking into account that the model was an idealized one with no density inhomogeneities in the interstellar medium, he found a satisfactory agreement. The density of the gas in the shell, estimated from its surface brightness and size, ranged from 1 cm^{-3} to 5 cm^{-3}, with an average of about 2 cm^{-3}, in harmony with the SNR model; likewise, the model radius, energy, and density for the ambient medium 115 pc, 3×10^{50} ergs, and 0.25 cm^{-3}, respectively, are consistent with the observed values: $R = 125 \pm 20$ pc, $\sim 5 \times 10^{50}$ ergs, and 0.1 to 0.2 cm^{-3}. He concluded that the Gum Nebula is a million-year-old cloud of expanding gas caused by a 5×10^{51} erg explosion, but which is now being ionized and excited by hot stars ζ Pup (spec = O4) and γ Vel (WC8 + O9 binary). Left to itself, the original SNR would have cooled so far as to be invisible. Thus, one possibility is that a supernova created the shell and two O-type stars now keep it excited, i.e., maintain it as a "classical" H II region! A stellar wind could not have produced a shell structure of this type nor an observed expansion velocity of the order of 10 to 30 km/sec. Such a wind might supply some extra energy.

It is important to know whether a Gum nebular-type scenario can occur elsewhere. Can filaments, arches, etc., be produced by stellar winds and supernova but the actual ionization and excitation be caused by hot associated stars? Apparently, IC 443 is one example of this phenomenon, but it is in a much different evolutionary state.

10D-3. 30 Doradus and NGC 604

In Chapter 1 we referred to the great 30 Doradus nebula, an H II region with a diameter exceeding 200 parsecs. The total mass of the involved stars is estimated as 400,000 $M_\odot$, but a total mass of 14,000,000 suns is involved in the gas, most of which is neutral. To maintain the current level of ionization in a steady state, the efforts of over one hundred O-type and WR stars is required.

*It has also been suggested that the Gum Nebula is not an SNR, but sheets of gas ejected by hot stars (Dickel).

The greatest single supplier of ionizing radiation seems to be an object denoted as R136. Visual and photographic observations and speckle interferometry show it to consist of at least three stars: M_V = -8.3, -7.7, and -7.3. Walborn (1973, 1983) regards R136 as similar to the six-fold multiple star HD 97950 which is at the center of the enormous H II region, NGC 3603. With a distance of 8.4 kpc, these stars fall in a region of diameter less than 1.0 pc!

In the evolution of the Tarantula nebula, winds from WR and O stars, as well as supernova explosions (each liberating 10^{50} ergs) could each play important roles. The kinematics of the ionized and neutral gas in 30 Doradus is extremely complicated. For example, using measurements of Hα in the neighborhood of R136, Meaburn (1981) found extensive sheets of ionized material with V(heliocentric) ranging from -80 to 410 km/sec, more or less symmetrically distributed with respect to the mean value of V(hel) ~ 270 km/sec for most of the ionized material. Large-scale sheets of ionized material flowing away from R136 are implied and there is some suggestion of interlocking shells. The wind and ionizing photons from R136 perhaps suffice to produce the motions. In addition to giant shells such as N70, with diameters ranging up to 300 pc, Meaburn has found supergiant shells in the LMC with diameters up to 1200 pc.

Analyses of the optical spectra of bright regions in 30 Doradus are consistent with a photoionization model, i.e., the energy comes primarily from ultraviolet spectra of imbedded stars or a supermassive object, not from the dissipation of shock-wave energy. It is possible that small filamentary wisps with strong forbidden lines of ionized sulfur [S II] are actually SNRs and owe their luminosity at least in part to the dissipation of mechanical energy.

NGC 604 is a giant H II region in the spiral galaxy, M33. It is somewhat similar in appearance to 30 Doradus. The core region whose diameter if about 220 pc is comparable in size. There is an outer smooth halo with a diameter of 450 pc. NGC 604 is a (90%) thermal source. Benvenuti _et al_. (1979) found evidence for only one supernova remnant and suggested that the ionization in this object was produced by at least one superluminous star of absolute magnitude -9 or brighter. Rosa and D'Odorico (1982) suggest that perhaps 50 Wolf-Rayet stars together with 50 O stars provide the ionization. Massive (~ 6×10^{-5} $M_\odot$/year) stellar winds blowing from WR stars with velocities of the order of 2000 km/sec are postulated to produce the dozen observed shells and structures in a gaseous mass of about a million suns. The shells are presumed to have mean radii of the order of 25 pc and move with velocities of about 25 km/sec. They conclude that the observed morphology and properties of the rims seen in the core and halo of NGC 604 can be interpreted via the interaction of stellar winds with the local interstellar medium. Furthermore, mass loss rates and time scales agree well with predictions based on a Wolf-Rayet phenomenon.

The spectra of the nebulosities in both 30 Doradus and NGC 604 fit expectations for pure radiative excitation processes as described in Chapters 3 and 4. There is no clear-cut evidence of any shock wave excitation. The graceful arms, rims, and shells seem most likely to have been set in their present positions by massive stellar winds described above or even by supernova detonations. Their present excitation and ionization seem to be primarily of thermal origin.

10E. Planetary Nebula

10E-1. The Problem of Distances

A fundamental problem in astronomy always is to determine the distances of the objects we are studying. H II regions are excited by hot stars whose spectroscopic parallaxes can often be inferred from their luminosity classes. For planetary nebulae, the situation is much more difficult. The absolute magnitudes of the central stars from which these objects derive their luminosity usually is not known. Planetary nebulae are so remote that trigonometric parallaxes cannot be measured, at least with present techniques. Other types of direct measurements are possible for a few objects:

1) In a few instances the central star is a binary, with a companion of some ordinary spectral class whose absolute magnitude is known.

2) Sometimes it is possible to measure both the radial velocity of expansion in km/sec and the angular expansion rate in ("/yr). Similar observations for novae and the Crab Nebula permit us to fix their distances, if we assume that the shell is expanding uniformly. It is unsafe to apply this method uncritically to planetary nebulae because their increase in angular size may arise partly because of a steady growth in the ionized volume, i.e., the ionization zone is penetrating into the neutral gas (as in the Orion Nebuloa), or the expansion rate is not uniform in all directions as appears to be true for the Crab Nebula.

3) Some planetary nebulae belong to stellar systems of known distance, but selection effects can be important since we necessarily detect the brightest and most compact objects. Thus, upper limits to PN luminosities and important statistical information may be found from studies of these objects in galaxies of the local group, particularly the Magellanic Clouds, and M31 and its companions.

One might anticipate that by studying objects in the galactic bulge, for example, we could get improved values of quantities such as the mean luminosity in Hβ, $\langle L(H\beta)\rangle$. By using radio frequency (including VLA) data we can eliminate effects of interstellar extinction and achieve good angular resolution, but selection effects favor again the brighter and more compact objects.

4) Interstellar extinction measurements in the field of a PN are often useful. We compare the color excess E_{B-V} of nearby field stars of known luminosity and measured V magnitude with the nebular color excess which we can calculate with the aid of the extinction factor C (see Eq. 8-5). The method is not applicable to objects at high galactic latitudes, nor does it work well if we have to use a mean $\langle E_{B-V}\rangle$ versus $\langle D\rangle$ relation, because of the patchiness of the interstellar medium.

5) Radio-frequency 21 cm absorption can be useful, as Pottasch (1983) pointed out, since the line displacement differs from one spiral arm to another. We can at least establish whether a given planetary is nearby, or is located at a considerable distance.

Distance estimates for most planetaries, however, are obtained by indirect statistical procedures:

Kinematical methods based on proper motions (e.g., Cudworth 1974) offer hope for determinations of mean parallaxes since their accuracy increases with time. The τ components which are perpendicular to the direction of solar motion are probably better than the υ components that lie in the direction of the solar motion. Galactic rotation has also been employed, but both procedures require careful calibration by other means.

A more fundamental problem is that all PN do not belong to the same population. They do not form a homogeneous set. Most PN apparently belong to an old flattened population type I system, but many appear to move in orbits of pronounced ellipticity (Kiosa and Khromov 1979; Purgathofer and Perinotto 1980). Thus, many may be associated with Baade's population type II.

Statistical methods are based usually on hypotheses concerning the evolution and development of PN. An underlying hope is that we can develop an evolutionary scenario that will give a quantitative connection between time-dependent absolute magnitudes, $M_s(t)$ and temperature $T_s(t)$ for the central star on the one hand, and nebular radius, $R_n(t)$, and total luminosity, $L_n(t)$, on the other. Once such an appropriate relationship has been developed, we can use PN of known distance to calibrate the system and establish a firm distance scale. The hope would appear to be overly optimistic, for we are likely to be frustrated by a spread in initial parameters. Traditionally, various simplistic assumptions have been employed. Suppose that the flux in Hα or Hβ and the angular diameter are available. One may set a limit on the distance by assuming i) that the mass of the fully ionized material falls in a well-defined range (Minkowski and Aller 1954) or ii) that it is constant (Skhlovsky 1956). Clearly, this method can be applied legitimately only to fully ionized nebulae. Also, actual variations in total shell masses can give substantial errors.

An alternative assumption (originally due to Zanstra 1931) and subsequently employed in various forms by many others is that during at least part of the life of a PN, L(Hβ) or L(Hα) is constant and equal to some fixed value. Minkowski (1964) suggested we calculatge a distance D_L on the hypothesis of a constant luminosity and a distance D_M on the Shklovsky hypothesis of a luminous shell of fixed mass. If the nebula is optically thick, D_M will certainly be too large, while D_L is presumed to approach the correct value; the reverse is true for an optically thin nebula.

Radio-frequency data (which are free from effects of interstellar extinction) offer a better opportunity for the Minkowski method. Milne (1982) assumed a constant nebular mass and a constant luminosity in the Lyman continuum. He supposed that the nebula moved along a line of fixed flux until full ionization occurs. Thereafter, a constant mass is postulated for the ionized material. The derived distance then depends on the assumed mass of the ionized shell. Hence, various distance estimates for these objects are often in good accord because various workers have tended to assume $\langle M \rangle \sim 0.2\ M_\odot$

For optically thick nebulae, however, serious and fundamental objections can be raised against the sssumption of a constant L(Hβ). These criticisms come from both observation and theory. We would expect that L(*) and L(neb) would depend on the stellar mass. Jacoby's (1980) studies of PN luminosities in the Magellanic Clouds and other galaxies of the local group do <u>not</u> support the hypothesis of a constant Hβ luminosity for optically thick planetaries. Actually, L(Hβ) may range over one or two orders of magnitude.

An alternative procedure, due to Daub (1982) uses radio-frequency data in a somewhat different way. He assumes $N_\varepsilon = 1.18\ N(H+)$, and approximates Eq. (8-7) as:

$$E(5\ \text{Ghz}) = 3.75 \times 10^{-39}\ N_\varepsilon^2 f(t) \qquad \text{erg cm}^{-3}\ \text{s}^{-1}\ \text{Hz}^{-1} \tag{11}$$

where:

$$f(t) = t^{-1/2}\left(1 + 0.110\ \ln t^{3/2}\right) \tag{12}$$

Then the flux received by an observer at distance, D, is:

$$f(5\ \text{GHz}) = \frac{3.75 \times 10^{-39}}{4\pi\ D^2}\ f(t) \int N_\varepsilon^2\ dV \qquad \left(\text{erg cm}^{-2}\ \text{s}^{-1}\right) \tag{13}$$

where dV is the element of volume. The corresponding nebular mass is:

$$M_{neb} = 1.67 \times 10^{-24} \int N(H+)dV = 1.32 \times 10^{-24} \int N_\varepsilon dV \ \text{grams} \tag{14}$$

From Eq. (11), (12), (13), and (14) Daub obtains:

$$\left(\frac{M^2 f(t)}{\varepsilon}\right) = 3.52 \times 10^{-9}\ F(5\ \text{GHz})\ D^5\ \Theta^3\ (\text{gram})^2 \tag{15}$$

where the filling factor, ε, is defined as

$$\varepsilon = \frac{[\langle N_\varepsilon \rangle]^2}{\langle N_\varepsilon^2 \rangle} \tag{16}$$

The angular diameter of the nebula is $\Theta = 2R/D$ where R is the radius of the ionized shell in parsecs and $S_D = 10^{23}\ F(5\ \text{GHz})$. To make further progress we must now bring in information for nebulae for which independent distance estimates are available from direct methods as described above. Daub finds satisfactory data for 14 PN and uses them to establish a relation between:

$$f_D(M) = \frac{M}{M_\odot}\left[\frac{f(t)}{\varepsilon}\right]^{1/2} \quad \text{and} \quad \frac{\Theta^2}{S_D} \quad (M_\odot = \text{mass of the sun})$$

For optically thick nebulae, with $\Theta^2/S_D < 3.65$,

$$\log f_D(M) = \log \Theta^2/S_D - 4.50, \quad D = 324\ \Theta^{1/5}\ S_D^{-3/5} \quad \text{(pc)}$$

while for optically thin nebulae, $\log \Theta^2/S_D > 3.65$.

$$\log f_D(M) = -\ 0.85, \quad D = 9300/(\Theta^3 S_D)^{1/5} \quad \text{(pc)}$$

Typically, $R \sim 0.12$ pc when a nebula first becomes optically thin. Distances calculated by Daub differ from those obtained by Milne or by Maciel and Pottasch who assumed an empirical relationship of the form M(H II) = 1.225 R - 0.0123. During the optically thick stage, the nebular luminosity increases rapidly with time, reaching a maximum quantum output of 5×10^{46} stellar Lyman continuum photons s^{-1}, just as the shell becomes optically thin. The expected range in L(Hβ) is still less than that found by Jacoby (1980).

In all of these methods we deal with statistical procedures. We cannot be sure that the ejected mass or filling factor is constant from object to object. In summary, we can expect not only random-type errors for individual objects but also systematic differences from one distance scale to another, depending on calibration procedures and initial assumptions. The fact that parent stars may have belonged to different population types exacerbates the difficulties.

10E-2. Some Statistics of Planetary Nebulae

The time interval over which a PN can be recognized as such seems to be about 20,000 to 30,000 years. Hence, at any epoch we should be able to see nebulae of all ages, some just emerging from their dusty cocoons, others such as the Abell objects that are fading away. In the solar neighborhood, the spatial density is about 45 to 52 PN kpc^{-3}. If we assume a density law of the form $n = n_o \exp(-z/z_o)$, where z is the height above the galactic plane, a scale height, z_o, of about 120 parsecs, is often quoted. This is appropriate to that found for stars of about two solar masses. Estimates of the local birth rate fall in the range $1.2 \rightarrow 5 \times 10^{-3}\ kpc^{-3}\ yr^{-1}$; the death rate of nearby main sequence stars or formation rate of white dwarfs is similar.

Good determinations of masses of PN shells are few because of uncertainties in the distance and because of the fact that the quantity measured is only the mass of the ionized portion of the shell which may be much less than the total mass of the shell. Jacoby (1981) made a direct determination of the shell mass of Abell 35 as $0.2 \pm 0.1\ M_\odot$, but this object is interacting with the interstellar medium. A number of investigators quote $\langle M \rangle \sim 0.5\ M_\odot$, although values of 0.1 to 0.2 have often been chosen. At the present time, we have neither a good value of $\langle M \rangle$, nor of the dispersion $\langle \Delta M \rangle$.

Within our local group of galaxies, the PN luminosity functions seem to be similar (Jacoby 1980). They fall close to the predictions of the Henize-Westerlund (1963) model. In this scenario, each PN is represented by a uniformly expanding homogeneous gaseous sphere that is ionized by a non-evolving central star. Hence, the number of PN found in a given luminosity interval, ΔL, varies as the fraction of the lifetime spent in that particular range. Daub (1982) finds the observed size distribution of PN to be similar to that predicted by a simple model of expanding shells. An exact fit is not to be expected, since it is difficult to allow for a spread in progenitor star masses, different nebular shell masses, different ejection velocities and possibly changes in T(*) and L(*) during a PN lifetime (Chapter 12).

Many investigators have estimated the total number of PN in our galaxy. Values ranging from 10,000 to 30,000 are suggested, the most likely number probably lies between 10,000 and 20,000. Most galactic PN are large, faint, and hidden by interstellar obscuration. Jacoby (1980) estimated there are about 300 to 1000 PN in the Small and Large Magellanic Clouds, respectively, about 750 in M33, while M31 has about 20,000.

These PN statistics, plus data on their motions, and chemical compositions of the ejected shells put some constraints on the types of stars that can evolve into PN. Most of these nebulae probably come from rather prosaic stars, although the immediate progenitors are not well established. Clearly, concepts and scenarios of evolution must be integrated with statistical methods, each endeavor assisting the other by an iterative procedure.

10E-3. Structures and Internal Motions of Planetary Nebulae

We now turn from the statistics of PN as a class of objects to a consideration of individual examples. In order to understand nebular evolutionary history and assess possible ejection scenarios, we must obtain both the three-dimensional structure and velocity patterns. Monochromatic images reveal the nebulae in projection. By measuring also the internal velocities at each point, we can separate the contributions of approach and receding shells, since we know that the gas is always streaming outwards from the star. Thus, a three-dimensional model can be constructed for a nebula that is expanding in a spherically symmetrical way. N.K. Reay and his associates of Imperial College London (1976) employ an imaging Fabry-Perot interferometer with an appropriate detector. The interferometer scans in wavelength and at each chosen λ we can construct a snapshot of the nebula. Detailed monochromatic models can be obtained for even rather complicated objects.

The earliest systematic studies of internal PN motions were those of Campbell and Moore (1918) who used a slit spectrograph to measure the splitting of the [O III] lines. Later, Wilson (1950) investigated other ionic lines and showed that the less ionized atoms had greater expansion velocities. For example, in NGC 2440, [Ne V] λ3426 shows no measurable expansion, [Ne III] and [O III] have expansion velocities of about 45 km/sec, but [N II] and [O II] show velocities of about 65 km/sec. Thus, the velocity in the shell increases with radius. Also, since the expansion velocity can vary from one direction to another, the nebular shape can be affected by the variation, $V(r, \Theta, \phi)$. Weedman (1968) suggested a formula: $R_n = aV_n + b$, where R_n is the nebular radius, a and b are constant, and V_n is the nebular expansion velocity. If V_n increases with nebular size, one might anticipate that every old PN would have large expansion velocities. Curiously, certain very large, old PN have low V_n values, $10 < V_n < 35$ km/sec (Bohuski and Smith 1974), while certain young objects, such as NGC 2392 have V_n = 60 to 86 km/sec. Presumably, radiation pressure accelerates the material (cf., e.g., Mathews 1966; Ferch and Salpeter 1975).

Although PN show a great variety of forms (see, e.g., the discussions by Greig 1971); Phillips and Reay (1977) assert that the forms of many of the objects they have studied can be interpreted via intrinsically simple structures. They used plausible ejection hypotheses and found that gravitational attraction, radiation pressure, and stellar rotation would suffice to represent the observed forms. They calculate the density distribution and emissivity along the line of sight, compute the surface brightness, and then convolve the theoretical image to allow for effects of finite resolution of the detection systems (seeing or antenna pattern). Within these various constraints many models are possible, but some variations are yet to be examined. For example, the shell may change from an optically thick configuration to an optically thin one as the nebula evolves. Additional physical processes may be significant. Material expelled along the magnetic axis of an oblique rotator could produce intricate structures (Münch 1968). If present, stellar magnetic fields could produce nonspherically symmetrical envelopes (Gurzadyan 1969). Interaction of the expanding shell with the interstellar medium could deform old planetaries close to the galactic plane, but not convincingly affect objects such as NGC 6543 that lie at high galactic latitudes. The structures of planetaries found in binaries might be expected to be substantially influenced by the presence of the companion (e.g., Morris 1981).

The Imperial College group has studied not only straightforward ring structures such as IC 418 but also more complicated objects such as NGC 6543 whose helical appearance led Münch (1968) to suggest a pair of coiled filaments moving symmetrically away from a central star. Phillips, Reay, and Worswick (1977) find that this nebula can be interpreted in terms of emission from a tilted, triaxial, elliptical shell.

Because of its large angular size and conspicuous stratification effects, the Ring Nebula in Lyra, NGC 6720, has been a favored object for structural studies. See Fig. 10-5. Attempts have been made to model this object with a toroidal structure (Osterbrock 1950; Minkowski and Osterbrock 1960; Hua and Louise 1972) and by an oblate spheroid (Louise 1974). From Fabry-Perot interferometer measurements of the velocity field and monochromatic images covering ions over a wide range in ionization potential, Atherton et al. (1978) found that the object could be represented by a closed spheroidal shell expanding radially from a central star; a toroidal structure is not required. The [O III] velocity field closely defines that of H+.

The remarkable nebula NGC 2392 consists of an inner, irregular, elliptical ring with dimensions 15" × 19" and a much fainter, relatively smooth, outer elliptical envelope of diameter 46", best seen in [O III]. Superposed on this structure are prominent condensations particularly conspicuous in Hα and [N II] (see Fig. 10-6). Within the inner ring is a hot, kinematically quiescent core where [Ne V] originates. The monochromatic and velocity data obtained by Reay, Atherton, and Taylor (1983) suggest a rather remarkable three-dimensional model, involving two concentric spheroidal shells viewed along the major axis. The inner shell appears to be of constant density. Its longest axis seems to be directed towards the earth. Its expansion velocity is about 65 km/sec but the broad line profiles suggest a velocity gradient within this shell with some material moving at a velocity of 120 km/sec, with respect to the central star. If the outer envelope is a prolate structure, the expansion velocities of the blobs which are about 75 km/sec can be proportional to the distance from the central star. What is surprising is that the smooth, underlying [O III] shells expands at a rate of only 30 km/sec. If the distance of NGC 2392 is 1 kpc, this smooth shell is 4000 to 7600 years old, but the blobs were ejected only 2500 years ago. Thus, these denser condensations are crashing through the smooth [O III] envelope. Furthermore, the inner shell with an expansion velocity of ~ 65 km/sec will overtake the outer envelope (see Fig. 10-7).

Possibly something similar may be occurring in the irregular filamentary, inhomogeneous nebula NGC 40, which resembles a truncated ring with a number of fainter wisps. Sabbadin and Hamzaoglu (1982) found here a smooth orderly expansion, but with the inner [O III] zone expanding more rapidly than the Hα shell that surrounds it. on the other hand, the irregular, hour glass-shaped inhomogeneous dumbbell nebula, NGC 6853 in Vulpecula, does not appear to be expanding symmetrically about a central star at all. Here a filamentary envelope, glowing in [N II] and [O II], surrounds a diffuse, more highly excited volume that emits [O III]. Goudis et al. (1978) concluded that the turbulent, bulk kinematics of this object can be explained by a cylindrical or elliptical nebula expanding radially with respect to its own axis!

Double shell planetaries have been studied for a number of years. In addition to NGC 2392, popular examples include NGC 1535, 2022, 3242, 6804, and 7354. There are three known examples of PN with triple shells: NGC 6826, NGC 7009, and NGC 7662. NGC 6826 has a small inner shell of diameter ~ 10", an outer shell ~ 25" diameter, and a giant halo of 130" diameter (Feibelman 1981). Outer haloes which have diameters ranging up to five times that of the main bright body of the nebula have been found for 27 of 44 nebulae observed with a CCD by Jewitt et al. (1983). These haloes often exhibit filamentary structures and have surface brightnesses ranging from 12 to 30% that of the main nebular body. The existence of double shells and haloes provide interesting insights on ejection processes in PN progenitors. Kaler (1974) suggested that in a number of instances the progenitor star may eject two or three shells of material, but the outer shells do not become visible until the evolving PN has become very bright and the

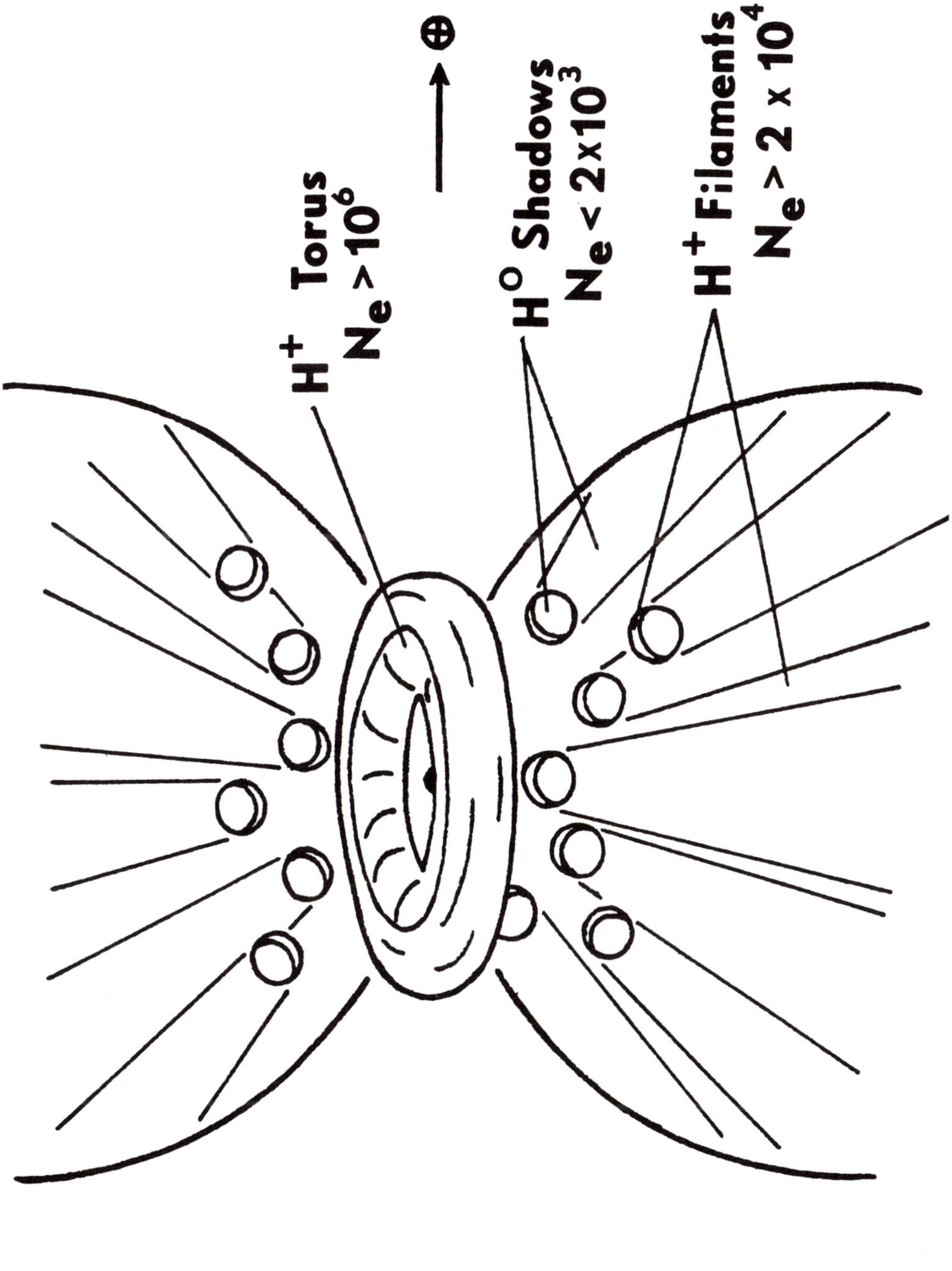
H^+ Torus
$N_e > 10^6$
H^0 Shadows
$N_e < 2 \times 10^3$
H^+ Filaments
$N_e > 2 \times 10^4$

Figure 10.8

inner shell optically thin. The haloes probably do not represent light scattered in surrounding dust clouds. The spectrum of the halo differs from that of the bright part of the nebula and polarization is very small. Jewitt et al. estimate the halo masses to be comparable with those of the bright portions of the nebula. They suggest that these masses are too large to be explained by winds from progenitor stars.

10F. Bipolar Nebulae; Transition Objects or Young Planetaries?

Compact, dense objects such as IC 4997 are easily recognized as young PN; the Abell objects as old dying nebulae, but what of the very youngest PN? The most likely candidates would appear to be found among the so-called "bipolar nebulae." This terminology has been used variously to denote a variety of objects from Minkowski's "Butterfly Nebula" M2-9 to conventional bilaterally symmetrical planetaries such as NGC 2346 or NGC 7026, and even to nebulae that are not related to PN at all, e.g., R. Mon! We follow Mark Morris (1981) in restricting the designation "bipolar nebula" to expanding envelopes around evolved stars -- objects which may be in fact incipient PN (Calvet and Cohen 1978; Zuckerman et al. 1976).

As a prototype BP nebula, we may choose GL 618 (see Fig. 10-8 due to Schmidt and Cohen 1981). They are typically symmetrical reflection nebulae with dust and both molecular and atomic gas expanding from an oft-hidden central star. This star may range from a relatively cool object to one of spectral class O. Molecular line widths indicate typical velocities of the order of 50 to 150 km/sec, although expansion velocities as high as 400 to 600 km/sec are sometimes found. Interferometer, mm and maser observations suggest mass loss rates in the range $10^{-5} < dM/dt < 10^{-4}$ $M_{\odot}$/yr, i.e., much larger than found in "standard" PN. Only about a dozen BP nebulae are known; they are rare because they probably represent but a moment in the lifetime of a small percentage of all stars.

Structures can be beautifully symmetric as in M2-9*. In the equatorial plane, there is a thick opaque ring that often conceals the star. The gas flow rate is markedly dependent on the latitude Θ, so that as Θ approaches 90°, the optical depth in the dust decreases. There are often dense blobs in the flow; much of the gas may be actually shadowed from the star. In GL 618, the optical radiation in the lobes is strongly polarized, showing much of it to be light from the central star that has been scattered by dust grains. There is also a contribution of forbidden lines from the neutral gas, [N], [C I], and [Mg I] and a small contribution of radiation from ionized gas that

*See, e.g., the photograph obtained by R. Minkowski and reproduce in Gaseous Nebulae and Interstellar Medium, Stars and Stellar Systems, 7, 483 (plate 11), 1968, ed. B. Middlehurst and L.H. Aller, Chicago, University of Chicago Press. The structures of M2-9 have been observed to change between 1952 and 1971. This object, whose age is estimated to be 5,000 years, is at a more advanced evolutionary stage than GL 618.

fills a small percentage of the volume in the unshadowed region. On the assumption that the density of the dust declines strongly with Θ, i.e., as $f(\Theta)r^{-2}$, Morris (1981) has been able to represent the observed nebular forms even with the "horns" that appear when density falls off strongly with Θ. At low latitudes where the optical depth in the dust is large, starlight does not penetrate very far, while at high latitudes there is not a sufficient amount of dust to scatter much light. Morris suggests that the BP nebulae originate from binary systems under rather special conditions. Only a fraction of PN would evolve from bipolar objects.

As the nebula evolves, the optical depth in the dust gradually declines. Finally, the star shines through and the former toroidal cloud now contributes most of the nebular light. The lobes become faint and the denser blobs in them appear as faint ansae. Ultimately, the structure may evolve into something like that of NGC 650-51 or NGC 2371-72.

References

Relations involving HII regions, excitation parameters, etc., and exciting stars (see Eq. 3) have been given by, for example:

Mezger, P.G., and Henderson, A.P. 1967, *Ap. J.* **147**, 471.
Schraml, J. and Mezger, P.G. 1969, *Ap. J.* **156**, 269.
Felli, M. 1979, in Stars and Stellar Systems, 195, ed. B. Westerlund, Dordrecht, Reidel Publ. Co., and references therein cited.
Georgelin, Y.M., Lortet-Zuckerman, M.C., Monnet, G. 1975, *Astron. Astrophys.*, **42**, 273.
Pottasch, S.R., Wasselius, P.R., and van Duinen, R.J. 1979, *Astron. Astrophys.*, **77**, 189.

Nebulosities of extremely low density are discussed, e.g., by:

Kirshner, R.P., Gull, T.R., and Parker, R.A.R. 1978, *Astron. Astron. Suppl.*, **31**, 261.
Reich, W. 1978, *Astron. Astrophys.*, **64**, 407.

The Orion Nebula. The standard, comprehensive account of modern work is Goudis, C., 1982, *The Orion Complex, A Case Study of Interstellar Matter*, Dordrecht, Reidel Publ. Co.

Our summary has been taken mainly from briefer review papers which also contain numerous references:

Zuckerman, B. 1973, *Ap. J.*, **183**, 863.
Balick, B., Gannon, R.H., and Hjellming, R.M. 1974, *P.A.S.P.*, **86**, 616.
Balick, B., Gull, T.R., and Smith, M.G. 1980, *P.A.S.P.*, **92**, 22.

Symposium on the Orion Nebula to Honor Henry Draper, 1982, ed. A.E. Glassgold, P.J. Huggins, and E.L. Shucking, Annals of the New York Academy of Sciences.

References (Continued)

Sharpless 155 and Cepheus 3 Association; the discussion is from:

Felli, M., Tofani, G., Harten, R.H., and Panagia, N. 1978, Astron. Astrophys., 69, 199, where references to numerous articles pertaining to this complex may be found. Other radiatively excited regions are discussed by, e.g.: Meaburn, J. 1978, Astrophys. Space Sci., 59, 193.

Wind-driven Shells and Interstellar Bubbles. For a summarizing review see:

McCray, R. 1983, Highlights of Astronomy 6, ed. R.M. West, Dordrecht, Reidel Publ. Co., and references therein cited. See also:

Castor, J., McCray, R., and Weaver, R. 1975, Ap. J. (Letters), 200, L107.

Dyson, J.E. 1977, Astron. Astrophys., 59, 161; 1978, 62, 269.

Weaver, R., McCray, R., Castor, J., Shapiro, P., and Moore, R. 1977, Ap. J., 218, 377.

Properties of ring nebulae associated with Of stars are examined by:

Lozinskaya, T.A. 1982, Astrophys. Space Sci., 87, 313, and references therein cited.

For discussions of expanding rings and shells around Wolf-Rayet stars, see:

Kwitter, K.B. 1981, Ap. J., 245, 154.

Chu, Y.H. 1982, Ap. J., 254, 578.

Heckathorn, J.M., Bruhweiler, F.C., and Gull, T.R. 1982, Ap. J., 252, 230.

Johnson, H.M. 1982, Ap. J., 256, 559.

Gum Nebula

Maran, S.P., Brandt, J.C., and Stecher, T.P. 1972, Gum Nebula and Related Problems. NASA SP-322.

Reynolds, R.J. 1976, Ap. J., 203, 151; 206, 679, and references contained therein.

30 Doradus

Numerous references to basic parameters for this nebula may be found in:

Cantó, J., et al. 1979, M.N.R.A.S., 187, 673, and in Blades, J.C., and Meaburn, J. 1980, M.N.R.A.S., 190, 59P.

References (Continued)

The spectrum of 30 Doradus is discussed, e.g., by:

Peimbert, M., and Torres-Peimbert, S. 1974, Ap. J., 193, 327.
Boeshaar, G.O. et al. 1980, Astrophys. Space Sci., 68, 335, and references cited therein.

The remarkable nature of R136 is discussed by:

Walborn N. 1973, Ap. J., 182, L21.
________ 1983, IAU Symposium No. 108 and references cited therein.
Meaburn, J. 1981, M.N.R.A.S., 196, 19P.

NGC 604: Supernova remnants, stellar winds, structure and morphology are reviewed by:

Benvenuti, P., D'Odorico, S., and Dumontel, M. 1979, Astron. Space Sci., 66, 39.
Israel, F.P., Gatley, I., Matthewes, K., and Neugebauer, G. 1982, Astron. Astrophys., 105, 229.
Rosa, M., and D'Odorico, S. 1982, Astron. Astrophys., 108, 339, and references contained therein.

The spectrum of NGC 604 has been studied by a number of observers; see, e.g., Kwitter, K.B., and Aller, L.H. 1981, M.N.R.A.S., 195, 938, and references cited therein.

Planetary Nebulae; General Surveys

Older studies and developments up to the mid-1960's are reviewed in Nebulae and Interstellar Matter; Stars and Stellar Systems, 7, 483, 1968, ed. B. Middlehurst and L.H. Aller, Chicago, University of Chicago Press. See also Planetary Nebulae, I.A.U. Symposium No. 34, 1968, ed. C.R. O'Dell and D. Osterbrock, Dordrecht, Reidel Publ. Co. More recent advances are described by Osterbrock (1973), AGN: Memoires Societe Royal des Sciences de Liege, 6th Series, 5, 1973; I.A.U. Symposium No. 76 (1977) and I.A.U. Symposium No. 103 (1983), Dordrecht, Reidel Publ. Co. An excellent comprehensive treatment is given by: Pottasch, S.R., Planetary Nebulae, 1983, Dordrecht, Reidel Publ. Co.

Of earlier work on PN distances we give as examples: Zanstra, H., 1931, Zeits. f. Astrofis., 2, 329; Berman, L., 1937, Lick Obs. Bull., 18, 73; Minkowski, R., and Aller, L.H., 1954, Ap. J., 120, 261; Shklovsky, I.S., 1956, Astr. J. (USSR), 33, 222, 315; Cahn, J.H., and Kaler, J.B., 1971, Ap. J. Suppl., 22, 319.

References (Continued)

Among more recent determinations, examples are:

Acker, A. 1978, Astron. Astrophys. Suppl., **33**, 367 (distance scales).
Cudworth, K.M. 1974, Astron. J., **73**, 1384 (proper motions).
Pottasch, S.R. 1982, I.A.U. Symposium No. 103 (21-cm data).
Milne, D.K. 1982, M.N.R.A.S., **200**, 51P.
Maciel, W.J., and Pottasch, S.R. 1980, Astron. Astrophys., **88**, 1.
Daub, C.T. 1982, Ap. J., **260**, 612.

For discussions of galactic orbits and statistics, see, e.g.,

Kiosa, M.N., and Khromov, G.S. 1979, Astrofizika, **15**, 105.
Purgathofer, A., and Perinotto, M. 1980, Astron. Astrophys., **81**, 215.
Maciel, W.J. 1982, Astron. Astrophys., **98**, 406.
Schneider, S.E., Terzian, Y., Purgathofer, A., and Perinotto, M. 1983, I.A.U. Symposium No. 103.

Masses of planetary nebular shells have been discussed by many authors. See, e.g.: Jacoby, G.H., 1981, Ap. J., **244**, 903, where a mass determination is given for Abell 35, a large PN deformed by interaction with the interstellar medium.

Internal motions in planetaries; the classical papers are:

Campbell, W.W., and Moore, J.H. 1918, Publ. Lick Obs., **13**, 77.
Wilson, O.C. 1950, Ap. J., **111**, 279; 1958, Rev. Mod. Phys., **30**, 1025.

For an assessment of more recent work, see Osterbrock AGN (1973); Weedman, D.W., 1968, Ap. J., **153**, 49, and Bohuski, J., and Smith, M., 1974, Ap. J., **193**, 197.

Structures of planetary nebulae have been investigated by many workers since the pioneering work of: Curtis, H.D., 1918, Lick Obs. Publ., **13**, 57. A frequently quoted classification of PN forms is that given by: Greig, W.E., 1971, Astron. Astrophys., **10**, 161. See also Khromov, G.S., and Kohoutek, L., 1968, I.A.U. Symposium No. 34, p. 227, Dordrecht, Reidel Publ. Co.

Attempts to calculate the shapes of planetary nebulae by theoretical means are numerous. A few examples are:

Gurzadyan, G.D. 1969, Planetary Nebulae, London, Gordon and Breach.
Louise, R. 1973, Mem. Roy. Soc. Liege, 6th series, **5**, 465.
Phillips, J.P., and Reay, N.K. 1977, Astron. Astrophys., **59**, 91.

The importance of radiation pressure in scouring out the inner regions of PN was pointed out by Matthews, W.G., 1966, Ap. J., **143**, 173, and further examined by Ferch, R.L., and Salpeter, E.E., 1975, Ap. J., **202**, 195, and by Pikel'ner, S.B., 1973, Astrophys. Lett., **15**, 91.

References (Continued)

Studies of Individual Nebulae involving both kinematical and structural measurements have greatly improved our understanding of these objects: A few examples:

NGC 6543: Phillips, J.P., Reay, N.K., and Worswick, S.P., 1977, Astron. Astrophys., 61, 695; but see also Münch, G. 1968, I.A.U. Symposium, 34, Planetary Nebulae, p. 259.

NGC 6720: Osterbrock, D.E., 1950, Ap. J., 131, 541; Minkowski, R., and Osterbrock, D.E., 1960, Ap. J., 131, 537; Hua, C.T., and Louise, R., 1972, Astron. Astrophys., 21, 193; Louise, R., 1974, Astron. Astrophys., 70, 189; Atherton, P.D., Hicks, T.R., Reay, N.K., Worswick, S.P., and Hayden-Smith, W. 1978, Astron. Astrophys., 66, 297.

NGC 2392: Reay, N.K., Atherton, P.D., and Taylor, K., 1983, M.N.R.A.S., 203, 1087, and references therein cited.

NGC 40 and NGC 7026: Sabbadin, E., and Hamzaoglu, E., 1982, Astron. Astrophys., 109, 131.

NGC 6853: Goudis, C., McMullan, D., Meaburn, J., Tebbutt, N.J., and Terret, D.L., 1978, M.N.R.A.S., 182, 13.

NGC 6302: Meaburn, J., and Walsh, J.R., 1980, M.N.R.A.S., 191, 5P; 193, 631; Phillips, J.P., Reay, N.K., and White, G.J., 1983, M.N.R.A.S., 203, 977.

Halo Nebulae: The first examples were found by Duncan, J.C., 1937, Ap. J., 86, 496. See: Kaler, J.B., 1974, Astron. J., 79, 594 (multiple shell PN); Feibelman, W.A., 1981, P.A.S.P., 93, 719 NGC 6826); Jewitt, D.C., Danielson, G.E., and Kupferman, P.N., 1983, Ap. J., xx, xxx.

Bipolar Nebulae: A useful summary is given by: Morris, M., 1961, Ap. J., 249, 572. See also: Allen, D.A., and Swings, J.P., 1972, Ap. J., 174, 583; Zuckerman, B., Gilra, D.R., Turner, B.E., Morris, M., and Palamer, P., 1976, Ap. J., 205, L15; Calvet, M., and Cohen, M., 1978, M.N.R.A.S., 182, 687; Schmidt, G., and Cohen, M., 1981, Ap. J., 246, 444.

List of Illustrations

Fig. 10-1. The Orion Nebula (Lick Observatory Photograph).

Fig. 10-2. A Model for the Structure of the Orion Nebula. Ultraviolet radiation from the Trapezium stars (particularly by the O6 component) has created a cavity, actually an HII region in the cool, dense, neutral gas cloud. As heated ionized gas escapes from the

List of Illustrations (Continued)

Chapter 11

CHEMICAL COMPOSITIONS OF GASEOUS NEBULAE

11A. Introduction

Ascertainment of the chemical compositions of gaseous nebulae has been a goal of many investigations since the pioneering studies of Bowen and Wyse (1939) and of Wyse (1942). At that epoch it was widely believed that the elemental abundances in all stars and nebulae were essentially the same. The concept of chemical evolution in galaxies was yet to be introduced.

Today we realize that knowledge of chemical compositions of gaseous nebulae help us to assess element building scenarios in advanced stages of stellar evolution and to chronicle the changing chemistry of galaxies. Emission line diffuse nebulae [H II regions] can be observed individually not only in our own galaxy but also in distant stellar systems where individual stars are simply lost in an amorphous glow. We can study point-by-point variations of the chemical compositions of diffuse nebulae from central regions to outermost spiral arms. Since this gaseous material presumably largely consists of debris from defunct stars, we can monitor the rate at which element building processes operate and thus get some information on the chemical evolution of a galaxy as a function of distance from its nucleus. The elements that can be studied are He, C, N, O, Ne, S, Cℓ, and Ar. Abundances of He, C, and N may yield clues to element building processes involving stars of modest mass. On the other hand, the manufacture of Ne, S, Ar and also O require massive stars and supernovae, as is also true for silicon and metals such as magnesium or iron.

The situation for planetary nebulae is more complex. A nebular shell can consist of: solely regurgitated material from the interstellar medium of the epoch of progenitor star formation or this original material plus products of nuclear burning in the core of the progenitor star. What happens depends on the initial mass and evolution of the progenitor star. Manufacture of helium, carbon and nitrogen is well-established for many planetary nebulae. For example, in the Magellanic Clouds, the n(C)/n(H) ratio in bright planetaries is much greater than in the H II regions; a similar effect is found in our own galaxy. An even more direct demonstration of the reality of element building in progenitor stars is the discovery of compact knots of highly processed material evidently ejected from the central star of the planetaries Abell 30 and 78 (Hazard *et al*. 1980; Jacoby and Ford 1983). Only in planetaries which come from progenitor stars of apparently low mass do we expect uncontaminated C and N to emerge. On the other hand, oxygen and heavier elements are probably not manufactured in PNP (planetary nebula progenitors).

11B. Observational and Theoretical Problems of Composition Determination; Diffuse Nebulae

H II regions are invariably of low excitation; hence, the number of observable ionization stages is limited so that much of the gas may exist in neutral or even molecular form. For example, often $n(He^+)/n(H^+) < n(He)/n(H)$ because much of the helium is neutral. It is particularly important to extend measurements to the infrared to obtain, e.g., [N III] λ57.3 μ and into the ultraviolet to get λ1909 C III.

In a nearby diffuse nebulae like M8 or Orion, or even 30 Doradus in the LMC where individual stars can be seen, one can observe gaseous clumps without heavy contamination by stars, but in galaxies such as M33 or M101, stars and nebulosity are superposed. Noise contributed by starlight obliterates weak lines. Even in objects such as the Orion Nebula, starlight scattered by dust can contribute most of the observed continuum. Absorption lines in scattered starlight can seriously affect measured nebular line intensities. A further difficulty is that the interstellar extinction law in the Magellanic Clouds and other galaxies may differ from that found in our galaxy (see Chap. 8).

Interpretation of line intensity data requires a knowledge of T_ε. Normally, we would use the 4363/5007 [O III] or 5755/6584 [N II] ratios but in H II regions with $T_\varepsilon \ll 10{,}000°K$, 4363 and 5755 are not observable and T_ε must be established by some other means. Theoretical models may be helpful but observations must supply guidance. In our own galaxy, radio-frequency line observations can provide the necessary temperature information, but in other galaxies radio data are not available, so other approaches must be used.

Monochromatic photographs show that throughout the brighter parts of many diffuse nebulae the ionization pattern is often similar, but in some objects such as N44 in the LMC, conspicuous differences between [O III] and Hα images can occur.*

Theoretical models have been computed for H II regions in the galaxy, the Magellanic Clouds, M33, M101 and other systems and attempts have been made to employ them to determine chemical abundances. Difficulties are imposed by a) lack of spherical symmetry; b) filamentary structure; c) nonuniformly-distributed dust; d) complex dynamical evolution with no steady state; e) excitation produced by a cluster of stars. Hence, it has been popular to employ *ad hoc* extrapolation recipes to allow for distribution of atoms among unobserved ionization stages. This procedure apparently works well for N and O, but it is of questionable value for Ne, S, Cℓ, and Ar; we must persevere in the development of theoretical models. By extending observations to the ultraviolet, we can also add the important element, carbon. Note that in both diffuse and planetary nebulae, substantial fractions of elements such as Mg, Si, Ca, and Fe can be locked in

*These excitation level variations can complicate chemical abundance determinations. See Czyzak and Aller, 1977, *Astrophys. Space Sci.*, **46**, 371.

grains. Carbon and oxygen also contribute to the formation of solid particles, although it is generally assumed that most atoms of these elements remain in a gaseous phase.

11C. Problems of Chemical Composition Determinations: Planetary Nebulae

Planetary nebulae generally have been more satisfactory targets for abundance determinations than have H II regions for a number of reasons: a) many have high surface brightness so that we can reach intrinsically faint lines and cover spectral ranges from the microwave region to the ultraviolet or alternately we can observe partially obscured planetaries where H II regions would be rendered too faint by the same amount of interstellar extinction; b) some show a wide range of excitation and ionization, so influence of ionization correction factors may be minimized, and c) some show fairly symmetrical structures, justifying theoretical models.

Accurate measurements of total spectral line fluxes can be made in the optical spectral region by photoelectric photometry, but in many instances it is essential to consider the small-scale angular structure, which we can observe with digital scanners. Much of the discordance between different spectrophotometric measurements does not necessarily imply serious errors in one system or the other but may arise from circumstances that one observer measured radiation from the entire image while another selected a small area. For practical reasons, intrinsically weak lines are observed in bright knots and filaments. Because of the low sensitivity of available detectors, infrared measurements frequently apply to the entire nebular image. For detailed analyses we need not only integrated nebular fluxes but also trustworthy spectrophotometry of individual clumps and local areas. Even such measurements normally yield only an integral along the line of sight through the nebula, but with high resolution spectra we may be able to separate approaching and receding shells (sec. 10E-3). Differential comparisons of bright clumps and their environs are essential to assess conditions in high-density versus low-density regions. A definitive solution to this problem by combining isophotic contours over the whole nebula with integrated monochromatic fluxes and velocity data is already technologically possible. Before we can proceed with a nebular analysis, we must assemble a consistent set of intensities. Some allowance must be made for the spatial distribution of the emissions. Even crude models of inhomogeneities may be useful in situations where we must compare emissions measured in integrated light with those measured only in the brightest portions of the nebular images. Information on isophotes is particularly important where we have to interpret observations that do not include the whole nebular image.

Having adopted a set of intensities we now employ methods described in Chapter 5 to obtain the nebular diagnostics, (N_ε, T_ε). With these data we may now evaluate $N(X_i)$ from collisionally excited lines of each observed ion i of element X.

The evaluation of $N(X_i)$ depends on whether or not we assume temperature fluctuations in the radiating strata. Following Peimbert (1967) we define a mean temperature T_o along the line of sight, ℓ, in a solid angle, Ω, and a mean square fractional temperature fluctuation $\langle\Delta\tau\rangle^2$ by the expression:

$$\langle\Delta\tau\rangle^2 = \frac{\int [T(\ell) - T_o]^2 \, N_\varepsilon(\ell)\, N_i(\ell)\, d\ell\, d\Omega}{T_o^2 \int N_\varepsilon(\ell)\, N_i(\ell)\, d\ell\, d\Omega} \qquad (1)$$

An effect of temperature fluctuations is that recombination line emission tends to be favored in cooler strata, since the emission is proportional to:

$$N_\varepsilon\, N(X^{i+1})\, T_\varepsilon^{3/2} \exp\left(\frac{hRZ^2}{n^2 kT_\varepsilon}\right)$$

On the other hand, collision rates go as $N_\varepsilon\, N(X^i)\, T_\varepsilon^{-1/2} \exp(-\chi_j/kT_\varepsilon)$ where χ_j is the excitation potential of the upper level of the transition. Hence, the emission of these lines is favored in the hotter regions. For example, Peimbert finds:

$$T(H\beta) \approx T_o[1 - 0.92\Delta\tau^2], \quad T\left(\frac{4363}{5007}\right) = T_o\left[1 + \frac{1}{2}\left(\frac{90{,}800}{T} - 3\right)\Delta\tau^2\right] \qquad (2)$$

Fortunately, it is possible to measure T_ε[O III], T_ε[N II], and $T(H_{nn'})$ (temperature determined from recombination lines in the radio-frequency range) in H II regions. Thus, Shaver et al. (1983) found mean values $\langle T_\varepsilon[\mathrm{O\ III}]/T_\varepsilon(H_{nn'})\rangle = 1.06 \pm 0.02$ and $\langle T_\varepsilon[\mathrm{N\ II}]/T_\varepsilon(H_{nn'})\rangle = 1.09 \pm 0.02$. In H II regions T_ε[N II] > T_ε[O III], apparently cooling is more efficient in the [O III] zones. They find no evidence in support of Peimbert's suggestion of a small-scale T fluctuation and suggest an upper limit $\langle\Delta\tau\rangle^2 < 0.015$.

Once the ionic concentrations have been found, we must next establish the ionization correction factor, ICF, i.e., $N(X)/N(X_i)$. Two methods are frequently used: 1) simple proportionalities or empirical formula; 2) theoretical models employed as interpolation devices. The first method has been used by numerous investigators; for example, by Peimbert and Costero (1969), by French (1981), and by Barker (1983), from whose compilation we quote the following formulae:

$$\frac{N(N)}{N(H)} = \frac{N(N^+)}{N(H^+)}\,\frac{N(O)}{N(O^+)} \qquad (3)$$

$$\frac{N(O)}{N(H)} = \frac{N(O^+) + N(O^{++})}{N(H^+)}\,\frac{N(He)}{N(He^+)} \qquad (4)$$

$$\frac{N(Ne)}{N(H)} = \frac{N(Ne^{++})}{N(H^+)}\,\frac{N(O)}{N(O^{++})} \qquad (5)$$

$$\frac{N(Ar)}{N(H)} = \frac{N(Ar^{2+}) + N(Ar^{+3}) + N(Ar^{4+})}{N(H^+)} \times \left[\frac{N(S+) + N(S++)}{N(S^{++})}\right] \qquad (6)$$

$$\frac{N(S)}{N(H)} = \frac{N(S+) + N(S++)}{N(H^+)} \left\{1 - \left[1 - \frac{N(O^+)}{N(O)}\right]\right\}^{-1/3} \qquad (7)$$

These approximate formulae appear to work well in a statistical sense, particularly for low-excitation nebulae. Hawley and Grandi (1978) and Kaler (1979) find that for gaseous nebulae without neutral helium, Eqs. (3) and (4) are reasonably valid, while Barker concludes they give satisfactory results for high-excitation nebulae NGC 6720 and NGC 7009. Their use may be recommended for nebulae where data are limited or complexities of structure preclude successful theoretical modeling.

The expression for neon is unsatisfactory for low-excitation objects and it also appears to fail at high excitation where we have data from Ne^{+3} and Ne^{+4}, however. For Ar and S a host of *ad hoc* expressions has been proposed. For S we quote Eq. (7) due to Stasińska; there is some indication that it may underestimate the S abundance.

11D. Examples of Abundance Determinations

As an example, consider the highly stratified Ring Nebula, NGC 6720. Two independent studies were carried out at Lick Observatory with the image tube scanner, the first by Hawley and Miller (1977) and the second reported by Aller (1976). Figure 11-1 shows positions selected in the latter program. A third set of measurements has been made by Barker (1981). Table 11-1 gives abundances derived by these simple extrapolations for several of the points depicted in Fig. 11-1, together with mean abundances found by Aller, by Hawley and Miller, and by Barker. From the small size of the chemical composition variations with position (less than inherent errors in observations and "acceptable" accuracy of the extrapolation formulae), Hawley and Miller and also Barker concluded that use of these simple expressions (Eqs. 3, 4, 5) was justified.

As an example of a procedure involving theoretical nebular models, let us consider a bright, high-excitation object, NGC 6818. The interstellar extinction is estimated by methods described in Chapter 8 to be $C = F_{obs}(H\beta)/F_{corr}(H\beta) = 0.15$. This value is used to correct all the observed line intensities with the aid of Eq. (8-1). To collate the ultraviolet and optical regions in this instance where total monochromatic fluxes are not available, so we use the theoretical ratio I(1640)/I(4686) = 6.81 from Seaton's calculation (interpolated for T_ε = 12,500, N_ε = 2500) and the observed ratio $I(4686)/i(H\beta)$ = 4.79. The next step is to use the observed intensities to obtain the nebular diagnostics (cf. Chapter 5). The diagnostic diagram suggests a range of densities and temperatures: [O II] 3729/3726 gives a higher density (N_ε = 2500 cm^{-3}) than does [S II] (N_ε = 1000 cm^{-3}), while T_ε [O III] = 12,800°K and T_ε [N II] = 11,300°K. In the hotter interior region of the nebula, T_ε [Ne IV] = 15,000°K. For most ions of intermediate excitation, e.g., Ne++, Si++, Cℓ++ and Ar++ we use T_ε = 12,800°K; for other ions we are guided by predictions of theoretical models. Table 11-2 lists details of the abundance calculation. Column 1 gives the wavelength in A, Column 2 gives the intensity on the scale $I(H\beta)$ = 100 as corrected for interstellar extinction. The ultraviolet line intensities were measured with the International Ultraviolet Explorer (IUE), by W.A. Feibelman. The optical region line intensities were measured mostly with an image tube

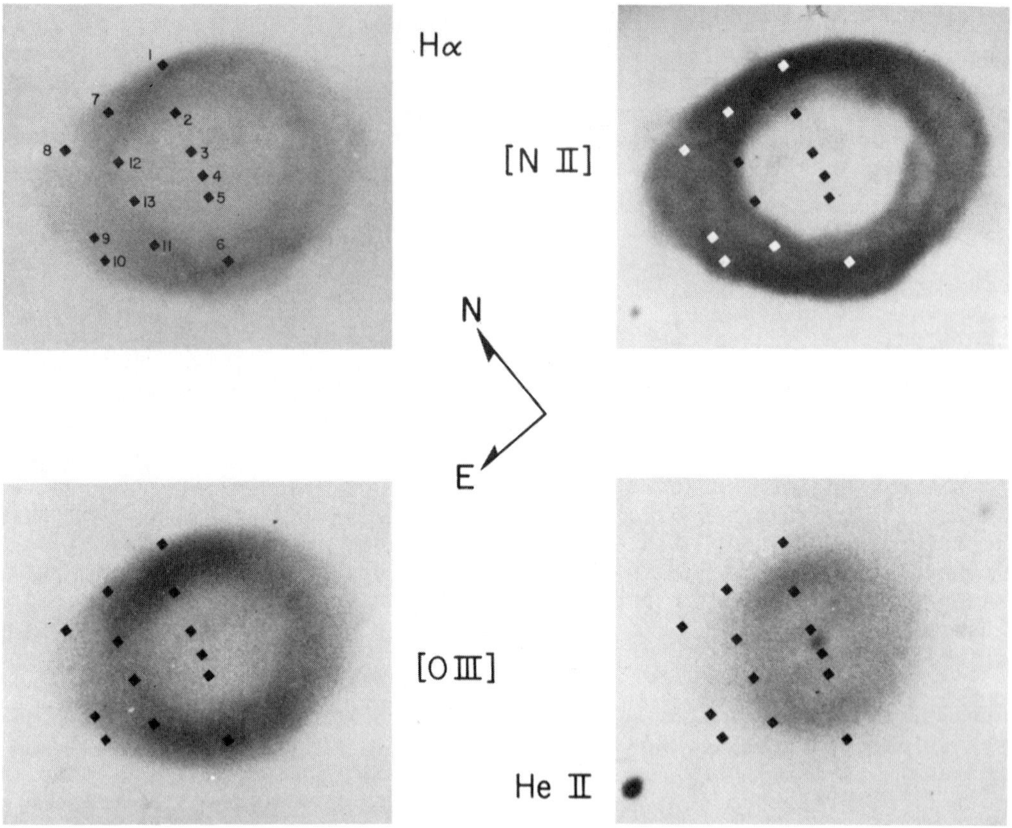

Figure 11.1

scanner at the Lick 3-meter Shane telescope. Column (3) gives the identifications.

The next step is to use the intensities and the chosen values of N_ε and T_ε to calculate $N(X_i)$ for various ions all in terms of $N(H^+)$. $N(He^+)$ and $N(He^{++})$ are calculated from the recombination lines of He I and He II, respectively, viz.:

	λ5876	λ4471	λ6678		λ4686	λ5411	λ4541
I	5.8	2.09	1.69		70.3	5.6	2.57
n(He+)/n(H+)	0.045	0.044	0.0455	N(He++)/n(H+) =	0.061	0.062	0.065

whence n(He)/n(H) = 0.107. Notice that the concentration of C III can be obtained both from λ4267 of C II interpreted as a recombination line, and the C III doublet 1907/1909 which is collisionally excited.

In Column 4 we tabulate the adopted values of $t = T_\varepsilon/10{,}000$, selected as described above, while Column 5 gives values of log $N(X_i)$ calculated by procedures described in Chapter 5. For S+ and N+, we assumed $N_\varepsilon = 10^3$ cm^{-3}, = 2500 cm^{-3}, otherwise. Column 6 gives logarithms of the sum of the $N(X_i)$ values for each element. The ionization correction factors, ICF, given in Column 7 come from theoretical models. The logarithm of the product $ICF(\Sigma N(X_i)$ = N(element), on the scale $N(H) = 10^{12}$ is given in Column 8.

We may also obtain N(element) directly from the theoretical model. One adjusts elemental abundances until a best fit with observed intensities is obtained. Some difficulty is encountered with $3p^n$ ions of S, Cℓ, and Ar. If we fit intensities of lines of ions of [S II], [Cℓ III] and [Ar III], those of [S III], [Cℓ IV], [Ar IV] and [Ar V] often do not agree. Some of this disparity may arise from errors in A-values, Ω-values, or charge-exchange cross sections, some from an inaccurate prediction of the distribution of atoms among different stages of ionization. The last column, Δ log N gives an estimate of the precision or internal consistency of the method. Note that in the extrapolation method we use the theoretical model only as a device for estimating the ICF factor. Slight adjustments in the theoretical model can produce marked variations in the predicted intensities, but they have only a small effect on the ionization correction factors.

Probably NGC 7662 has been the most favored object of model builders; Harrington (1969, 1970), Kirkpatrick (1972), Pequignot (1980), Seaton, Adams, Harrington, and Lutz (1982). The most elaborate, detailed, satisfactory model is that of Harrington et al. NGC 7662 consists of two rings, an inner, denser brighter incomplete one which has the appearance of a "C" and average radius of about 8" and an outer fainter one with a radius of about 15" containing many condensations. These structures are illustrated, for example, in O.C. Wilson's slitless spectrum and isophotes derived therefrom (Aller 1956). For purposes of the theoretical model, the nebula may be treated as a symmetrical spherical structure. This object has been studied in the normal optical range in the ultraviolet, in the infrared, and in the radio frequency range. The model by Harrington et al. takes into account irregularities that can complicate treatment of the radiation field. They also found it necessary to consider the influence of dust, whose presence is indicated by an infrared excess. If $n(C^{3+})/n(H^+)$ is obtained from the C III λ2297 dielectronic recombination line,* the intensity of C IV λ1549 is predicted to be about three times the observed value. This C IV emission undergoes repeated resonance scattering in the course of which it becomes attenuated by dust absorption. Harrington et al. estimate that an optical depth, $\tau_D \sim 0.1$, with solid carbon grains of radii slightly larger than 0.1 μm will suffice to reconcile predicted and observed intensities. They find the oxygen abundance to be about half the solar value, while C/O and N/O are respectively 3 and 1.4 times solar values. The neon and sulfur values appear to be essentially solar, but gaseous magnesium and silicon are respectively 0.02 and 0.14 solar values. They are able to reproduce the intensities of most of the nebular lines; an exception being [Ne V] where it does not appear that observational errors can explain the discordance.

Such detailed modeling is worthwhile only for nebulae for which good observational data are available. When it can be applied it probably yields the best information we can get on nebular chemical compositions. Alas, for most PN detailed models are not yet available. The ionization correction factors (Eqs. 3-7) can be applied but may give errors of the order of a factor of 2 or even more for elements such as S, Cℓ, and Ar. Some nebulae can be analyzed by methods described for NGC 6818: a) construct a model and determine abundances needed to reproduce the line intensities; and b) calculate the ionic abundances in the usual way with the aid of the nebular plasma diagnostics, use the model to improve the T_ε estimate and obtain the ionization correction factors, and calculate the abundances. Comparison of the two sets of results gives some kind of check on the method (see the last column of Table 11-2). For many elements the discordances $\Delta \log A \lesssim 0.1$, but this does not mean that the abundance is found with corresponding accuracy, since errors in atomic data, particularly Ω's and A's, can still enter with devastating effect. For most nebulae analyzed by this method, we estimate the following levels of accuracy: He, $\Delta \lesssim 0.02$, C, N, $\Delta \lesssim 0.2$; O, Ne, $\Delta \lesssim 0.1$; S, Cℓ $\Delta \lesssim 0.2$, Ar $\Delta \lesssim 0.1$. For rarer elements line intensity errors play a dominating role. The larger uncertainties for C and N arise from the fact we depend most on UV lines where uncertainties in T_ε and dust absorption can be important. Furthermore, the observed UV permitted lines often contain substantial contributions from the central star, particularly in objects of small angular size such as IC 2165 and NGC 6741. For S, Cℓ, and Ar, errors in atomic parameters for important ions can be significant.

11E. Compositions of Planetary Nebulae

Table 11-3 compiles some abundance data for PN. As previously remarked, the chemical composition of a given nebula depends on the original mix in the progenitor star and also on modifications subsequently produced by nucleogenesis (see Chapter 12). Here we describe some proposed groupings and try to and assess the accuracy attainable for various elements.

*Note that abundances derived from dielectronic recombination lines depend weakly on T_ε, while those found from collisionally-excited lines are sensitive to T_ε.

The third column of the table gives log N(element)/N(H) + 12 or the ratio N(element/N(oxygen). Since these ratios may vary with distance from the galactic center, the tabulated "STD" values are intended to apply to the solar neighborhood (r = 10 kpc). The values in this column apply to samples taken over a large number of objects; they may be regarded essentially as "control values." The fourth, sixth, eighth, tenth, twelfth and fourteenth columns give the references as listed at the bottom of the table. The fifth and seventh columns give data for nebulae belonging to Population Type I and II as listed by Kaler (1978). The ninth column refers to objects that show λ4686 He II in their spectra. The N-rich objects are here defined as those with log N(N)/N(H) > 8.0 on the scale log N(H) = 12. The carbon-rich, high-excitation objects are those for which n(C)/n(O) > 1. The last column gives the solar abundances. The grouping differs from that proposed by Peimbert in 1977 at IAU Symposium No. 76.

At the onset it is important to recognize that PN compositions may vary with the mass of the progenitor star (and thus the population type), and location in the galaxy. We must distinguish between the central bulge, the disk and the halo. Noting that the input data in Table 11-3 refer mostly to relatively bright, nearby PN of both population types, we can comment: 1) The He/H ratio seems well established; a comparison with the sun is not easy because the solar helium abundance is poorly known; 2) For heavier elements there seem to be systematic differences between Population Type I and II PN in the sense that the "metal"/H ratio is smaller in the latter but they scale by about the same factors, so N(element)/N(O) remains approximately constant; 3) The carbon and nitrogen abundances show marked variations from one type of object to another. Results from different investigators show a large spread in both C and N determinations, arising partly from diverse selections of objects and different analytical techniques employed. Carbon and nitrogen can be involved in the element-building process in progenitor stars; 4) On the other hand, Ne, S, Cℓ and Ar whose nuclides must originate in massive stars appear to have similar abundances in Population I, high-excitation, carbon-rich and nitrogen-rich PN; 5) A comparison with solar abundances shows a number of close similarities and well-defined differences. Carbon tends to be enhanced (but compare Kaler 1981), while nitrogen shows a considerable spread in abundance. Oxygen definitely appears to be less abundant in PN than in the sun (Peimbert and Serrano's 1980 value of logN(O) = 8.83 corresponds to an adopted temperature fluctuation $\langle \Delta\tau^2 \rangle = 0.035$). In Population Type I planetaries, the abundances of Ne, S, Cℓ and Ar, and perhaps also F, Na, and K appears to be essentially solar; 6) Some elements such as Mg, Si, Ca and Fe appear to be depleted by grain formation (Shields 1978; Aller 1978; Harrington <u>et al</u>. 1982).

11F. Planetary Nebular Compositions and Galactic Structure

Both planetary and diffuse nebulae may be used as probes for galactic abundance gradients. Because of their high surface brightnesses and the fact they are not closely confined to the galactic plane as are H II regions (so they can be seen more frequently through windows in the interstellar extinction) planetaries can be used as probes of chemical composition from the galactic bulge to distant spiral arms and through the galactic halo.

A number of investigators, including Peimbert and Kaler and their respective associates, has tried to sort out planetaries according to the population groups of their progenitors, spiral arm population (Baade's Type I), old disk population, and halo population although finer subgroups had been attempted. In the spiral arm population, we try to distinguish between those where pronounced enhancements have occurred and those where this has not happened. Thus, Kaler (1980) noted that planetaries at large distances, z, above the galactic plane and with large radial velocities had lower abundances of N, O, Ne, etc., than do planetaries in the galactic disk.

The most distinctive abundance anomalies are those found in halo planetaries. Table 11-4 gives data for several planetaries as taken from work of several authors. Notice that in this compilation the He/H ratio is about the same as that found in an average planetary nebulae, i.e., essentially "cosmic." This result suggests that the He/H ratio may have been already established at the time the halo was formed. On the other hand, Kaler concludes from his analyses that the He/H ratio of extreme planetaries tends towards 0.085, the value found by Faulkner and Aller (1965) and by the Peimberts (1974) for the Small Magellanic Cloud and 30 Doradus in the large cloud. The carbon abundance deduced from $\lambda 4267$ interpreted as a recombination line has a high value in Hu 4-1, suggesting that carbon may be manufactured in these objects as well as in normal planetaries. The abundance of nitrogen, although uncertain, is generally smaller than in the sun. Oxygen shows a fluctuation up to a factor of 8, approaching within a factor of 2 of the solar value in Ha 4-1. The neon abundance is less than solar or "cosmic" by a factor of 12 to 40. Also, sulfur and argon are less than solar or "normal" planetary nebular values. In particular, Barker notes that argon is two orders of magnitude less abundant than in general field nebulae, and suggests that these objects may have been formed from a galactic halo depleted in argon. Examples of such greatly reduced ratios are rare.

Chemical compositions of planetaries in the central bulge of the galaxy have not yet been studied adequately. Webster (1976) suggested they had "normal" compositions, but Price (1981) has proposed a subpopulation of bulge nebulae with abnormal abundances. For one of these objects, H 1-55, he finds log N values as follows: He = 11.3; N = 8.87 ± 0.2; O = 9.50 ± 0.4; S = 7.4 and Ar = 7.4. Possibly it originated from an oxygen-rich star in the bulge. Since no reliable method of distance determination exists for planetaries, it is difficult to know whether any particular object is actually located in the bulge or is simply a foreground object.

Abundance gradients in the disks of galaxies are well known from studies of H II regions (see Chapter 11-G). Attempts have been made to employ planetary nebulae for the same purpose, e.g., D'Odorico <u>et al</u>. (1976), Aller (1976), Barker (1978), Peimbert (1978), Kaler (1978), and Peimbert and Serrano (1980). Peimbert (IAU Symposium No.

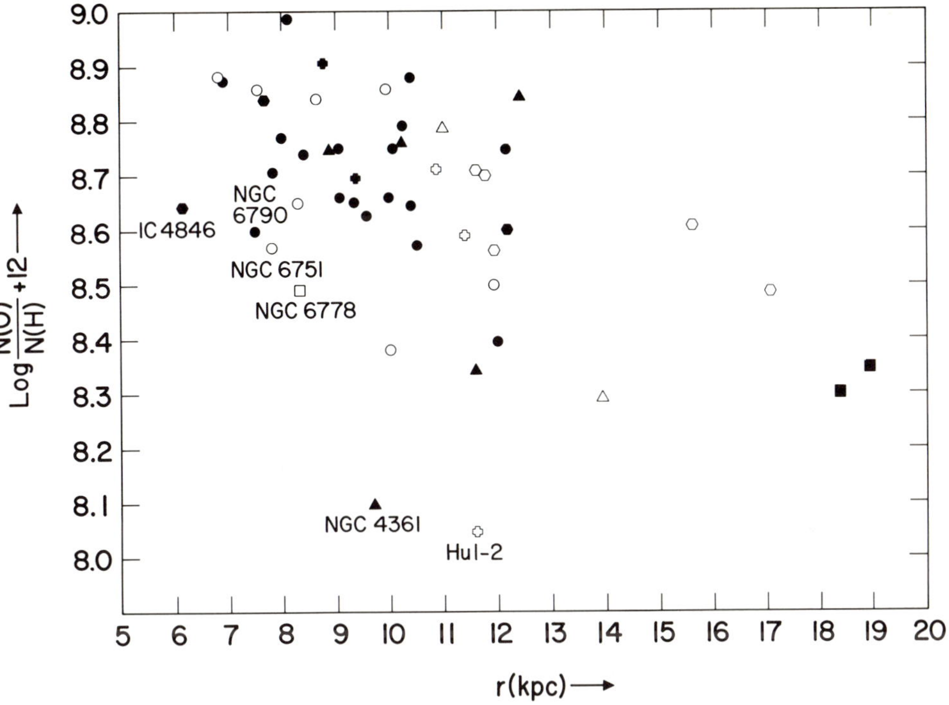

Figure 11.2.a

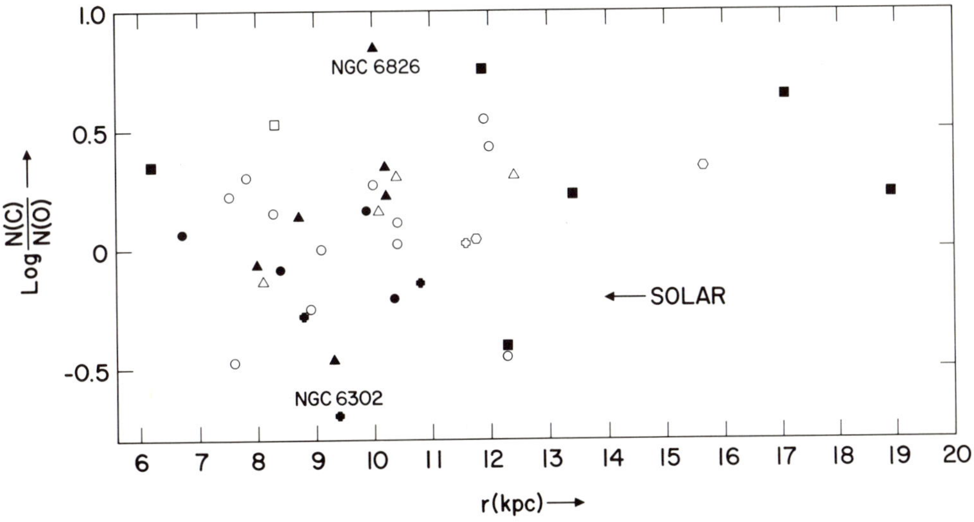

Figure 11.2.b

76, 1977) and Peimbert and Serrano employ the following composition types: Type IP, N(He)/N(H) > 0.14 [or N(N) > N(O)]; Type IP - IIP, 0.140 > N(He)/N(H) > 0.125; IIP and IIIP, N(He)/N(H) < 0.125 logN/O < -0.2. Their Type IIP planetaries are intermediate populations with $|z| < 1$ kpc and peculiar radial velocities $V < 60$ km/sec. Their Type IIIP constitutes Population II objects with $|z| > 1$ kpc and V(peculiar) > 60 km/sec. Type IVP comprises halo objects. For Types IIP and IIIP, they find a tight correlation of N/O and He/H, but for Type IP they find no such relationship. Low N/O and He/H ratios are found in the direction of the anticenter, while higher values are found towards the galactic center. We use the notation IP, IIP, etc., to avoid confusion between the Peimbert composition types and usual population types.

Note that the values of the gradients $\Delta_{gal} = d/dr_{gal} \log X$ depend on whether or not we assume temperature fluctuations $(\Delta\tau^2)$. Peimbert and Serrano found an O/H gradient in good agreement with that obtained by Talent and Dufour (1978) for H II regions (see Table 11-6). Carbon and nitrogen, however, are manufactured in planetary nebulae themselves so that they cannot be used to deduce meaningful galactic composition gradients unless one has a means of clearly separating those planetaries whose progenitor stars have manufactured C and N in their interiors from those that did not.

Kaler (1979) has pointed out that one must be cautious in interpreting galactic abundance gradients derived from planetaries as distinct from those obtained from H II regions. The former refer to ejecta of individual stars while the latter refers to the general interstellar medium, which is made up of contributions from many defunct stars and supernova together with the original material. Thus, if massive stars tend to occur near the galactic center and then thin out, possible planetaries originating from them would show effects of nuclear processing and thereby yield a gradient that had a special meaning in that it reflected only this effect. Likewise, if the ratio of halo-to-disk stars increases outward from the center of the galaxy, the resultant planetary nebular composition would reflect this variation.

Different investigators draw somewhat different conclusions for the carbon abundance. The Peimberts (1977) found a large enhancement from λ4267; French (1983) who used also other C ions likewise conclude C is enhanced. Kaler (1981) noted that C/O does not correlate with He/H as theories of nucleogenesis by Iben and Becker would suggest. Absolute scaling of his C/O ratio is ultimately dependent on UV lines. Kaler (1981) used 4267 (interpreted as a recombination line) to get $n(C^{++})/n(H^{+})$ for 53 planetaries and then employed an empirical argument to get a final C/H ratio. He concludes that carbon is not enhanced in most planetaries. The carbon abundances presented in Table 11-3 are mostly from ultraviolet lines. Recall, however, the observed UV lines of C and N may be affected sometimes by a contribution from the central star, especially for objects of small angular size.

Nitrogen offers a number of engaging problems. Variations in the N/H ratio were first clearly suggested by the work of Wyse (1942), further substantiated by Minkowski's and Aller's observations in the late 1940s (Aller 1957) and confirmed by extensive studies in the 1970s, e.g., the Peimberts (1971), Boeshaar (1975), Kaler (1974, 1978), Barker (1978). Peimbert has estimated that 10 to 30% of planetaries are in the nitrogen-enriched class.

Perhaps the most extreme example of nitrogen enhancement is shown by a very bright planetary in the irregular galaxy NGC 6822. Dufour and Talent (1980) find the following logarithmic abundances vis log N(He) = 11.27 ± 0.08; log N(N) = 8.8 ± 0.3, log N(O) = 8.1 ± 0.1, log N(Ne) = 6.9 ± 0.2 which are to be compared with the corresponding values for NGC 6302 (Aller et al. 1981), He = 11.26; N = 8.92, O = 8.7; Ne = 8.0. The NGC 6822 planetary has the highest N/O ratio, although the absolute abundance of nitrogen may be slightly greater in NGC 6302. The planetary nebula N67 in the Small Magellanic Cloud (studied by Dufour and Killen 1977) is also nitrogen-rich.

Oxygen and heavier elements do not seem to be made in the progenitor stars of planetary nebulae, even of those showing great enhancements of N, C, etc. Thus, the O/H ratio is determined by the composition of the interstellar material from which these antecedent stars were formed, and should offer a good clue to the population to which the planetary nebula belongs. Over a wide range of excitation, $n(O)/n(H) = [n(O^+) + n(O^{++})]/n(H^+)$. Kaler's (1980) survey covered many more objects than had been included in most studies, so he was able to subdivide his material into finer population subclasses: young disk -- log N(O) = 8.72; old disk -- log N(O) = 8.55; intermediate disk: log N(O) = 8.65; old halo -- N(O) = 8.05.

Kaler (1978) also studied the Ne/O, Ar/O, and Cℓ/O ratios throughout the galaxy and found them to be essentially solar (cf. Table 11-3). Furthermore, he found no S/O dependence on r_{gal}, the distance from the galactic center, or on the height, z, from the galactic plane. As distinct from procedures involving theoretical models, the method is a highly empirical one that compares ratios of ionic numbers in different ionization stages as a function of a parameter related to the temperature of the central star.

The data shown in Figs. 11-3 through -5 are based on a homogeneous set of observations, involving both optical and ultraviolet observations where these are available. The abundances are derived primarily by a procedure in which theoretical models are used to obtain ionization correction factors that are then employed to correct observed ionic concentrations, $N(X_i)$, to get elemental abundances. The method was illustrated in our example for NGC 6818. The nebulae are sorted into three groups, Population type I (as classified by Kaler), Population type II (similarly classified), and nitrogen-rich objects. In each group the nebulae are further designated according to the quality of the analysis, as indicated by the code. Certain nebulae that fall far from the "mainstream" are explicitly indicated.

Consider first the gradient in log N(O)/N(H). For the entire group of nebulae, including both those of stellar populations I and II, we find $\Delta_{gal} \sim -0.06$ to -0.07, but if we consider only Kaler's Population I objects for which good data are available (quality groups A and B), the gradient is somewhat smaller.

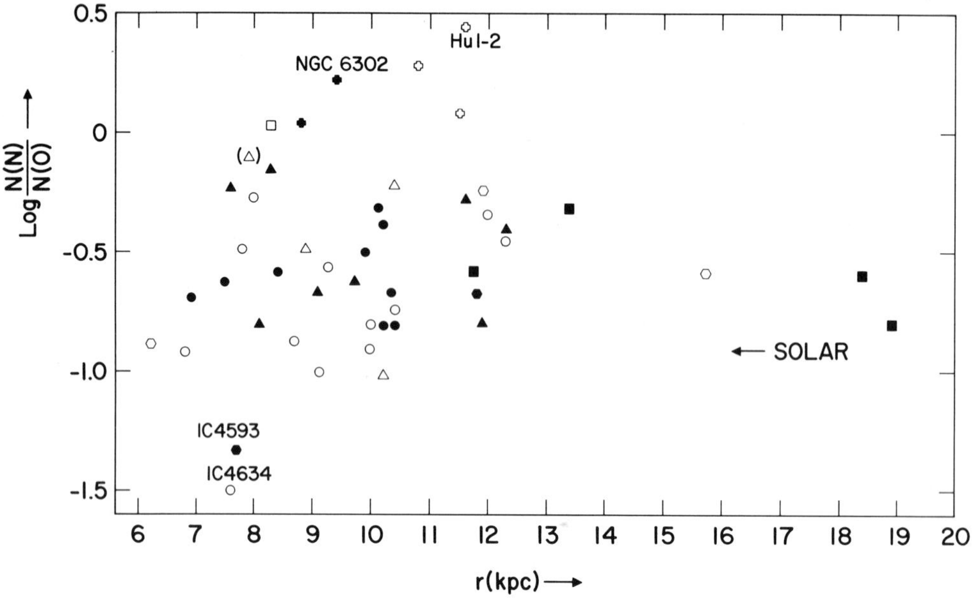

Figure 11.3.a

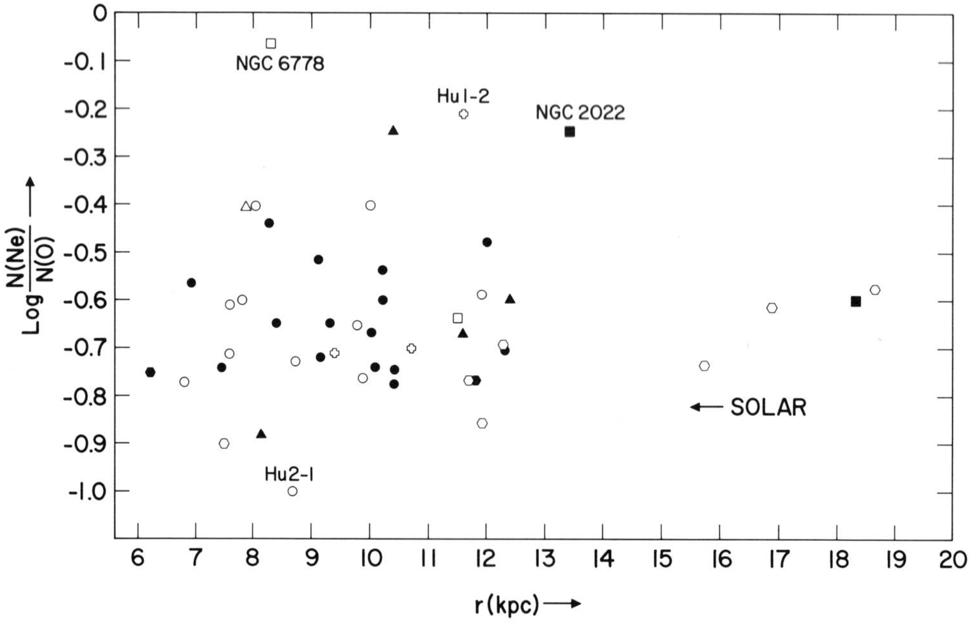

Figure 11.3.b

Figures 11-4 and -5 show log N(element)/N(O) as a function of distance from the galactic center. The solar value of the ratio is also indicated but we must emphasize that because oxygen is deficient in the planetaries a nebular Ne/H ratio, equal to the solar value for example, corresponds to a nebular Ne/O ratio that exceeds the adopted solar one. But beware! Solar neon abundances are not as well-established as we would like. Individual plots show a considerable scatter but there is no evidence that the ratios of Ne, S, Cℓ, and Ar with respect to oxygen change with distance from the sun. Much of the spread must be real. It may represent differences in chemical compositions of progenitor stars and differences in their evolutionary histories. We defer an interpretation of these abundance differences within the framework of stellar evolutionary processes to the following chapter.

11G. Abundance Determinations in H II Regions and Composition Gradients in Galaxies

Chemical composition gradients in galaxies offer challenging theoretical and observational problems both in our own and in other galaxies. The impetus to such investigations is to obtain an insight into element building rates in suceeding generations of stars since the formation of the galaxy, to test possible and alternate scenarios of nucleogenesis, and track the chemical evolution of the galaxy itself. Can we obtain any information on star formation rates and initial mass functions? For example, was there a tendency for very massive stars to be formed in greater numbers in the very early years of our galaxy's history? Our own galaxy offers special advantages in that we can study a variety of objects -- planetary nebulae (§ 11F), stars of various ages, origins and luminosities, molecular clouds (which permit us to assess isotope ratios) and H II regions. Of these various types of objects, only H II regions and supernova remnants (SNRs) are readily accessible in distant galaxies. Note that abundance gradients measured in H II regions refer to the current composition of the interstellar medium, whereas planetary nebular data, for example, involve material from progenitors with a great range in ages, with frequent additions of freshly synthesized carbon and nitrogen.

Let us consider first some data from H II regions in other galaxies, where the dependence of excitation on distance from the nucleus was first found. When such systems lie nearly in the plane of the sky, they offer a number of distinct advantages. Intrinsic interstellar extinction is usually small, while that due to foreground dust in our own galaxy is usually fairly constant over the solid angle subtended by the system. In particular, we shall consider analyses of H II regions that are carried out with the aid of theoretical nebular models. Stasinska et al. (1981, 1982) have carefully examined the uncertainties in some of these calculations. In addition to the aforementioned difficulties arirising from geometrical irregularities, inhomogeneties, irregular dust distribution and time-dependent effects, all of which can occur in planetaries, there are other considerations that merit attention.

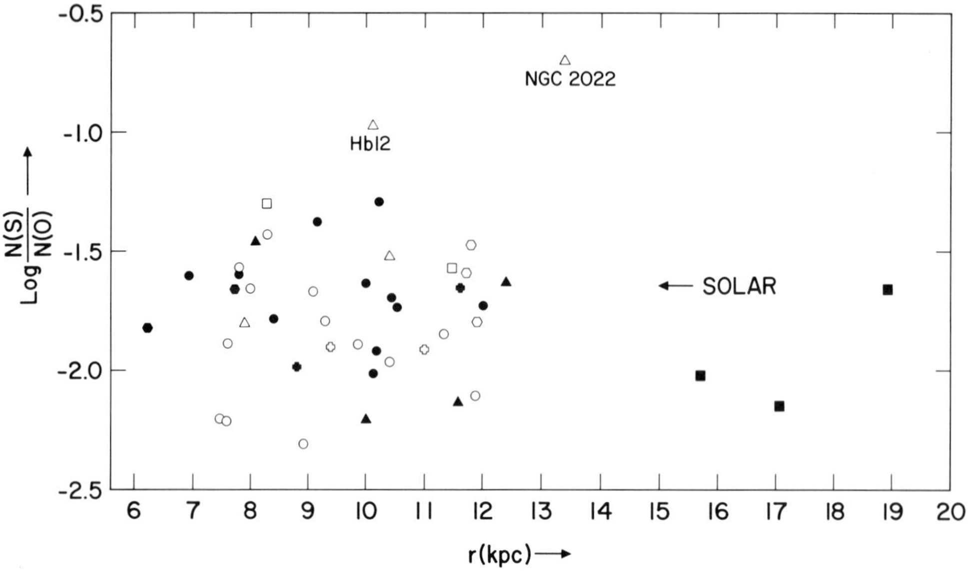

Figure 11.4.a

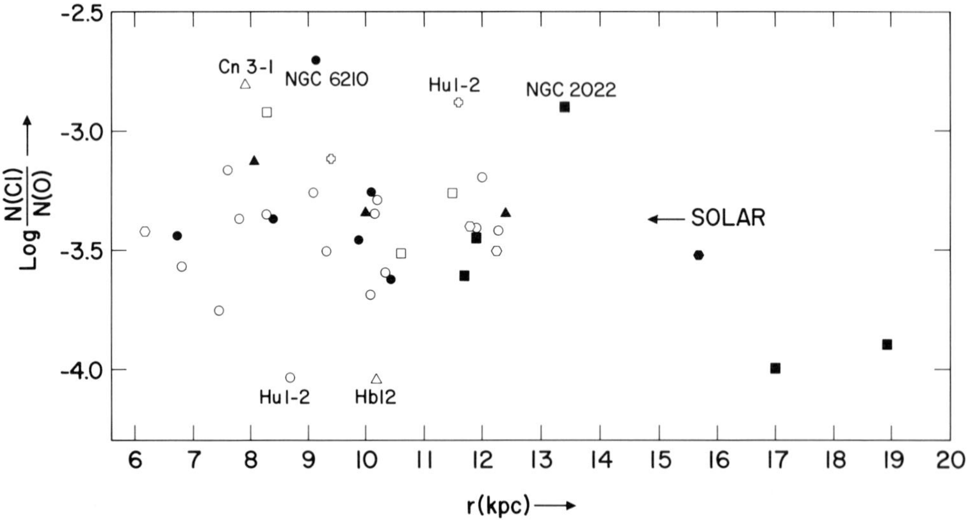

Figure 11.4.b

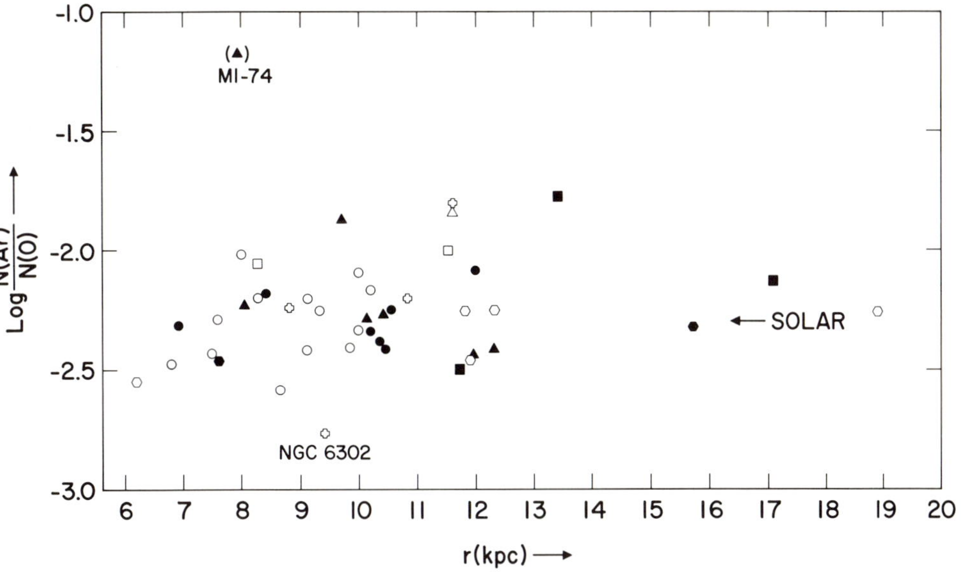

Figure 11.4.c

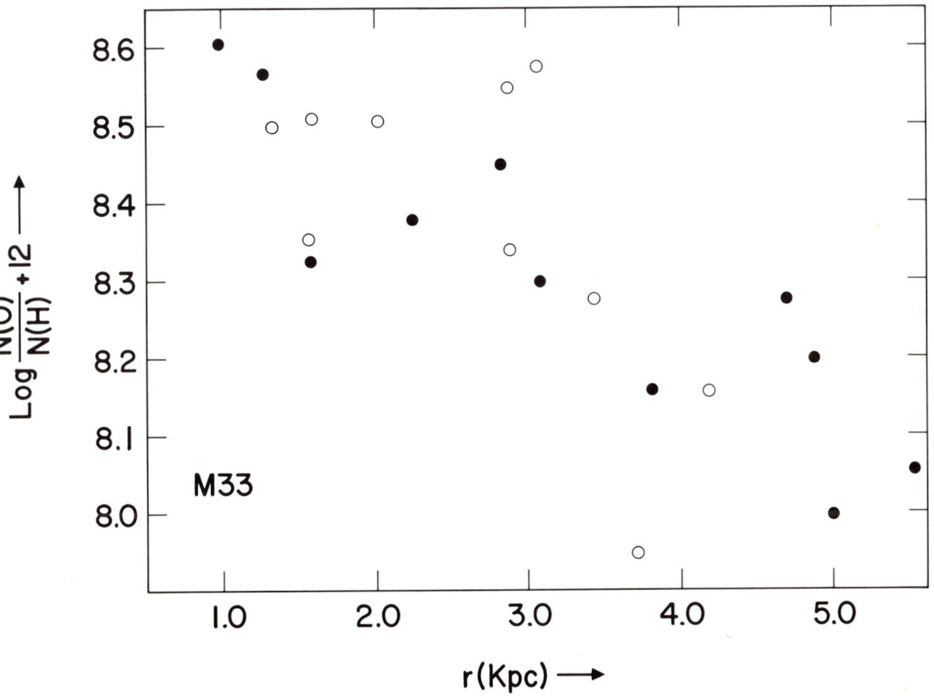

Figure 11.5

all of which can occur in planetaries, there are other considerations that merit attention.

Excitation appears to be produced by very luminous O stars. Stellar atmospheric fluxes have been calculated for objects with a variety of effective temperatures, surface gravities, and limited ranges of chemical compositions. Both LTE and non-LTE atmospheres have been considered. Shields and Searle (1978) in their discussion of M101 proposed an internal consistency argument. The stellar ionizing flux distribution should correspond to that of a star having the same chemical composition as the surrounding H II region, from whose material it had been formed. Further, unlike in planetary nebulae, the ionizing flux in a giant H II region may arise from a cluster of stars. In H II regions in our own galaxy (e.g., Orion), the ionization is often dominated by a single star, but in H II regions with diameters of hundreds of parsecs (e.g., M101), several sources may be involved.

One of the most serious difficulties encountered is that an important diagnostic tool, [O III] λ4363, is often not observable in H II regions. If T_ε is low, this line becomes very weak and is lost completely in the noise. This situation is often encountered in inner regions of galaxies and is correlated with the O/H ratio in the sense that the greater the abundance of oxygen, the lower the temperature. This result is easily understood because oxygen is a powerful coolant (see Chapter 6). Various workers have proposed procedures for estimating T_ε where λ4363 is not observable. The best criterion is the [O III] + [O II]/Hβ ratio, proposed by Pagel _et al._ 1979, 1980, and calibrated for our own galaxy by Shaver _et al._ (1983). The relationship between this line ratio and T_ε, viz.:

$$T_\varepsilon = 3570 \pm 130 + (7940 \pm 880) \log\left\{\frac{[\mathrm{O\ II}] + [\mathrm{O\ III}]}{\mathrm{H}\beta}\right\} \qquad (8)$$

is firmly established from a combination of optical data for line intensities and T_ε obtained from radio frequency data. It reflects a connection between T_ε and the O abundances:

$$\log\frac{n(\mathrm{O})}{n(\mathrm{H})} + 12 = (9.82 \pm 0.02) - (1.49 \pm 0.11)\left(\frac{T_\varepsilon}{10{,}000}\right) \qquad (9)$$

There would be some scatter about the mean relationship arising from the influence of geometrical irregularities and dust.

At least for M33 an independent check on the composition gradient is provided by analyses of supernova remnants (SNR) which consist almost entirely of swept-up interstellar medium material (Dopita _et al._ 1980). Shock wave theory (of Chapter 9) may be employed to interpret the spectroscopic data. In particular, abundances of nitrogen and oxygen may be obtained. Figure 11-6 shows the O/H ratio as a function of the distance from the center of this galaxy. Note that the SNR and H II region data are in good accord. A well-defined gradient exists both in O/H and in N/H (compare Pagel and Edmunds 1981), as well as in other elements. The concentration of these elements is greater near the nucleus.

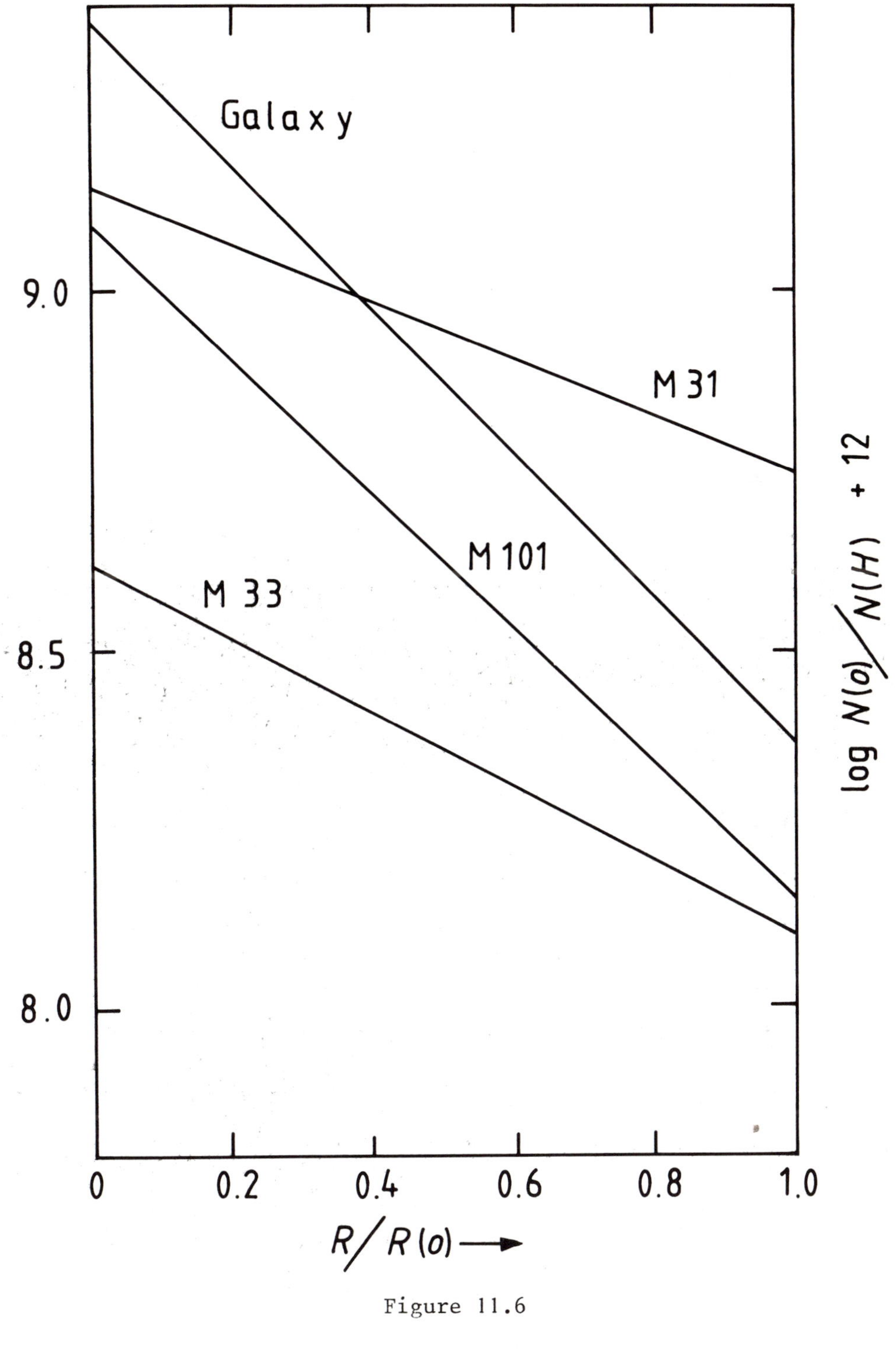

Figure 11.6

As a second example, consider the giant spiral M101, which has been studied spectroscopically by many workers (Searle 1971; Smith 1975; Shields and Searle 1978; Hawley 1978; Pagel and Edmunds 1980; Rayo et al. 1982). This is a galaxy on the grand scale. If we adopt a distance of 7.4 megaparsecs, we find that the largest H II regions have dimensions of the order of 50 to 100 parsecs.

An interpretation of H II regions by a method based on theoretical models illustrates some of the difficulties involved. The models are used to reproduce as closely as possible the observed line intensities and are then used to get ionization correction factors. If we assume that the ionization is produced by a single star and apply the Shields and Searle procedure one tends to find very low densities and an unlikely large Ne abundance. Alternatively, we can suppose that the excitation as produced by two stellar groups of effective temperatures T_1 and T_2; hence, $F_\nu(\text{total}) = xF_\nu(T_1) + (1-x)F_\nu(T_2)$ where x is an adjustable parameter. A number of problems emerge: 1) uniqueness -- two models that give an equally good representation of line intensities may give ionization correction factors (ICFs) differing by a factor of 1.25 for nitrogen but they may be in good accord for argon; 2) the direct model calculations for homogeneous nebulae tend to underestimate n(O); 3) to achieve energy balance, additional coolants are required. The models suggest that the He^+ zones are smaller than the H^+ zones so that the "observed" He/H ratio is probably a lower limit. Figure 11-7 shows the N/H gradient.

Table 11-5 compares some elemental abundance ratios in M33, M101, the Magellanic Clouds, an "average" planetary, the Orion nebula, the sun and a typical H II region at 10 kpc from the galactic center. Although O/H shows a pronounced gradient in M33, the ratios of N, Ne, S, and Ar to O are remarkably insensitive to distance from the core, a situation that appears to hold to some exctent in other galaxies. A number of investigations (Dufour 1975; the Peimberts 1974, 1976; Aller et al. 1974, 1977; Pagel et al. 1978) indicate that in the Magellanic Clouds there does not seem to be any dependence of chemical composition on position. A similar result seems to hold in barred spirals and in stellar systems that have complex internal motions. (Pagel and Edmunds 1981 and references cited therein).

Note the striking similarity in the N/O, Ne/O, S/O, and Ar/O in the typical H II region at r = 10 kpc and in the sun. In fact, the Ne/O ratio is remarkably similar in the objects listed in Table 11-5. Likewise, the ratios for S/O, Cℓ/O and Ar/O are nearly similar but there is a larger error arising from uncertainties in the data and possibly in atomic parameters. Evidently, element building processes proceeded in similar fashions in parent material of all these objects insofar as O and heavier elements are concerned (see Chapter 12). The N/O ratio shows considerable variation. In M101 and M33, it is perhaps a factor of 1.5 to 3, respectively, lower than in the sun. This ratio is substantially lower in the Magellanic Clouds where carbon also is notably less abundant.

Stellar compositions are more difficult to obtain in extragalactic systems except for the brightest stars in the Magellanic Clouds. Unfortunately, the atmospheres of these objects teeter on the brink of instability so that theoretical model atmosphere methods are probably of limited usefulness. Furthermore, highly evolved supergiants may display products of nucleosynthesis from their own interiors. We can avoid some of these difficulties by going to somewhat fainter stars as is now technically possible. It would appear that metal deficiencies consistent with those expected from H II region analyses do occur in these supergiants. In other galaxies, one must deal with the integrated colors and line intensities of stellar populations or even mixtures of stellar populations. The data for spiral galaxies do suggest that the metal-to-hydrogen ratio decreases with increasing distance from the nucleus, in qualitative agreement with results from H II regions. See also Hunter (1982).

11H. Abundance Gradients in the Disk of Our Galaxy

Turning to our own galaxy we find both advantages and disadvantages when we seek to establish abundance gradients. One can study stars, SNRs, H II regions and molecular clouds.

Stellar abundance determinations entail the following difficulties: 1) At great distances, it is possible to study only very luminous objects, i.e., bright main-sequence O and B stars or supergiants. O and B stars show lines of H, He, C, N, O, Ne (all interesting elements) but they are often formed under conditions of non-local thermodynamic equilibrium (non-LTE). Abundances are often difficult to determine even when fairly detailed information on stellar energy distributions, line shapes and intensities is at hand; 2) Supergiant atmospheres are sometimes difficult to model and interpret. As remarked above, the atmospheric compositions may have been modified by nuclear processes within the stars. Sometimes open star clusters could be useful in providing a check since effects would differ for main sequence and supergiant stars; 3) Interstellar extinction puts a severe constraint on how far we can see in the galaxy. Of course, it limits not only stellar but also H II region investigations.

From an analysis of kinematical and photometric properties of about 600 dF and 600 gb to gK stars, M. Mayor (1976) found a mean radial chemical gradient of $d(\mathrm{metal/H})/dr = -0.05 \pm 0.01\ \mathrm{kpc}^{-1}$ with some indication that the metallicity gradient could be larger for the youngest stars, viz: $-0.10 \pm 0.02\ \mathrm{kpc}^{-1}$. She concluded that an important fraction of heavy elements actually present in the solar neighborhood seems to have been synthesized during the life of the galactic disk. K.A. Janes (1979) from a study of CN in gG and gK stars likewise concludes that the metallicity gradient in the solar neighborhood is about $-0.05\ \mathrm{kpc}^{-1}$ and he finds no indication it is different for younger objects.

The most comprehensive study of abundance gradients as derived from H II regions is that by Shaver *et al.* (1983). They employed a judicious combination of radio and optical data. A critical limitation on the analysis of H II regions has been the uncertainty in T_ε. In our galaxy one can employ both radio frequency and optical data; the former give us T_ε, so that from the optical line measurements we can then obtain the ionic concentrations.

They were able to measure both the thermal background continuum and also radio frequency lines such as H 137α, H 109α, and He 109α. By fitting the line profiles with Gaussian curves they could obtain $\Delta\nu_L$ = line width at half maximum and find an estimate of T^c, i.e., the LTE value of T_ε from $\Delta\nu_L T_L/T^c$. Here T^c is the brightness temperature of the continuum (with the background subtracted), T_L is the line temperature from the Gaussian fitting program (see Eq. [4-78]). The next problem is to calculate the true, nonLTE electron temperature T_ε from T_ε^c. Shaver had found that there existed a relationship between the frequency ν and the emission measure, EM, such that by using radio frequency data at a frequency $\nu \sim 0.08\,(EM)^{0.36}$, $\Delta T_\varepsilon = T_\varepsilon^c - T_\varepsilon$ would be a minimum. The remaining small correction could easily be calculated by Eq. (4-79). The results clearly show that galactic H II regions can have gas kinetic temperatures as low as 4300°K. An independent check is provided for some of these cool, low-density regions where the turbulent velocity is very small. The Doppler temperature found from the line profiles approaches the electron temperature found by other means. Gas kinetic temperatures can be established to an accuracy of a few percent.

From this analysis emerged a clear-cut relationship between T_ε(H II) and the distance from the galactic center, viz.:

$$T_\varepsilon = (3150 \pm 110) + (433 \pm 40) r_{gal}$$

where r_{gal} is the distance in kpc from the galactic center. The rise in T_ε as r_{gal} increases is a consequence of an abundance gradient in the galaxy. The beauty of this investigation was that it gave: "a self-consistent picture in which the derived abundances and abundance gradient agree closely with that which is required to explain the observed temperatures and temperature gradients." The scatter in T_ε at any value of r_{gal} can be attributed to variations in N_ε and $T_{eff}(s)$ in individual H II regions.

Table 11-6 compares results previously obtained by several workers with those found by Shaver et al. There appears to be no abundance gradient in helium. The nitrogen gradient appears to be larger than the oxygen one. Gradients obtained for other elements are less certain, partly because of less accuracy in the observations and partly because the ionization correction factors are poor. Figure 11-7 compares gradients of O/H for M31, M33, M101, and our galaxy. The distance from the nucleus is expressed in terms of the characteristic radius, R(0), as defined by de Vaucouleurs.

A knowledge of not only chemical abundances but also of isotope ratios is important for the formulation of element building scenarios. Such ratios as D/H, $^{13}C/^{12}C$ can be measured in molecular clouds in our own galaxy. Eventually, some of them can be studied in other galaxies. We summarize briefly some of the principal results (see Penzias 1980; Wannier 1980).

Let us follow the conventional notation of indicating the atomic weight only for the rarer isotopes; thus, for example, $^{13}C/^{12}C$ is denoted simply as $^{13}C/C$; $^1H_2{}^{12}C^{16}O$ is written simply as H_2CO but $^1H_2{}^{13}C^{16}O$ is called $H_2{}^{13}CO$. Observations of molecules are useful for obtaining isotope ratios but not elemental abundances since it is impossible to handle the complicated, "dirty test-tube" chemistry of the interstellar medium. A further complication is that fractionation is important for hydrogen and sometimes can be significant for carbon. In particular, deuterium tends to be concentrated in molecules but by measurements with the Copernicus satellite of the ultraviolet interstellar absorption lines of H, D, H_2 and HD in the spectra of selected stars, all of the deuterium and hydrogen can be accounted for, and a D/H ratio of 1.8×10^{-5} established. Carbon fractionation is unimportant in dense clouds. Another problem is saturation which occurs because of large abundance ratios, e.g., $^{16}O/^{18}O = 486$ in the sun. The weak line is nicely visible; the strong line is saturated. Hence, double isotope ratios are often measured. That is, we compare pairs of lines each of which involves a rare isotope. Thus, $[C^{18}O]/[^{13}CO]$ gives $(C/^{13}C)/(O/^{18}O)$ and if we can obtain $C/^{13}C$ by some other means we can find $O/^{18}O$. One thing that is surprising is that, in contrast to elemental abundance ratios, the isotope ratios for C, N, O, Si and S (where they can be measured) seem to be independent of distance from the center of the galaxy, once you get outside the central bulge. The second point is that isotope ratios such as $^{17}O/^{18}O$, $N/^{15}N$, and $Si/^{29}Si$ often differ from the solar value, although curiously for the rare isotope ratios $^{29}Si/^{28}Si$ and $^{34}S/^{33}S$ the interstellar and solar system values agree closely. Different isotopes of a given element arise from different nuclear processes, and should give us some clues pertinent to the history of element building in stars. As Wannier (1980) notes, the apparent absence of isotope ratio radial gradients in the galaxy, while chemical abundance gradients exist, poses some engaging questions. The problem is far from solved. Herkel, Wilson, and Bieging (1982) found a gradient of $(^{12}C/^{13}C) = 0.04$ dex per kpc over the range from 4 to 12 kpc. They used observations of H_2CO.

The problem is to try to get a reliable, comprehensive picture of present-day abundance gradients in the galactic system and relate this to well-defined scenarios of stellar evolution and nucleogenesis. Chemical compositions of gaseous nebulae supply powerful tools for such studies of the evolution of galaxies, but they are most useful when supplemented with corroborative data from stellar sources and isotope ratio data.

Table 11-1.

Derived Abundances for Five Typical Regions in the Ring and Interior of NGC 6720, and Mean Composition

Position	(1) N. Ring	(3) Interior	(4) Interior	(6) S. Ring	(10) E. Ring	A Mean	H,M Mean	B 1981
Adopted								
T_ε[O III]	10,200	10,200	11,500	10,200	10,200			
T_ε[N II]	10,200	9,800	9.500	10,200	10,200			
x	0.13	0.13	0.08	0.25	0.10			
log N_ε	3.12	3.12	2.93	3.40	3.00			
He	11.00	11.02	11.01	11.02	10.98	11.00	11.06	11.04
N	8.53	8.21	8.21	8.36	8.31	8.32	8.40	8.34
O	8.90	8.86	8.87	8.83	8.87	8.87	8.98	8.80
Ne	8.39	8.23	8.10	8.39	8.74	8.28	8.40	
S	7.13	7.09	6.69	7.01	7.46	7.15	6.65	6.99
Cℓ	5.26			> 5.4		> 5.2		
Ar	6.57	6.65	> 6.19	6.55		6.59		6.57

The numbers (1), (3), etc., denote observed regions as indicated on the chart given by Aller, Epps, and Czyzak (1976) (see Fig. 1). The electron temperatures for the [O III] and [N II] lines are obtained from the ITS data for λλ4363, 4959, 5007, 5755, 6548, and 6584, respectively. The x-values are adopted from the [S II] λλ6717/6730 and the [Cℓ III] λλ5517/5537 ratios, influenced by the [O II] results by Minkowski and Osterbrock (1960). The results in columns 2-7 are from Aller (1976, Pub. Astron. Soc. Pac., 88, 574), except for S which is estimated with the aid of an improved ionization correction procedure. The mean value for neon excludes Region 10. Hawley and Miller's (1977 Ap. J., 212, 94) results are given in the column labeled H, M; Barker's (1981 Ap. J., 240, 99) results are given in the last column.

Table 11-2
Analysis of NGC 6818*

λ	I	Ion	t	log $N(X_i)$	log Σ	ICF	log N(element) Direct	log N(element) Model	Δ
		He I	1.28	10.65		1.0	11.03	11.05	-0.01
		He II	1.28	10.79	11.03				
4267	0.44	C III	1.28	8.65					
1907/1909	696	C III	1.28	8.50	8.56	1.30	8.67	8.67	0.00
1548/1559	173	C IV	1.35	7.59					
6583	34.9	N II	1.13	6.58					
1747	43.2	N III	1.28	7.95	8.20	1.03	8.21	8.32	-0.11
1487	45	N IV	1.4	7.73					
1238/1241	21.2	N V	1.56	6.95					
3727	52.2	[O II]	1.28	7.04					
4959/5007	2310	[O III]	1.28	8.41	8.43	1.45	8.60	8.72	-0.12
1661/1666	49.8	O III	1.28	8.54					
3868	110	[Ne III]	1.28	7.64					
4725	1.33	[Ne IV]	1.5	7.67	7.94	1.01	7.94	8.06	-0.12
2423	165	[Ne IV]	1.5	7.60					
3346/3426	122	[Ne V]	1.55	7.32					
1892	17.5	Si III	1.28	6.18	6.18	22	7.53	7.66	-0.13
3241	1.8	[Na IV]	1.35	5.89	5.89	2.8	6.33	6.33	0.00
6717/6730	6.71	[S II]	1.13	5.17	6.58	3.0	7.05	7.06	-0.0
6312	2.77	[S III]	1.14	6.56					
5517/5537	1.4	[Cℓ III]	1.28	4.70	4.70	3.0	5.18	5.14	-0.04
7135	15	[Ar III]	1.28	5.85					
4740	8.4	[Ar IV]	1.28	6.24	6.42	1.18	6.49	6.57	-0.08
6435/5007	3.0	[Ar V]	1.37	5.19					
6101	0.41	[K IV]	1.28	4.75	5.07	1.53	5.26	5.10	+0.16
5603	0.081	[K VI]	1.35	4.80					
5309	0.13	[Ca V]	1.35	4.47	4.47	4.3	5.10	5.26	-0.16

*After W.A. Feibelman and L.H. Aller (1983).

Notes to Table 11-2

$\log N(He^+)/N(H^+) = 0.294 + 0.16 \log t + 0.016/t + \log I(\lambda 4471)/I(H\beta)$
$= -0.133 + 0.235 \log t + \log I(\lambda 5876)/I(H\beta)$
$= 0.851 + 0.216 \log t + 0.0158/t + \log I(6678)/I(H\alpha)$
$\log N(He{+}{+})/N(H{+}) = -1.077 + 0.135 \log t + 0.135/t + \log I(4686)/I(H\beta)$

For other He^+ lines; see Table 4-5.

$I(\lambda 5411)/I(4686) = 0.0776\ (t = 1.0) = 0.0809\ (t = 2.0)$
$I(\lambda 4541)/I(4686) = 0.0333\ (t = 1.0) = 0.0358\ (t = 2.0)$

Carbon:

$\log N(C^{++})/N(H^+) = -0.963 + 0.14\ t + \log I(4267)/I(H\beta)$

Table 11-3

Mean Chemical Composition of Planetary Nebulae

Element	Quantity	STD	R	Population I	R	Population II	R	High Excitation	R	N-Rich	R	C-Rich	R	Solar
Helium	log He/H +12	11.03	(1)	11.06(*)		11.03	(5)	11.04	(7)	11.13	(7)	11.03	(7)	11.0
		11.02	(2)	11.03(+)										
Carbon	log C/H +12	8.92	(6)	8.82	(5)	8.72	(5)	8.89	(7)	8.50	(5)	8.99	(7)	8.66
Nitrogen	log N/H +12	8.21	(1)	8.12	(5)	7.94	(7)	8.39	(7)	8.88	(5)	8.15	(7)	7.98
		7.79	(3)											
		7.72	(4)											
Oxygen	log O/H +12	8.83	(1)	8.68	(5)	8.58	(2)	8.66	(7)	8.71	(5)	8.69	(7)	8.91
		8.67	(1)	8.65	(12)	8.55	(12)							
		8.51	(3)	8.72										
		8.44	(4)											
Neon	log Ne/H +12	8.03	(8)	8.08	(5)	7.88	(5)	8.02	(7)	8.05	(5)	8.05	(7)	8.05
		7.83	(3)											
		7.84	(4)											
	Ne/O	0.225	(11)	0.247	(5)	0.192	(5)	0.23	(7)	0.27	(7)	0.23	(7)	
Sulfur	log S/H +12	7.08	(8)	6.96	(5)	6.88	(2)	7.03	(7)	6.98	(5)	7.09	(7)	7.23
		7.32	(9)											
		6.8	(10)											
	S/O			0.0208	(5)	0.0177	(5)	0.023	(7)	0.0169	(5)	0.025	(7)	
Chlorine	log Cℓ/H+12			5.28	(5)	5.13	(5)	5.27	(7)	5.4	(5)	5.27	(1)	(5.5)
	Cℓ/O × 10^4	3.3	(11)	4.46	(5)	4.4	(5)	4.1	(7)	5.5	(7)	3.8	(7)	

Table 11-3 (Continued)

Mean Chemical Composition of Planetary Nebulae (Continued)

		STD	R	Population I	R	Population II	R	High Excitation		N-Rich		C-Rich		Solar
Argon	log Ar/H +12	6.44	(8)	6.42	(5)	6.22	(5)	6.48	(7)	6.65	(5)	6.46	(7)	6.57
	$Ar/O \times 10^3$	7.0	(11)	5.57	(5)	4.48	(5)	6.6	(7)	5.33	(5)	5.9	(7)	
Fluorine	logF/H +12	5:	(13)					4.6	(7)					(4.6)
Sodium	logNa/H +12	6.23	(13)					6.18	(7)					6.31
Potassium	logK/H +12	4.90	(13)					5.0	(7)					5.15
Calcium	logCa/H +12	5.10	(13)					5.0	(7)					6.34

Mg, Si, Fe see text.

References to Table 11-3

1. Peimbert, M., and Serrano, A. 1980, Rev. Mexicana Astron. and Astrophys., 5, 9 (r = 10 kpc).
2. Kaler, J.B. 1978, Ap. J., 226, 947.
3. Barker, T. 1978, Ap. J., 221, 145, Group I.
4. Barker, T. 1978, Ap. J., 221, 145, Group II.
5. Present survey, r = 10 kpc.
6. French, H. 1980, Bull. Amer. Astr. Soc., 12, 842; IAU Symposium No. 103, 1983.
7. Aller, L.H., and Czyzak, S.J. 1981, Proc. Natl. Acad. Sci. USA, 78, 5266.
8. Beck, S.C., Lacy, J.H., Townes, C.H., Aller, L.H., Geballe, T.R., and Baas, F. 1981, Ap. J., 249, 592.
9. Dinerstein, H. 1980, Ap. J., 237, 486.
10. Natta, A., Panagia, N., and Preite-Martinez, A. 1980, Ap. J., 242, 596.
11. Kaler, J.B. 1978, Ap. J., 225, 527.
12. ________ 1980, Ap. J., 239, 78.
13. Aller, L.H. 1978, Proc. Astron. Soc. Australia, 3, 213; IAU Symposium No. 76, 225.

Helium (*) includes all Population Type I objects in Study (5).
(+) Population Type I objects, excluding NGC 6302, NGC 6445, and Hu 1-2, which have He/H ratios exceeding 0.15.

Solar values are from a compilation prepared for a 1980 Santa Cruz workshop and revised for "Spectroscopy of Astrophysical Plasmas," ed. D.L. Layzer and A. Dalgarno, Cambridge Universilty Press (in press).

Table 11-4

Adopted Compositions for Halo Planetaries

	He	C	N	O	Ne	S	Ar	Refs.
108 - 76°1	11.03	(8.48)	(7.3:)	7.8	7.72	6.48	4.59	1,2
49 + 88°1	11.03	8.60	7.5:	8.3	6.75	5.30	4.69	1,2,4
K 648	10.99	8.45	6.3:	7.60	6.40	5.60	4.26	1,2,3
Hu 4 - 1	10.99	9.39:	7.87	8.50	6.80		< 5.2	3
Sun	11.0	8.66	7.98	8.91	8.05	7.23	6.57	5
Aver. P.N.	11.02	(8.8)	7.96	8.66	8.03	7.08	6.44	5

References to Table 11-4

1. Barker, T. 1980, Ap. J., 237, 482 (see also I.A.U. Symposium No. 103, 1983).
2. Aldrovandi, S.M.V. 1980, Ap. Space Sci., 71, 393.
3. Peimbert, M., and Torres-Peimbert, S. 1979, Rev. Mex. Astr. Ap., 4, 341.
4. Hawley, S.A., and Miller, J.S. 1977, Ap. J., 212, 94.
5. Adopted from data in Table 3.

Table 11-5

Comparison of Elemental Abundance Ratios With Respect to log N(element)/N(O)

	M33	M101	SMC	LMC	Galactic Sources PN	Sun	Orion	H II r_{gal} = 10.0 kpc
C			-0.91	-0.51	0.17:	-0.25	-0.10	
N	-1.25	-1.07	-1.55	-1.43	(-0.7)	-0.93	-1.12	-0.93
Ne	-0.77	-0.74	-0.60	-0.68	-0.63	-0.86:	-0.81	-0.80
S	-1.55	-1.68	-1.59	-1.45	-1.58	-1.68	-1.48	-1.64
Cℓ	-3.56				-3.40	-3.4	-3.50	
Ar	-2.40	-2.47	-2.26	-2.17	-2.22	-2.34	-2.33	-2.28

References to Table 11-5

M33: Kwitter and Aller (1981), theoretical models.

M101: Sedwick and Aller (1981), theoretical models. Rayo et al. (1982) derive a smaller N/O and Ne/O ratio but comparable S/O and Ar/O ratios.

Magellanic Clouds: The carbon abundance is from Dufour, Shields, and Talbot (1982). For the other elements we tabulate mean values from this work and from Aller, Keyes, and Czyzak (1977). Theoretical models are employed here to derive ICFs. Extensive investigations were carried out previously by Dufour (1975), the Peimberts (1974-1976), Aller et al. (1974), Dufour and Harlow (1977), and by Pagel et al. (1978).

Galactic Sources: Planetary Nebulae (Tables 3 and 4); Sun (see Table 3).

Orion Nebula: Data by the Peimberts (1977) as revised and extended by Dufour et al. (1982); H II regions (at r_{gal} = 10 kpc), Shaver et al. (1983).

Table 11-6

Galactic Abundance Gradients per kpc Derived From H II Regions

$\frac{d\ \log(He/H)}{dr_{gal}}$	$\frac{d\ \log N/H}{dr_{gal}}$	$\frac{d\ \log O/H}{dr_{gal}}$	References
0 ± 0.02	-0.10 ± 0.02	-0.04 ± 0.03	1
-0.02 ± 0.01	-0.23 ± 0.06	-0.13 ± 0.04	2
-0.008 ± 0.008	-0.083 ± 0.018	-0.059 ± 0.0171	3
0.00	-0.09 ± 0.015	-0.07 ± 0.015	4

References to Table 11-6

1. Hawley, S.A. 1978, Ap. J., 224, 417.
2. Peimbert, M., Torres-Peimbert, S., and Rayo, J.F. 1978, Ap. J., 220, 516.
3. Talent, D.L., and Dufour, R.J. 1979, Ap. J., 233, 888.
4. Shaver, P.A., McGee, R.X., Newton, L.M., Danks, A.C., and Pottasch, S.R. 1984, M.N.R.A.S., in press.

Table 11-7

Comparison of Isotope Ratios

Isotope Ratio Measured	Central Zone	Disk	Solar
$C/^{13}C$	26	60 ± 8	89
$^{18}O/^{17}O$	3.7	3.2	5.5
$O/^{18}O$	270	500	500
$N/^{15}N$	> 600	320	272
$Si/^{29}Si$	10	10	20

(Computed from data by: Wannier, P.G. 1980, Ann. Rev. Astron. Astrophys., 18, 399.)

References

(Note that references included as footnotes to tables or in captions to figures are not necessarily repeated here. Extensive bibliographies may be found in review articles cited below.)

The pioneering studies of the chemical compositions of gaseous nebulae were those of:

Bowen, I.S., and Wyse, A.B. 1939, Lick Observatory Bulletin, **19**, 1, and Wyse, A.B. 1942, Ap. J., **95**, 356.

The first detailed investigation explicitly employing modern-type expressions for rates of atomic processes was contained in the last paper of the Harvard series on Physical Processes in Gaseous Nebulae, 1945, Ap. J., **102**, 239, but substantial progress was possible only with improved observational data and Seaton's collisional cross sections (Chapter 5), see, e.g., 1954, Ap. J., **120**, 401 (where importance of density fluctuations in NGC 7027 was emphasized); 1957, Ap. J., **125**, 84.

The influence of small-scale temperature fluctuations on nebular spectra was examined by M. Peimbert, 1967, Ap. J., **150**, 825.

Examples of ionization correction formulae are given by: Peimbert, M., and Costero, R., 1969, Bol. Obs. Tonanzintla y Tacubaya, **5**, 3; Barker, T., 1983, Ap. J., **267**, 630; French, H., 1981, Ap. J., **246**, 434; Stasińska, G., 1978, Astron. Astrophys., **66**, 257.

Use of theoretical models in analyses of H II regions is discussed, e.g., by Hawley, S.A., and Grandi S.A., 1978, P.A.S.P., 90, 125; Stasińska, G., 1978, Astron. Astrophys. Suppl., **32**, 429; 1980, Astron. Astrophys., **84**, 320 (see Chapter 7).

Discussions of compositions of planetary nebulae are given in the references to Tables 1, 3, and 4, in IAU Symposia No. 76, 1978, and No. 103, 1983. See also:

Torres-Peimbert, S., and Peimbert, M., 1977, Rev. Mex. Astron. Astrofis., **2**, 181.
Barker, T. 1978, Ap. J., **220**, 193.
Kaler, J.B. 1970, Ap. J., **160**, 887; Ap. J., **226**, 947; 1979, Ap. J., **228**, 163; 1981, Ap. J., **244**, 54; 1981, Ap. J., **249**, 201.
Boeshaar, G.O. 1975, Ap. J., **195**, 695.
Webster, L.B. 1976, M.N.R.A.S., **174**, 513.
Price, C.M. 1981, Ap. J., **247**, 540.
Aller, L.H., and Czyzak, S.J. 1983, Ap. J. Suppl., **51**, 211.

Direct evidence for nuclear processed material is found in the shell ejected from Abell 30. See:
Hazard, C., Terlevich, R., Morton, D.C., Sargent, W.L.W., and Ferland, G. 1980, Nature, **285**, 453., and also Jacoby G.H., and Ford, H.C. 1983, Ap. J., **266**, 298.

References (Continued)

Entrapment of certain elements in solid grains can be an important factor in influencing the composition of the gaseous phase. In the context of planetary nebulae, a basic paper is:

Shields, G.A. 1978, Ap. J., 219, 559.

A Few Examples of Analyses of Individual Objects

NGC 7662: Péquignot, D. 1980, Astron. Astrophys., 83, 52.
Harrington, J.P., Seaton, M.J., Adams, S., and Lutz, J.H. 1982, M.N.R.A.S., 199, 517.

IC 3568: Harrington, J.P., and Feibelman, W.A. 1983, Ap. J., 265, 258.

NGC 7009: Perinotto, M., and Benvenuti, P. 1981, Astron. Astrophys., 101, 88.

NGC 7027: Shields, G.A. 1978, Ap. J., 219, 565.
Perinotto, M., Panagia, N., and Benevuti, P. 1980, Astron. Astrophys., 85, 332.
Péquignot, D., and Stasińska, G. 1980, Astron. Astrophys., 81, 121.

NGC 2440: Shields, G.A., Aller, L.H., Keyes, C.D., and Czyzak, S.J. 1981, Ap. J., 248, 569.
Condal, A.R. 1982, Astron. Astrophys., 112, 124.

NGC 6720: See references to Table 1.

NGC 6302: Aller, L.H., Keyes, C.D., Ross, J.E., and O'Mara, B.J. 1981, M.N.R.A.S., 197, 95.

Analyses of Planetary Nebulae in Various Galaxies

The field of endeavor is attracting additional attention as very efficient radiation detection systems become available. A general review with references to earlier work is given by:

Ford, H.C. 1983, IAU Symposium No. 103, p. 443.

A few examples of individual investigations:

Magellanic Clouds: Osmer, P. 1976, Ap. J., 203, 352.
Webster, B.L. 1978, IAU Symposium No. 76, 11.
Dufour, R., and Killen R. 1979, Ap. J., 211, 68.
Aller, L.H. 1983, Ap. J., 273, 590.
Aller, L.H., Keyes, C.D., Ross, J.E., and O'Mara, B.J. 1980, M.N.R.A.S., 194, 613.
Maran, S., Stecher, T., Gull, T., and Aller, L.H. 1982, Ap. J., 253, L43.

References (Continued)

NGC 6822: Dufour, R., and Talent, D. 1980, Ap. J., 235, 22.

M31: Jenner, D., Ford, H., and Jacoby, G. 1979, Ap. J., 227, 392.

Fornax: Danziger, E.J., Dopita, M.A., Hawarden T.G., and Webster, B.L. 1978, Ap. J., 220, 458. See also Maran, S.P. 1984, Ap. J. (in press).

H II Regions and Abundance Gradients in Galaxies:

Although a change in the level of excitation of the H II regions in M33 with a distance from the center was noticed long ago (1942, Aller, L.H., Ap. J., 95, 52), a systematic study was first undertaken by Searle, L., 1971, Ap. J., 168, 327. Further investigations were carried out by Smith, H.E., 1975, Ap. J., 199, 591, who extended the study to several spirals, by Shields, G., and Searle, L., 1978, Ap. J., 222, 821, by Sedwick, K.E., and Aller, L.H., 1981, Proc. Nat'l. Acad. Sci. USA, 78, 1994, and by Rayo, J.F., Peimbert, M., and Torres-Peimbert, S., 1982, Ap. J., 255, 1, who investigated M101. Similar studies have been carried out by many investigators. See also Alloin, D., Collin-Souffrin, S., Joly, M. 1979, Astron. Astrophys. Suppl., 37, 361; Jensen, E.B., and Strom, K.M. and S.E. 1976, Ap. J., 209, 748, Webster, B.L., and Smith M.G. 1983, M.N.R.A.S., 204, 743. A useful summarizing article with an extensive bibliography is: Pagel, B.E.J., and Edmunds, M.G. 1981, Annual Reviews Astron. Astrophys., 19, 77. See also Peimbert, M. 1975, Ann. Rev. Astron. Astrophys., 13, 113.

The relation between H II regions and star formation in irregular galaxies is discussed by D.A. Hunter, 1982, Ap. J., 260, 81.

Magellanic Clouds: Extensive investigations were carried out by Peimbert, M., and Torres-Peimbert, S., 1974, Ap. J., 193, 327; 1976, Ap. J., 204, 581; Aller, L.H., Czyzak, S.J., Keyes, C.D., and Boeshaar, G., 1974, Proc. Nat'l. Acad. Sci. USA, 71, 4496, Dufour, R.J., 1975, Ap. J., 195, 315; Dufour, R.J., and Harlow, W., 1977, Ap. J., 216, 706, Pagel, B.E.J., Edmunds, M.G., Fosbury, R., and Webster, B.L., 1978, M.N.R.A.S., 184, 569.

Theoretical models were calculated by: Dufour, R.J., Shields, G.A., and Talbot, R.J., 1982, Ap. J., 252, 461. Aller, L.H., Keyes, C.D., and Czyzak, S.J., 1979, Proc. Nat'l. Acad. Sci. USA, 76, 1525.

Abundance Gradients in Our Galaxy: See particularly Shaver et al., 1984, M.N.R.A.S., in press. For a discussion of abundance gradients obtained for our galaxy from planetary nebulae, see Peimbert, M., and Serrano, A., 1980, Rev. Mex. Astron. Astrofis., 5, 9, and references therein.

References (Continued)

Two representative examples describing abundance gradients derived from stars are: Mayor, M., 1976, Astron. Astrophys., 48, 301; Janes, K.A., 1979, Ap. J. Suppl., 39, 135. Some abundance gradients derived from H II regions are listed in Table 6; special attention is directed to the article by Shaver et al. (1982).

Isotope ratios are discussed by: Wannier, P.G., 1980, Ann. Rev. Astron. Astrophys., 18, 399; Penzias, A.A., 1980, Science, 208, 663.

List of Illustrations

Fig. 11-1. Monochromatic Images of NGC 6720: The numbered squares give the positions at which spectra scans were obtained. The distance between Squares (1) and (8) is 35 arcsec. (Aller, L.H., Epps, H.W., and Czyzak, S.J. 1976, Ap. J., 205, 798.) Note huge stratification effects.

Fig. 11-2a. Dependence of Oxygen Abundance on Distance From Galactic Center: Symbols are used to indicate population types and quality of analyses on a scale A, B, C, D, corresponding to uncertainties in abundance determination Δ log N $\leqslant$ 0.1, 0.1 - 0.2, 0.2 $\rightarrow$ 0.3 and 0.3 $\rightarrow$ 0.5. Nebulae belonging to Population I are indicated as follows: A, solid dots; B, open circle; C, solid triangle; D, open triangle. Population II: A, solid hexagon; B, open hexagon; C, solid squares. Nitrogen-rich nebulae: A, filled crosses; B, open crosses; C and D, open squares.

Fig. 11-2b. The Carbon-to-Oxygen Ratio as a Function of Distance From the Galactic Center: The same symbols are used as in Fig. 11-2a. The solar carbon/oxygen ratio is indicated by the arrow.

Fig. 11-3a. The Nitrogen/Oxygen Ratio as a Function of Distance From the Galactic Center. The same symbols are used as in Fig. 11-2a. The solar nitrogen/oxygen ratio is indicated by the arrow.

Fig. 11-3b. The Neon/Oxygen Ratio as a Function of Distance From the Galactic Center: The same symbols are used as in Fig. 3a. The solar neon/oxygen ratio is indicated by the arrow. Although the solar Ne/O ratio is lower in the nebulae, the average Ne/O ratios are similar.

Fig. 11-4. The Abundance Ratios of Sulfur, Chlorine, and Argon With Respect to Oxygen, as a Function of Distance From the Galactic Center.

Fig. 11-5. Abundance Gradient of Oxygen in M33: Open circles: data from supernova remnants (Dopita, M.A., D'Odorico, S., and Benvenuti, P. 1980, Ap. J., 236, 628); Filled circles: data from H II regions (Kwitter, K.B., and Aller, L.H. 1981, M.N.R.A.S., 195, 939). Compare also Pagel and Edmunds (1980).

Fig. 11-6. Comparison of O/H Gradients in M31, M33, M101, and the Galaxy: The logarithmic O/H gradient from H II regions is plotted against R/R(0) where R(0) is the characteristic radius of a galaxy as defined by the de Vaucouleurs (1976 Reference Catalogue of Bright Galaxies, University of Texas Press). Data for our galaxy are from Shaver et al. (1983), for M31 from Blair, Kirshner, and Chavalier (1982, Ap. J., 254, 50), for M33 from Fig. 5, and for M101 from Shields and Searle (1978), Rayo et al. (1982) and Sedwick and Aller (unpublished).

Chapter 12

NEBULAR COMPOSITIONS, ELEMENT BUILDING, AND STELLAR EVOLUTION

The relationship between chemical abundances in gaseous nebulae and element building processes in stars remains one of the most challenging problems in astrophysics. There are several aspects that merit examination:

1) The linkage between nucleogenesis and nebular chemical compositions is best defined for planetaries and a few other objects where we study the ejecta from a single star. How can theories of stellar evolution help us understand these objects?

2) Although a large fraction of the stellar population probably evolves into planetaries, the role of very massive O, Of, and Wolf-Rayet (WR) stars is extremely important because these objects can return highly processed material, either through a stellar wind, through demise as a supernova, or both, and have a possible role in the origin of cosmic rays.

3) In particular, supernovae (although few in number) have a great effect on the chemistry of the ambient interstellar medium. Especially enlightening are a few objects such as Cas A where the chemical composition of the individual fragments of the detonation can be studied.

4) Given the composition of a sample of the interstellar medium, as illustrated by an H II region, what can we infer about the stellar or primordial origins of individual elements? For example, from what type of objects did the bulk of the nitrogen originate?

5) To what extent are composition studies of H II regions helpful for investigations for studies of chemical evolution of galaxies, rates of star formation, etc.?

Thus, we are concerned with nuclear processes in advanced stages of stellar evolution and the subsequent migration to surface layers and possible mixture of processed matter with the original material from which the star was formed.

If it is massive enough, the star evolves, through giant and supergiant stages of its evolution, and finally the outer layers are ejected, perhaps gently as in stellar winds, perhaps violently as in supernovae, or perhaps by a wind at one stage and violent pulsations and ejection at another. Let us first examine some characteristics of the stellar winds. In Chapter 7 we discussed them from the point of view of their influence on observable radiation from the nebular shell. We must now examine their role in advanced stages of stellar evolution and the part they may play in supplying material to the interstellar medium.

12A. Stellar Winds

Main sequence stars of moderate luminosity appear to lose negligible mass by a stellar wind unless they spin rapidly (v sini $\gtrsim$ 300 km/sec^{-1}. For example, the solar wind carries off a mass $\dot{M} \sim 10^{-13}$ $M_\odot$/yr. The situation is otherwise for very luminous main sequence and highly evolved giant and supergiant stars. Their masses can decrease substantially, often involving the loss of envelopes that have been "contaminated" with the products of nuclear processing in their interiors.

Direct observational evidence for stellar winds is found from:

a) narrow circumstellar lines in the spectrum of α Herculis indicating the escape of the tenuous outer envelope of the highly evolved M-type component;

b) P Cygni-type lines, prominent particularly in the ultraviolet;

c) extensive chromospheres of late-type stars (revealed by ultraviolet observations);

d) high spatial resolution radio frequency techniques that have demonstrated directly the existence of circumstellar shells;

e) some of the ejected material condensing as dust whose thermal emission is detected in the infrared;

f) microwave measurements that reveal the presence of circumstellar molecules, e.g., OH at 1612 Mhz.

Mass loss occurs from stars more massive than about 12 $M_\odot$ throughout their existence, while less massive stars that evolve into red giants and supergiants develop winds during their final evolutionary stages. The wind speeds up from about 10 km/sec near the photosphere to about 1000 km/sec within about 10 R(*), reaching a terminal velocity v(∞) at about 100 $R(_\odot)$. Lamers (1983) writes:

$$v(\infty) = A\, v_{esc} = A[2\, GM\, K_{rad}/R(*)]^{1/2}$$

where K_{rad} is a correction factor to account for the levitational effects of the mechanical force exerted by radiation scattered by electrons in the atmosphere. Lamers estimates the A factor to be 2.5, 3.5 and 1.0 for O stars, early B stars, and A stars, respectively.

Although Garmany et al. and Abbot et al. (1981) concluded that the mass loss rate, $\dot{M} = dM/dt$, depends only on stellar luminosity, L, Lamers (1981) and Chiosi (1981) suggested that the mass-loss rate also depends on surface gravity, viz.:

$$\dot{M} = 1.5 \times 10^{-5}\, [L/10^6 L_{(sun)}]^{1.42}\, (g/10^3)^{-0.31} \quad (2)$$

This formula predicts a lower mass loss rate for main sequence stars than does the expression involving L only. For Wolf-Rayet stars, see §12G. In very massive stars, $120 > M(^*) \gtrsim 60\ M_\odot$, mass loss rate as high as 0.01 solar masses per year, may persist for a time and produce profound effects on the evolution of the star involved.

Although $\dot{M}$ increases as a star evolves for the last time along the giant branch, the predicted wind rate never suffices to produce a bona fide planetary nebula. One must invoke either a much more violent wind (sometimes called a superwind) or perhaps suppose that the outer envelope is wrenched away by some type of dynamical instability, e.g., a pulsation or a thermal pulse (Section 12-B).

12B. Late Stages of Stellar Evolution for Objects of Low-to-Intermediate Mass

Red giant stars have long been regarded as precursors of planetary nebulae, although the detailed scenario is far from understood. We note that red giant stars often involve non-solar C/O abundance ratios, abnormal $^{12}C/^{13}C$ ratios, enhancements of elements produced by slow neutron capture, i.e., s-process elements (including the appearance of the radioactive nuclide Tc) and an overabundance of lithium. Sometimes all of these effects occur in one star. They have been assessed by stellar evolution models which can be used to estimate the contributions that stars of moderate mass can make to galactic nucleogenesis of ^3He, various isotopes of C, N, O, Ne and s-process nuclides.

We shall now consider stellar post-main-sequence evolution of stars of moderate mass (roughly one to eight solar masses; see Iben and Rezini 1983). From the standpoint of the stellar or nebular spectroscopist, what counts is the composition of the material that reaches the outermost strata. Sometimes this outer envelope that gets ejected into space has been heavily "salted" with products of nuclear burning in the stellar core. In other stars, very little of the nuclear processed material ever reaches the surface. When such stars die, the substance returned to the interstellar medium will have a chemical composition identical with that from which the star was originally formed. The single important factor here is the stellar mass; the greater the mass, the greater the amount of processed material.

While it remains on the main sequence, the star burns hydrogen in its core. When the hydrogen is exhausted, there is an inert, nearly isothermal, core of helium which is surrounded by a thin shell in which hydrogen is still being converted to helium, mostly by the C-N cycle (see Figure 12Ia). Surrounding this shell is a steadily expanding envelope where energy is conveyed outwards by a vigorous convection. The star quickly leaves the main sequence and ascends the giant branch for the first time. At this stage, what is called the "first dredge-up" of processed material can occur. Convection currents penetrate into the region where the C-N cycle had been operating during the main sequence life of the star. As a consequence, ^{12}C is converted

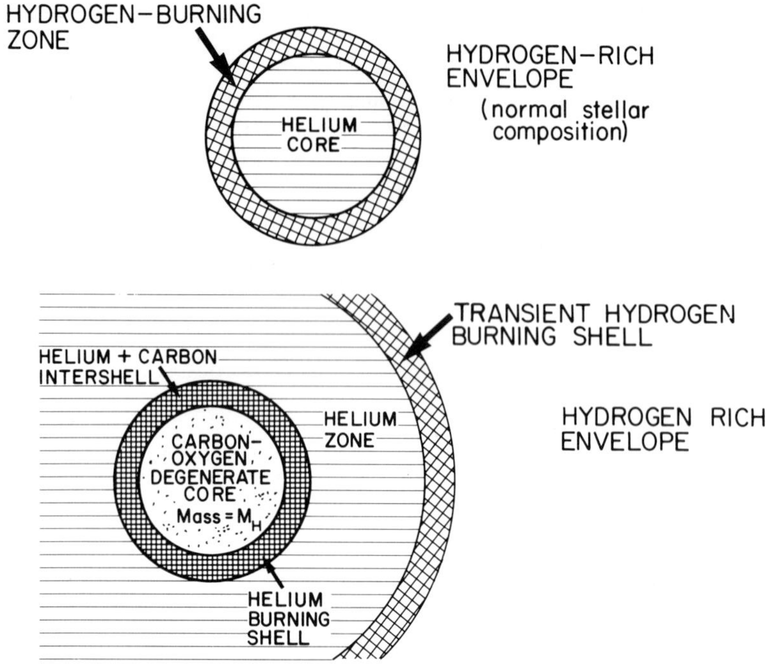

Figure 12.1

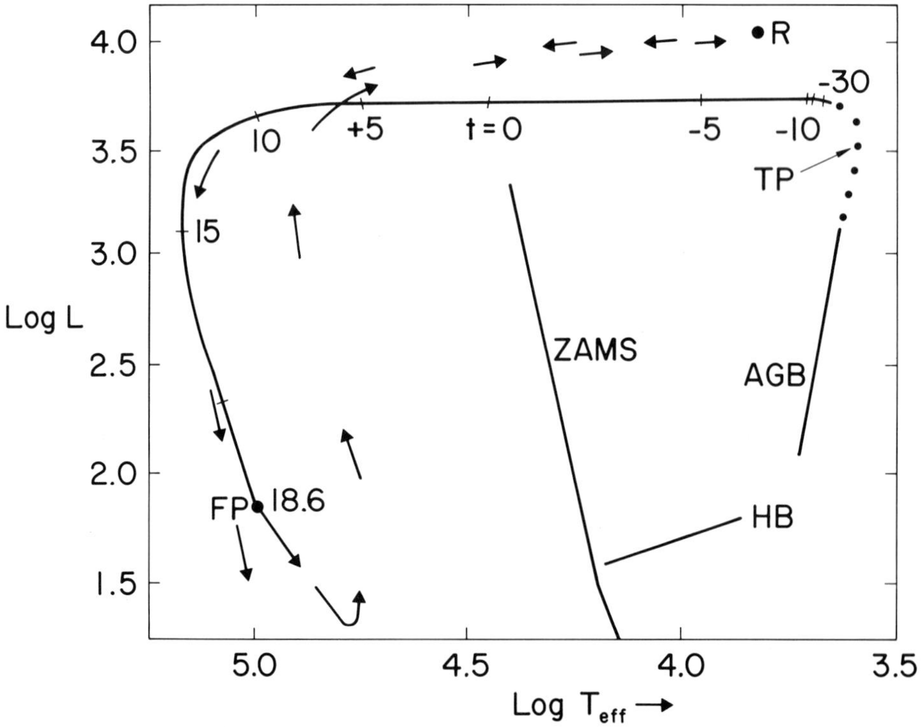

Figure 12.2

substantially to ^{13}C and ^{14}N; these isotopes are convected to the surface where ^{12}C is lowered in abundance but O is unchanged. Iben and Renzini (1982) suggest that if one starts with C, N, and O in the ratios 1/2, 1/6, and 1, one can then obtain ratios 1/3, 1/3, and 1 over a wide range in masses. The $^{12}C/^{13}C$ ratio is lowered from 90 to about 20 or 30 in the stellar atmosphere. Furthermore, mass loss can occur. It is during this first ascent up the giant branch that low-mass stars are presumed to lose an important fraction of their mass. Massive stars lose relatively little at this epoch. Presumably, a planetary nebula is not formed on the first ascent.

Subsequent evolutionary developments are complicated and become the more so the more massive the star. As the He core contracts, its temperature as well as its density rise until ^4He burns into ^{12}C by the triple-α process $3^4He \rightarrow {}^{12}C$. The star now has a central energy source and leaves the giant branch. It goes to a lower luminosity depending on its mass. Population II stars enter the horizontal branch. The energy release depends on a very high power of the temperature. Eventually, there appears a carbon core which becomes degenerate. Energy is now produced in a helium-burning shell and the star rises in luminosity as it becomes a giant for the second time. Its evolution is now said to follow what is commonly called the asymptotic giant branch (AGB), since for the less massive stars it lies slightly to the blue of the T_{eff} - L relationship followed on the first giant branch. In the early stages of AGB evolution, the H-burning shell is turned off and helium burning supplies most of the energy required to expand the envelope and provide the star's luminosity. This is a particularly complicated but interesting phase of a star's life, for when it goes up the giant branch this time it may become a Mira type, very red, long-period variable with a period of the order of 200 to 400 days. The variability presumably arises from pulsations in an extended envelope. It is conjectured that sometimes the pulsations may become so severe that the outer envelope is ejected.

Convection currents in an extended envelope are presumed to penetrate into deeper layers in which the material has been processed by H-burning reactions during an earlier phase. Such a mechanism is called a "dredge-up." At this stage ^{12}C and even some ^{16}O may have been converted to ^{14}N. The second dredge-up occurs in stars initially more massive than 3 to 5 M(sun). Following the ignition of the helium-burning shell, the convective envelope penetrates into the region where all H has been converted to helium. Since ^4He and ^{14}N are dredged up, the surface abundance of these nuclides are enhanced, while C and O are depleted.

Detailed theoretical calculations show that as the star moves up the AGB, H becomes reignited for a time in a thin shell above the He zone. The nuclear engine runs in an "off and on" routine with the consequence that the star now begins to pulse thermally. What happens is that the energy now liberated in the triple-α reaction raises the temperature and density in the inner shell zone (Fig. 12Ib). Expansion and cooling both occur, the He burning slows down as expansion turns off the hydrogen burning. The star then contracts until the transient H shell is again ignited, the interior gradually heats up until the triple-α process again increases significantly and a new thermal pulse is generated.

During the hot part of the cycle, much ^{12}C is produced, along with smaller amounts of ^{16}O. The ^{12}C and ^{16}O that get into the H-burning shell are largely reduced to ^{14}N, while ^{14}N may interact with α particles. Products of the ^{14}N-α reactions include some interesting nuclides. Calculations (of Iben and Renzini 1982 and references therein cited) show that if the mass of the core is $0.6\ M_{\odot} \rightarrow 0.9\ M_{\odot}$, the following sequence of events can occur:

$$^{14}N + \alpha \rightarrow {}^{18}F + \gamma,$$
$$^{18}F \rightarrow {}^{18}O + \varepsilon^{+} + \gamma, \quad {}^{18}O + \alpha \rightarrow {}^{22}Ne + \gamma \qquad (3)$$

The ^{22}Ne may serve as an important source of neutrons and thus the production of s-process elements. If $M_{core} \sim 0.9 \rightarrow 1.0$, solar masses, then $^{22}Ne + \alpha \rightarrow {}^{25}Mg + n$ can occur. With the thus-supplied neutrons, solar system abundances of s-process isotopes can be obtained. Note that the reaction $^{13}C + \alpha \rightarrow {}^{16}O + n$ could be important as a source of neutrons but only if no ^{14}N nuclei were present.

A third dredge-up process may occur during AGB evolution of fairly massive stars. Following each of the He shell flashes, the base of the convective envelope may penetrate through the edge of the H-depleted zone into a region where incomplete He burning previously has occurred with the result that appreciable amounts of ^{4}He and ^{12}C can be convected outwards. At the peak of thermal gradient, convection "plumes" can penetrate into the intershell zone and carry out products of helium burning and neutron capture (Scalo and Ulrich 1973). Thus, in the third dredge-up process, these materials are eventually conveyed to the surface.

There are further complications. Just because a particular nuclide has escaped the helium-burning core, it does not mean that it is safe from further transformations. Nuclear processes involving protons can occur at the base of the convective envelope. These envelope-burning processes can play complicated and important roles. For example, in massive stars, the ^{12}C dredge-up from the core gets converted mostly to ^{14}N by the C-N cycle, so very massive objects can never become carbon stars (see Section 12D). Whether a particular star evolving up the AGB becomes a carbon star depends on its mass and the ratio of mixing length to scale height.

In situations where both the third dredge-up and envelope burning processes are significant, a great variety of surface abundances can ensue. Note that substantial amounts of primary ^{13}C and ^{14}N can be produced. Some material produced by the CN cycle during the main sequence life of the star also can get moved to the surface. Big differences in abundance patterns can occur for stars of different initial masses. For example, the C/O and N/O ratios can vary by a factor of 60. We compare predictions of theory and observation below. Several stars in this evolutionary phase become lithium-rich; ^{7}Li is produced by the mixing of ^{3}He with H in the convection zone leading to ^{7}Be, which undergoes K capture; $^{7}Be + \varepsilon^{-} \rightarrow {}^{7}Li + \gamma + \nu$.

In summary, the theory of stellar structure appears to show quite convincingly how nucleogenesis can occur in stars and the products mixed to the surface, explaining the principal abundance anomalies found in red giants and supergiants. The well-observed phenomena of stellar winds and planetary nebula production assure us that these heavier element-enriched materials can be supplied to the interstellar medium. The details of the process, however, are incompletely understood, and annoying discrepancies between theory and observation appear. Studies of the compositions of stars in globular star clusters show that mixing of nuclear material with the surface layers occurs much earlier in the evolution of a star than is predicted. Substantial effects seem to be manifest as the star starts to move up the giant branch for the first time.

12C. Origin of Planetary Nebulae; Some Theoretical Insights and Observational Clues

As the star evolves up the AGB, the mass of the envelope, M_e, above the hydrogen-burning shell decreases for two reasons: At the base, hydrogen is converted to helium and sinks into the core. At the top of the envelope, mass is lost into interstellar space. At some point it is envisaged that the bulk of the outer envelope is torn off to create the planetary nebula. The core then settles down to form a white dwarf. If the mass of the core exceeds 1.4 $M_\odot$, carbon is ignited in the degenerate core and a supernova event follows.

Just what happens at the end of the AGB evolution is obscure. It has been suggested that Mira stars (long-period variables) may be immediate antecedents of planetary nebulae. Bessell and Wood (1983) suggest that as these variables evolve they shift from oscillations in the first overtone to the fundamental. At this point a shell is ejected, perhaps in what Renzini calls a superwind. This material may form a dense, dusty shell that temporarily hides the star. The dying star settles down, retaining for a while perhaps a thin remnant of its original hydrogen-rich envelope. Its surface temperature rises and as it reaches $T_{eff} \sim 30{,}000°K$ the surrounding shell may be turned on as a planetary nebula. Whether or not this will actually occur depends on the time that elapses between the cessation of the superwind and the rise of the core temperature to 30,000°K. If the required time is too long, the shell will disperse before a planetary nebula can be turned on. Paczyński's theory (1971) predicts that the more massive the core, the more rapidly it will fade. A core as massive as $M_\odot$ will fade 2.5 magnitudes in 2,000 years, while if $M(*) = 0.6\ M_\odot$ the star will be virtually unchanged during the lifetime of the nebula. Stars with initial masses between one and two solar masses would be expected to produce core masses between 0.55 and 0.7 solar masses. Although Schönberner (1981) concluded that planetary nebular cores all fell in a narrow mass range around $M = 0.58\ M_\odot$, investigations by Kaler (1983a) and by Stecher et al. (1982) for planetaries in the Magellanic Clouds suggest that M_{core} can extend from 0.55 $M_\odot$ to $M_\odot$ with many falling in the interval 0.7 to 0.8 $M_\odot$. The core masses still tend to be smaller than those predicted by Iben and Truran.

In these scenarios the evolution of the planetary nebula would be closely linked to that of the nucleus through the time variation of the ultraviolet flux, the mass loss rate, and the wind velocity. If the central star varies over the evolutionary time scale, clearly we must follow the development of the star plus nebula within an integrated picture.

In Fig. 12-2, which is taken from the work of Iben and his associates, we depict the evolution of a star of residual mass $M(*)_{res.} = 0.6\ M_{\odot}$. Of course, the original star had a much larger mass, M(*). The star moves up the asymptotic giant branch, AGB, and then undergoes a number of thermal pulses (see Section B), the temperature and luminosity at the beginning of each pulse being indicated on the diagram by a dot. The epoch at which the nebular shell is ejected and details of the evolution at this time are not precisely known, but the residual core gets off to a relatively slow start on the horizontal path. The tick marks indicate its location at times 30,000, 20,000, 10,000, and 5000 years before the time t = 0 when T(*) = 30,000°K. It moves then relatively rapidly to the domain of high surface temperature and then fades as a hot dwarf, finally with an almost constant radius. At time t = 18,600 years, a final thermal pulse indicated by FP may occur (see Section D).

The theoretical problem of shell detachment is an extremely difficult one. Sun Kwok (1981) has proposed a somewhat simpler scenario. In his model, one envisages a slow (V ~ 10 km/sec) wind from a red giant star that carries away a mass ~ $10^{-5}\ M_{\odot}$/year for some tens of thousands of years. Many solid grains are formed by condensation. Then the outer stellar envelope dissipates and the hot core is exposed. From this hot source there now blows a tenuous wind ~ $10^{-6}\ M_{\odot}$/year with a velocity of about 1,000 km/sec. This high-velocity wind now overtakes and interacts with the slowly moving envelope so that the mixture moves with a velocity of 10 to 20 km/sec. In this picture, the planetary nebula consist of a shell of red giant ejecta, bulldozed by the fast wind from the nucleus and surrounded by an envelope of low density.

In the conventional picture, as a red giant star nears the end of its AGB evolution, it surrounds itself with an optically thick, circumstellar, dusty cloud of about 70 astronomical units, which may obscure the star, but can be detected by infrared measurements. The temperature of the dust may be in the neighborhood of 100 to 150°K. Furthermore, molecular envelopes are observed around such objects. Zuckerman (1977, 1978) has mentioned several red giant stars that are suspected progenitors of planetary nebula. The redder the object, the greater the rate of mass loss. He suggests that carbon-rich stars, which have greater outflow rates than oxygen-rich stars, will evolve into C-rich planetaries. Zuckerman concludes that oxygen-rich planetaries tend to originate from less massive stars.

CW Leo (= IRC + 10216) is an infrared source with a mass loss rate of $10^{-4}\ M_{\odot}$/year. It appears to be pulsating in its fundamental mode. Iben and Renzini (1981) siggest it is in a superwind phase. Other possible progenitors mentioned by them are IR sources in the Air Force Cambridge Laboratory Catalogue without optical counterparts and OH/IR sources.

Bipolar nebulae such as GL 618, M 1-92, MZ-3, and M 2-9 (see Chapter 10) seem to represent a very early phase in the life of certain types of planetary nebulae, perhaps binaries.

Among the youngest bona fide planetaries are dense objects such as IC 4997 which has shown pronounced variations in the (T_ε, N_ε)-sensitive line ratio 4363/[5007 + 4959], and possibly V1016 Cyg, HM Sge, and HBV 475.

12D. Carbon Stars and Their Relationship to Planetary Nebulae

Several types of evidence indicate that many planetary nebulae must originate from carbon stars:

1) Excesses of C are frequently found (see Chapter 11, especially the discussion of NGC 7662 by Harrington *et al.* 1982);
2) Properties of absorbing dust in A30 (Greenstein 1981) suggest carbon particles;
3) Presence of CO in NGC 7027 and some other young planetaries while other compounds of O seem to be weak or missing;
4) Galactic spatial distribution and kinematics of carbon stars resemble those of planetary nebulae;
5) Although $n(C)/n(O) < 1$ in the interstellar media in the Magellanic clouds (Dufour *et al.* 1982), the planetary nebulae so far investigated show $n(C)/n(O) > 1$ (Maran *et al.* 1982) (see Chapter 11).

The ratio of the number of carbon to M stars, $n^*(C)/n^*(M)$, varies strikingly from one part of our galaxy to another and particularly between our galaxy and the Magellanic clouds. In the galactic bulge, this ratio is ~ 0.002, while in the solar neighborhood it is about 0.01. Compare these values with those found for the Large and Small Magellanic Clouds which are 1.7 and 25, respectively! Differences in the ratio of heavy elements to hydrogen (popularly called the metallicity) and the age of the respective stellar populations can explain many of the differences. With decreasing metallicity, the percentage of AGB stars that become carbon stars increases; likewise, with decreasing metallicity, the AGB tracks shift to higher temperatures and the percentage of stars later than M5 falls. Comparison of the predicted and observed C, N, and O abundances in red giants is difficult because there are uncertainties in stellar composition determinations for C, N, and O. Furthermore, effects of binaries are significant. Different components may have different evolutionary histories.

12E. Comparison of Theory and Observations

Some theories of stellar evolution are sufficiently detailed to allow definite predictions of the relation between the initial stellar mass, the mass of the residual core, the envelope mass at ejection (which may exceed the mass of the observed nebula), and elemental abundances that can be compared with observations.

Elaborate calculations of late stages of AGB evolution, taking into account successive dredge-up stages, have been made by Becker and Iben (1979, 1980) and by Renzini and Voli (1981). These calculations not only yield L, T_{eff}, and $\dot{M}$ as a function of time and initial mass, M_i, but also surface abundances of He, C, N, O and

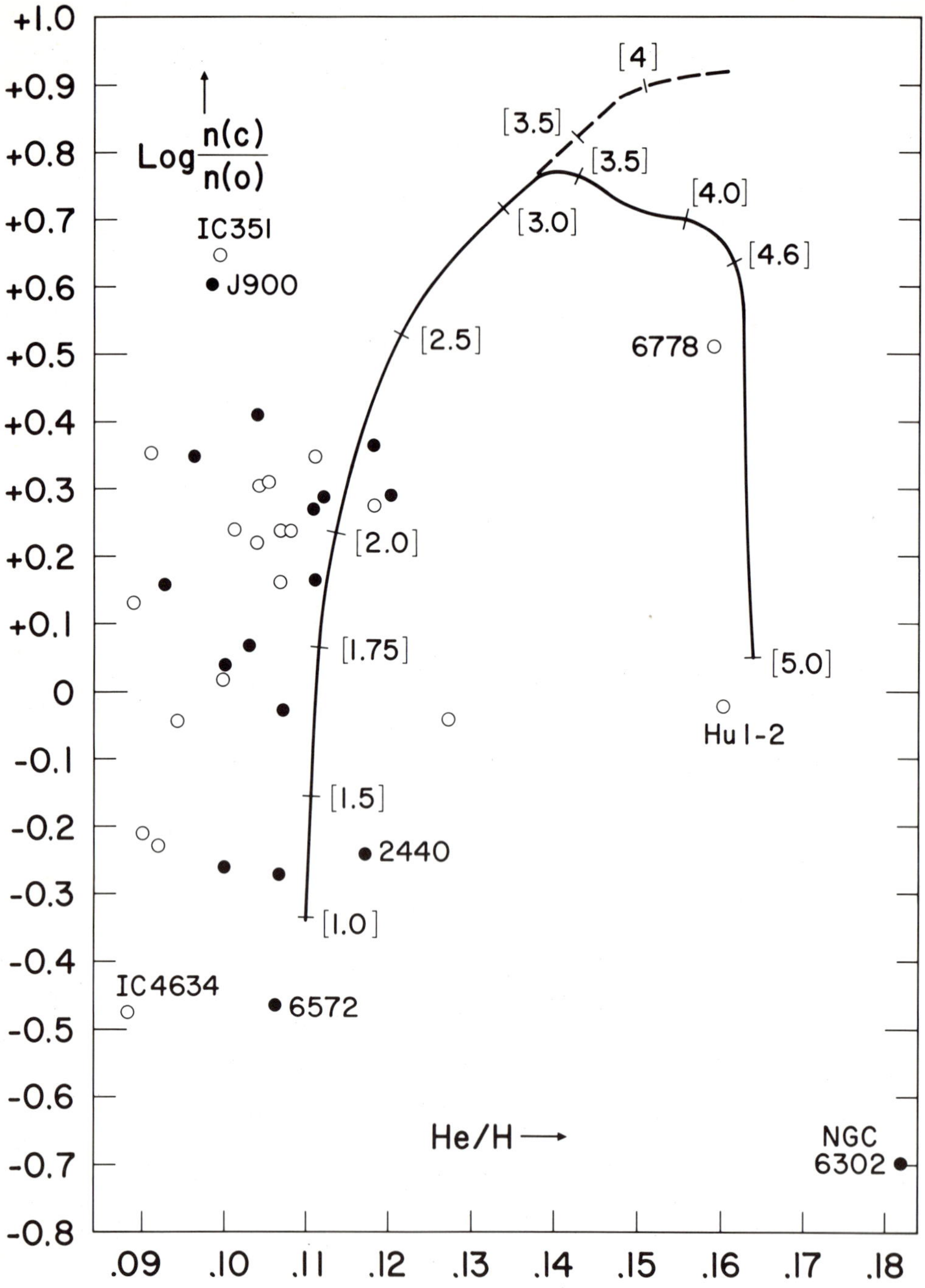

+1.0
+0.9
+0.8
+0.7
+0.6
+0.5
+0.4
+0.3
+0.2
+0.1
0
-0.1
-0.2
-0.3
-0.4
-0.5
-0.6
-0.7
-0.8
Log n(c)/n(o)
IC351
J900
IC4634
6572
2440
6778
Hu 1-2
NGC 6302
[1.0]
[1.5]
[1.75]
[2.0]
[2.5]
[3.0]
[3.5]
[3.5]
[4]
[4.0]
[4.6]
[5.0]
He/H
.09
.10
.11
.12
.13
.14
.15
.16
.17
.18

Figure 12.3

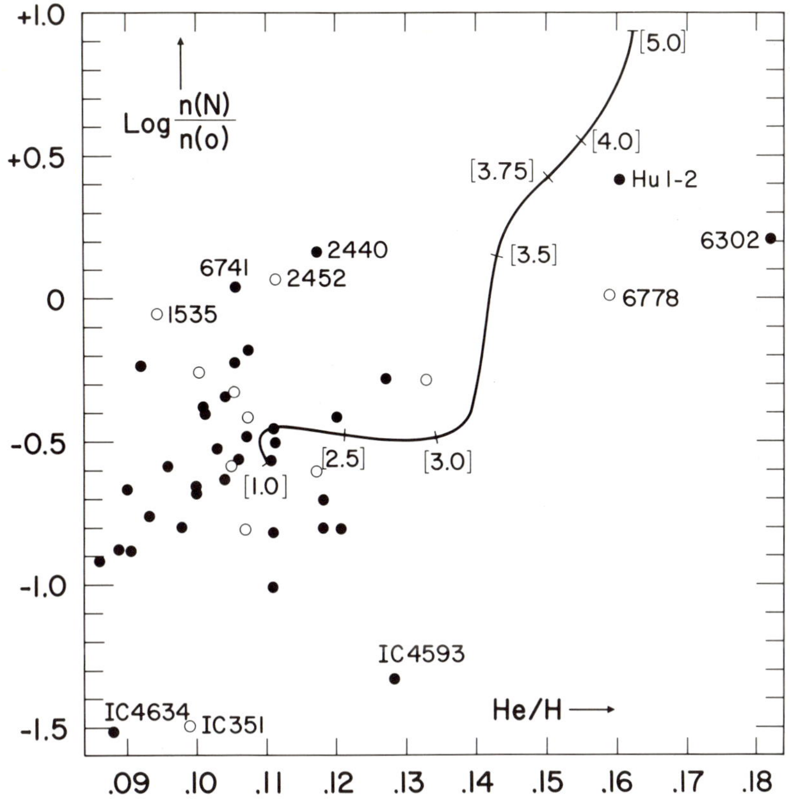

Figure 12.4

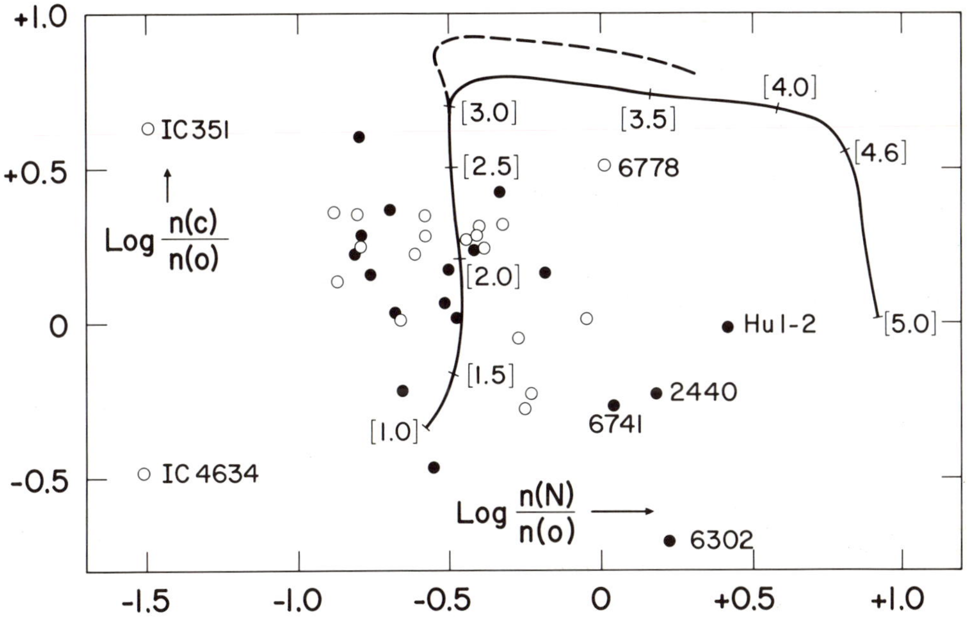

Figure 12.5

s-process elements. Figures 12-3, -4, and -5 compared observed C/O versus He/H, N/O versus He/H, and C/O versus N/O ratios with the Renzini-Voli (1981) predictions which appear to depend on assumptions about the parameter $\alpha = \ell/H$ (the ratio of mixing length to scale height). The observed values are taken mostly from Aller and Czyzak (1983). Similar comparisons have been made by Peimbert (1981) and by Kaler (1983b).

A horizontal shift of the theoretical curves in Figs. 12-3 and -4 can be made if we assume different initial He/H ratios. The masses assigned for the progenitor stars are uncertain, although the ordering is correct. In particular, we cannot conclude from Fig. 12-3 that progenitor masses lie between 1.0 and 2.5 for most planetaries since the mass scale is uncertain. The He/H ratios should be accurate to ~ 0.01 but the log C/O ratios may have uncertainties of the order of 0.2 to 0.3. Note that the nitrogen-rich, carbon-depleted NGC 6302 departs markedly from the curve by a large amount compared to observational uncertainties. Hu 1-2 and NGC 6778 would appear to have originated from massive stars.

The observed n(N)/n(O) ratios (Fig. 12-4) should be more accurate than the n(C)/n(O) ratios. The Renzini-Voli theoretical relationship shows a weak mass dependence for small masses and He/H ratios and then a rapid rise, while the observations suggest a gentle rise with the He/H ratio. The apparently N-depleted objects, IC 4634, IC 351, and IC 4593 seem to form a separate group.

Again, Fig. 12-5 illustrates a tendency of the Renzini-Voli theory to predict N/O ratios that are too high. They assumed a mechanism of envelope burning that efficiently converted carbon to nitrogen. On the other hand, both the Iben-Becker and Renzini-Voli treatment of nucleogenesis through the third dredge-up gave too much carbon. Kaler (1983b), who utilized also additional observational data on low surface brightness planetaries, finds that a satisfactory fit can be obtained if it is assumed that carbon is reconverted to nitrogen at a less efficient rate than by envelope burning. Figure 12-6 gives his results. The theoretical work required to give an adequate quantitative treatment of late stages of stellar evolution will be extremely difficult. Further, we will need improved nebular chemical compositions.

Direct evidence for nucleogenesis in progenitors of planetary nebulae is provided by Abell 30 and Abell 78 in which the inner regions are hydrogen-deficient, showing that the ejected material must have been ejected by a highly evolved star. One might ask why we do not often see hydrogen-depleted material in planetary nebular envelopes. It would appear that in most instances the outermost ejected layers are still hydrogen-rich; the star has not yet squeezed the last bit of nuclear energy from the protons in the thin outer shell. Iben, Kaler, Truran, and Renzini (1983) have proposed an interesting scenario for the evolution of a planetary nebula nucleus (PNN) that experiences a final helium shell pulse as it evolves into a white dwarf along a path of nearly constant radius. In its final convulsion (thermal pulse), most of the hydrogen that remained in the outer envelope is incorporated into a heliun-burning shell and burned. Following this pulse, the star swells up over an interval of about 100 years to become a red giant. The track is indicated by the arrows in Fig. 12-2. The

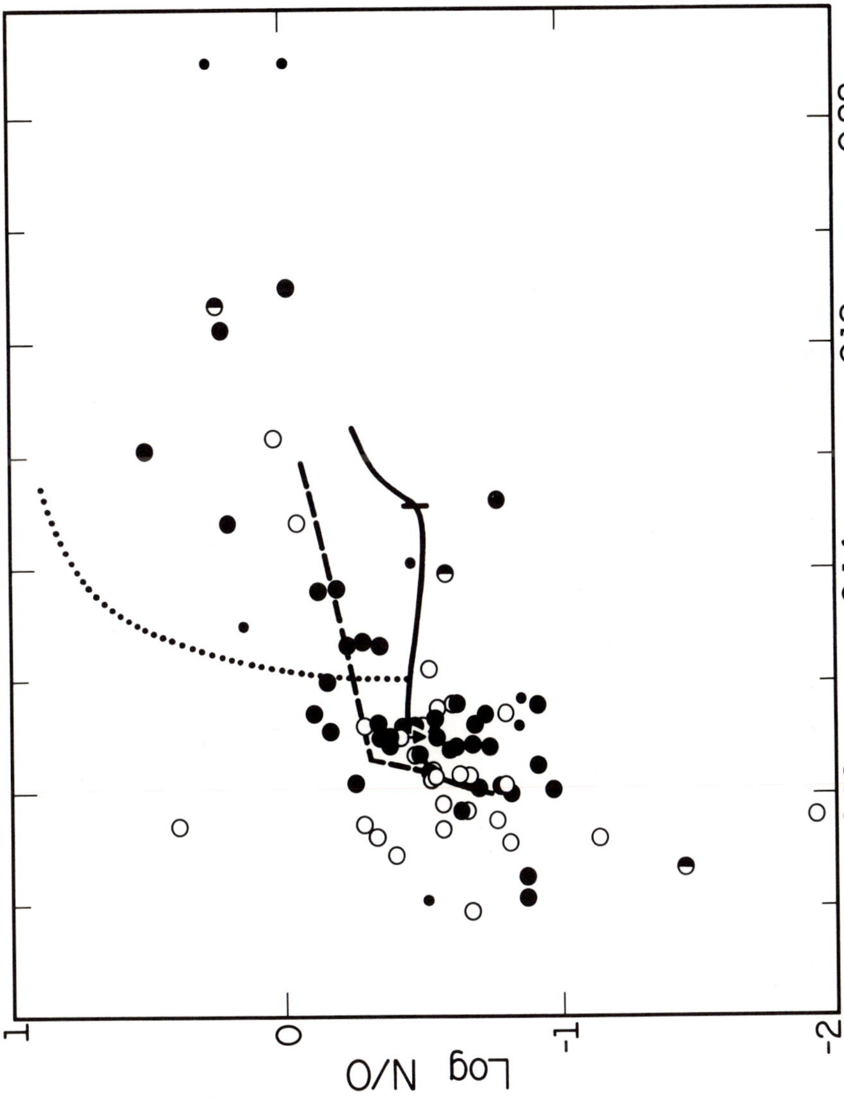

Figure 12.6

final pulse starts at the point FP. Thereafter, the star moves along the dotted line on a rapid time-scale. They identify A30 and A78 and some high-excitation planetaries as post "final-pulse" planetaries. The absence of hydrogen seems to be due to a final H-burning episode during the peak of the final thermal pulse, or to mixing that occurred in a deep surface convective zone during the giant phase after the last thermal pulse. They estimate that about 10 percent of planetaries experience such a scenario. Iben et al. identify the giant star that follows the pulse with an R Cor Bor star which has no H, a great quantity of He, and C, a luminosity up to 10^4 $L_\odot$, $T_{eff} \sim 6000$°K, $R \sim 100$ $R_\odot$, and a high rate of mass loss. Most R Cor Bor stars are presumed to be formed when the star has faded beyond the point FP and the planetary nebula has long since disappeared.

Some PNN have Wolf-Rayet-type spectra (see Sec. 12G). Others show a featureless continuum in the optical region and P Cygni profiles in the ultraviolet. Some, such as the nuclei of NGC 1535, 6058, 6804, 6853, 7009, IC 2149, 3568, show O-type absorption spectra of H and He, or like the nuclei of NGC 2392, 6210, 6891, or IC 4593, show Of spectra that closely resemble "classical" Of stars. These are objects in which a thin lay of H overlies the hot core, PN ejection having occurred before a final mixing of surface layers. It is important to establish accurate chemical compositions of the surrounding nebulae. Can we have enriched PN while H-rich material still forms the stellar atmosphere? There is some indication that PN surrounding such stars may tend to have more nearly "normal" compositions. (See, e.g., Aller 1976, ref. Chap. 11.)

12F. Notes on the Evolution of Very Massive Stars

Since massive stars are so few in number, one might wonder at first glance why they are important. Their relevance lies in the quantity and types of nuclear processed material returned to the interstellar medium. Up to about 20 solar masses, the evolutionary tracks resemble those for stars of moderate mass. The star evolves into a red giant or supergiant, the outer envelope escapes, and the core becomes a white dwarf, a neutron star, or a supernova. A star in the mass range, $20\ M_\odot < M\ 60\ M_\odot$ will evolve into a red supergiant, but then it will lose its outer envelope and return to the high temperature range as a Wolf-Rayet (WR) star. WR stars have strong, broad, emission lines indicative of rapid outward mass flow. On the violet edges of many emission lines, sharp absorption features often occur; these are produced by absorbing atoms in the line of sight. WR stars fall into two sequences, an N sequence with lines of He, N and rather weak carbon, and a carbon sequence characterized by lines of He, C, and O, but no N.

Mass loss by winds increases with the stellar mass. When a star more than 60 times as massive as the sun leaves the main sequence and starts to evolve towards the cool side of the HR diagram, the wind becomes so severe that it actually never reaches the red supergiant (RSG) stage. It becomes, for a time a superluminous Hubble-Sandage variable. The theory of stellar evolution suggests that when an exceedingly massive star reaches $\log T_{eff} = 4.0$, and $\log L/L_\odot = 6.0$, there appears a deep zone in vigorous convection. The mechanical stability of these layers is marginal, at best. As T_{eff} decreases, one must take into account the gradient of the turbulent pressure with depth. This gradient is related to the dissipation of mechanical energy flow, and associated with it is a turbulent acceleration, g_{turb}. When turbulent and gravitational accelerations become comparable in size, the stellar envelope becomes unstable. Oscillations develope as the star dissipates energy and moves back and forth from one side to the other of the limit of stability. There can occur a quasicatastrophic phase, lasting just a few scores of years or centuries during which annual mass loss rates of ~ 0.01 $M_{\odot}$ occur. Can we expect to catch a star or stars in such a phase?

Perhaps η Carinae is such an object! It has been variously interpreted as a massive binary, a contracting star, or an incipient SN (which indeed it may be) but perhaps the most likely explanation is that it is a massive post-main sequence supergiant with strong mass loss (~ 0.01 to 0.08 $M_{\odot}$/year) and L_{bol} ~ 4,000,000 $L_{\odot}$! Outside the bright disk of the material ejected in the 1843 outburst, there are a few scattered condensations which show emission line spectra that are easier to interpret than is the intricate spectrum of the 2" shell or homunculus. Davidson, Walborn, and Gull (1982) found the spectrum of the brightest of these condensations to be truly remarkable. Nitrogen was represented by five stages of ionization, but none of carbon or oxygen were observed. They find C/N < 0.05, O/N < 0.5, which is to be compared with Maeder's prediction (1983) of C/N ≅ 0.025 and O/N = 0.17 - 0.9, values very different from commonly accepted "standard" cosmic values.

12G. The Role of Wolf-Rayet Stars

The upper limit to stellar luminosity for red supergiants appears to be fixed by considerations of mechanical stability that depend on T_{eff} and L rather than on the ratio of heavy atoms to hydrogen (so-called metallicity). Subsequently, as the supermassive star loses material, the fractional core mass, q(He, C, O), increases and when it becomes large enough, the star moves to the high-temperature side of the diagram. By this time, it has been stripped to the strata which reflect effects of nucleogenesis and we observe a Wolf-Rayet star. The mass loss rates are of the order of $\dot{M} = 3 \times 10^{-5}$ M(sun)/yr^{-1} and do not depend on luminosity (Willis 1982).

As high mass loss rates continue, Maeder finds that WR stars become chemically quasihomogeneous and surface compositions can be calculated as a function of time. The star evolves from a late WN-type (WNL) to an early (WNE) type and then quickly into a WNC star. In the early stages, the composition is dominated by strata reflecting action of the CNO cycle. Note that since the formation of ^{14}N is the bottleneck in the cycle, the proportion of ^{14}N rises as ^{12}C and ^{16}O are consumed. An initial "cosmic" ratio of $^{14}N/^{16}O$ of 0.11 is transformed to 12.5. The abundances of ^{17}O rises as ^{16}O and ^{18}O are destroyed until the ^{17}O and ^{16}O isotopes are comparable in number at an advanced CNO stage. During this phase the star evolves from WNL to WNE as hotter regions become exposed. In the He-burning zones of these massive stars, Maeder (1983) showed that ^{14}N is transformed to ^{22}Ne by reaction (3). Since galactic cosmic rays show some significant overabundance of ^{22}Ne, WR stars may supply about two percent of interstellar neon.

When products of helium burning ($3He \rightarrow {}^{12}C$) reach the stellar surface, the chemical composition of the latter abruptly changes. Now carbon and oxygen, which is made by the $^{12}C(\alpha,\gamma)^{16}O$ reaction, may reach the surface as ^{14}N is subducted and destroyed, e.g., by reactions: $^{14}N(\alpha,\gamma)^{18}O$; $^{18}O(\alpha,\gamma)^{22}Ne$. In stars as massive as the order of 100 suns, Mg can be built by $^{22}Ne(\alpha,\gamma)^{26}Mg$ or $^{22}Ne(\alpha,\gamma)^{25}Mg$; ^{20}Ne can be built by $^{16}O(\alpha,\gamma)^{20}Ne$. Because of the constant stripping off of the outer layers, isotopes such as ^{17}O and ^{22}Ne escape the destruction that would await them in normal stellar interiors.

Although the "classical" or population type I WR stars are very massive, planetary nebulae nuclei (PNN) often show spectra of the O, Of, and WR types. All WR central stars belong either to the C or O VI sequences; none are members of the N group. Large masses are not involved in PNN, but evidently heavy mass loss rates occur. The CNO cycle products may be ejected, as in the N-rich PN, or they may be reprocessed in the core by the time we see the stellar spectrum. The O VI stars are super-hot C stars.

12H. Supernovae

Both theory and observation suggest that WR stars may evolve into supernovae. Supernova remnants occur frequently in stellar systems containing many WR stars, e.g., M33, and in early-type stellar associations in our own galaxy. The WR stage need not necessarily precede an SN outburst; Jura and Morris (1982) suggest that the supergiant α Orionis may be destined to become a supernova within a few millenia. Analyses of young SNR's by methods described in Chapter 9 give some clue to the type of material supplied to the interstellar medium. Several factors are to be noted: 1) As seen in Chapters 9 and 11, many SNR's consist mostly of material swept up from the interstellar medium, but occasionally one finds a mix of SN ejecta and interstellar medium material. For example, Pup A shows an excess of O compared with the standard cosmic mix, enabling us to estimate the mass fraction of the new material supplied by the detonation; 2) From observed abundances in an SNR not yet appreciably altered by mingling with the interstellar medium, one can use theoretical model predictions to estimate the mass of the progenitor star. From observations of the Crab nebula, Davidson *et al.* (1982) found a large He abundance, but small C and O abundances from which they deduced a mass of 8 $M_\odot$ for the precursor star; 3) In many SNR's it is possible to get only average abundances over the shell, but sometimes individual, more detailed information can be obtained. The Cass A supernova remnants have supplied some particularly instructive results since the moving blobs of material appear to be largely untained ejecta from the original detonation which occurred about 1657. Baade and Minkowski (1954) who made the first optical identification described two types of ejecta, quasistationary floculli (QSF), and fast moving knots (FMK). The QSF are nitrogen-rich. The FMK are overabundant in products of advanced stages of nucleogenesis. Detailed studies have been carried out by Peimbert (1971), by Peimbert and van den Bergh (1971), and by Chevalier and Kirshner (1977, 1978, 1979).

The envelope of the star (represented by the QSF) appears to have been removed prior to the detonation that ejected the fast moving knots. Different knots show different chemical compositions but hydrogen makes up a very small fraction of the knot material. Chevalier and Kirshner (1979) derived abundances by comparing observed line intensities with predictions of shock models. It is possible to correlate compositions of individual (FMK) knots with those of successive layers in the model for the pre-supernova outburst. For example, the abundances of S, Ar, and Ca (which are products of O burning) as deduced from [S II], [Ar III], and [Ca II] are correlated with each other. This group of elements shows large variations with respect to O. In the deeper strata, O is burned to Si, and eventually Si is converted to even heavier elements. Hence, the ejected strata represent a smörgåsbord of nuclear cooking. Ejected parcels with substantial amounts of both O and Si appear to have undergone incomplete explosive oxygen burning. Among the FMK there appears to be no correlation between blob ejection velocity and composition as would be required for an orderly explosion. The core collapse appears to have been asymmetrical, whereas in other SN such as Tycho, the expanding shell seems to have been uniform. The mass of the Cas A shell appears to be about 8-12 m(sun). For a number of young SNR's, X-ray spectral observations have been secured. Tycho's supernova produced significant amounts of Si, S, and Ar, but there was no excess of Mg and Fe above the solar value. The X-ray spectrum of the Kepler SNR as observed by the Einstein satellite was similar to that of Cas A, while the ejecta of the SN, G292.0+1.8 is oxygen-rich and showed an overabundance of S and an underabundance of Fe relative to the sun.

Supernova ejecta appear to have played an important role in the material from which the solar system was formed. Clues to the nature of the SN events that must have seeded the relevant primordial interstellar clouds of dust and gas and to the time interval from the last blast to the condensation of solid bodies are provided by meteorites. A discussion of this topic lies outside the scope of our review.

12I. Principal Sources of Enrichment to the H II Regions

We now examine the question "Can we identify the principal sources of some of the more abundant elements in the interstellar medium, particularly those observed in H II regions such as C, N, O, Ne, S, and Ar?" We have seen that stellar envelopes, often heavily affected by results of nucleogenesis, are ejected into space by:

a) Planetary nebulae;
b) Winds from cool giant and supergiant stars in advanced evolutionary stages, including objects such as η Carinae;
c) Wolf-Rayet stars;
d) Novae and nova-like variables;
e) Supernovae.

Helium appears as the immediate product of H burning and is supplied to the interstellar medium by many stars. We might expect He to be enriched in PN; as we have seen, indeed it is in many objects, but PN are not important suppliers of helium. Although a large percentage of the material ejected by WR stars is helium, these objects probably do not play an important role in building up the galactic He because they are rare. In fact, most helium is believed to have come from the Big Bang.

The origin of interstellar carbon has been discussed in the context of C stars and planetary nebular ejecta. If planetary nebulae were the prime source of material for the interstellar medium, we might expect it to be carbon-rich, at least in Magellanic clouds, but this is not the situation (Dufour et al. 1982, Chapter 11). Evidently, supernovae and other objects which supply copious amounts of oxygen must be dominant.

Several sources have been proposed for interstellar nitrogen. Classical novae may play a role. The abundances in the ejected shells depend on their masses. Gallagher and Starrfield (1978) suggest that if a conversion of ^{3}He to ^{7}Li plays an important role in the outburst, novae might be able to contribute significant amounts of ^{7}Li, ^{13}C, and ^{15}N to the interstellar medium. In the LMC, Walborn (1982) has identified four Of-like objects with extended nitrogen-rich envelopes. Other examples are the ring nebulae associated with WN stars such as NGC 6888, and possibly nebula NGC 6164-6165 ejected by the Ofp star and that associated with AG Carinae. Walborn has suggested that most O and early B supergiants are moderately nitrogen-enhanced. Some (the OB IV stars) are markedly enhanced in nitrogen. Nitrogen enhancement may be a common occurrence in massive stars.

One might expect WN stars to be important suppliers of N to the interstellar medium, but Maeder concludes ^{14}N yields by WN stars are not large because the excess of ^{14}N in WN stars is only one-tenth the excess of ^{12}C in carbon stars. The carbon WC stars are copious suppliers of C, of course. The rare isotopes ^{17}O and ^{22}Ne are manufactured in WN and WC stars respectively. Maeder notes that in galactic cosmic rays, the excess of ^{12}C, of the isotope ratios, ^{22}Ne/^{20}Ne, ^{25}Mg/^{24}Mg, and ^{26}Mg/^{24}Mg can be explained by supposing that one particle out of 50 originates in WR stars.

Quantitative data are lacking but it would appear that in our galaxy and possibly in the Magellanic clouds and elsewhere, massive stars evolving off the main sequence may supply significant amounts of nitrogen. Oxygen and heavier elements can be furnished only by very massive stars and supernovae which succeed in mixing results of nuclear processing into their surface layers. Studies of "transient gaseous nebulae" in the context of novae, supernovae and nebulous shells surrounding supergiants can give us important tools for probing late stages of stellar evolution. Lambert and his coworkers have suggested that O may have been made in more massive stars than those that produced Fe and s elements. Audouze (1983) proposed that N comes from low-mass stars, C from intermediate mass stars, and Fe from more massive stars. Stars more massive that 120 m(sun) would supply predominantly O. Among PN progenitors, however, it appears that the copious suppliers of N seem to be more massive than those that supply mostly C.

12J. H II Regions and the Chemical Evolution of Galaxies

Chemical compositions of galaxies have been investigated by many procedures. So far, individual bright stars can be studied only in the Magellanic Clouds. For more distant systems, we have to rely on colors, integrated spectra, and variations thereof with distance from the nucleus, plus, of course, H II regions. To what extent does a study of abundances in diffuse nebulae offer clues to the history of the chemical evolution of a galaxy?

A popular scenario is that galaxies started out as roughly spheroidal clouds of gas, some of which later collapsed into disk forms. If essentially all stars were formed while the primordial gas ball was spheroidal, an elliptical galaxy resulted. The stars in the central bulge and halo of our own galaxy presumably were formed at an early epoch, while the gas was still distributed spheroidally. In spiral galaxies, gas became concentrated to the plane where star formation continues to the present. Some spirals show prominent central bulges; others such as M33 appear to have little or no central condensations.

Stars in the halo or galactic bulge and high-velocity stars in the solar neighborhood, which were formed largely from primordial material, are deficient in elements that are produced in stellar nucleogenesis. Also, there seems to be some evidence for a rough correlation between these heavier elements and the epoch of formation of disk stars.

One of the most striking phenomena in ordinary spiral galaxies is a correlation between the (O-group/H) ratio and the distance from the nucleus. This suggests that the rate of stellar evolution and consequent element formation proceeded at a faster pace in denser central regions than in the outer spiral arms. The rate of element formation in these outer regions of objects such as M33 and M101 seems comparable with that found in the Magellanic clouds (see, e.g., discussion by Pagel 1979). In irregular galaxies and barred spirals, it appears that abundance gradients may be smoothed out by noncircular motions. The work of Hunter *et al.* (1982) on noninteracting irregular galaxies suggests that processes of star formation and element building may be stochastic and quite complicated.

Scenarios proposed for the chemical evolution of galaxies assume we start with gases left over from the Big Bang: hydrogen, deuterium, and helium, with a trace of lithium. Much of this material is presumed to have condensed into short-lived stars of moderate-to-high mass, which manufactured carbon by the triple-α process, oxygen, and some neon by successive α-particle capture.

Primary elements are defined as those produced directly in massive stars by converting H successively to He, C, O and perhaps heavier elements such as iron. No initial content of heavy elements is required. Secondary elements are those made in less massive objects by processes that transform previously present elements. Thus, C, O, and Fe are primary elements, while ^{13}C, ^{22}Ne, and s-process elements are secondary. Some N may be primary. It is produced by the C, N, O cycle as H is introduced into very hot regions containing C. Some N is certainly secondary. Nuclides produced either by primary or secondary processes eventually find their way into the interstellar medium by stellar winds and the formation of planetaries. The most massive objects manufactured iron and heavier elements and ejected them into the interstellar medium by supernova detonations. Some s-process nuclides were released by stellar winds.

If this scenario is correct, we might expect to find some dim, ancient stars with no trace of any element except H or He. Such relics have never been detected. Even the very oldest, most markedly metal-deficient objects still contain iron and other heavy elements, clearly indicating they are second-generation stars.

There is considerable interest in establishing the primordial abundances of ^1H, ^2H and ^4He in order to fix the parameters of the Big Bang. Deuterium and lithium are destroyed in the stars, while helium is enhanced. In certain red giant stars, however, ^7Li is produced, evidently by the $^3He\,(\alpha,\gamma)\,^7Be$ reaction, which is followed by a quick dredge-up. The original abundance of deuterium in the solar system as deduced from meteorites, Jovian atmosphere, and cometary data is of the order $^2H/^1H \sim 2.5 \times 10^{-5}$, which is in good agreement with best estimates for the interstellar medium. This must be a lower limit on the primordial value which easily could be two or three times larger.

Much interest has been focussed on the helium abundance which can be obtained not only from H II regions but also from hot stars by indirect, less accurate methods such as comparison of the empirical and theoretical mass luminosity correlation and the zero-age main sequence. Big Bang cosmology (Yang et al. 1979; Olive et al. 1981) suggests a primordial helium mass fraction of $0.20 < Y_p \lesssim 0.25$. At any given place and epoch in the interstellar medium, the local value of Y depends on the initial helium abundance, Y_p, and the contribution from element building processes in stars. The latter should be proportional to Z, the fraction of heavy elements by mass. Thus:

$$Y = Y_p + \left(\frac{dY}{dZ}\right) Z$$

In principle, by a comparison of Y values for different galaxies with different Z-values or even from a comparison of Y-values in different parts of certain spiral galaxies, it should be possible to find Y_p and dY/dZ, although the dependence of Y on Z is very weak. For example, from a study of emission nebulosities in M101, Rayo et al. (1982, Ref., Chapter 11) find $Y_p = 0.22$. A detailed study of metal-poor objects by Kunth and Sargent (1983) Yields $Y_p = 0.245 \pm 0.003$ with no clear evidence for a dependence of Y_p on Z among highly-depleted objects. An accurate Y_p determination is of great cosmological importance and of great difficulty.

The simplest element-building model supposes that stars are formed in isolated, well-mixed zones. In a given galaxy these could be concentric cylinders. The ratio of mass in stars to mass in gas, M_s/M_g, plays an important role in element building. As Searle and Sargent (1972) noted, a large-scale abundance gradient can occur because the M_s/M_g ratio falls steeply away from the nuclear region. This simple model explains qualitatively the O/H ratio behavior discussed in Chapter 11, but gives no adequate quantitative interpretation. The problem is extremely complex. For example, we must adopt an hypothesis of star formation rate as a function of mean gas density, $\bar{\rho}$, and fluctuations thereof, $\langle\Delta\rho\rangle$. During the long history of the galaxy the density must have changed in a complicated way. The star formation rate must be enhanced if the density is increased, for example, when the interstellar medium is shocked as it passes through a spiral density wave.

A second important point is the distribution of masses among newly formed stars -- Salpeter's initial mass function, IMF. For each stellar mass, the theory of stellar structure should enable us to assess the fraction, ζ, that is returned to the interstellar medium when the star dies. Then $(1 - \zeta)$ represents the mass fraction that is locked into "condensed bodies" -- white dwarf or neutron stars. Another important parameter is what fraction ξ of the ejected mass consists of products of stellar nucleosynthesis. Both ζ and ξ increase with the stellar mass. Furthermore, the initial mass function must depend in some intricate way on the total mass of the pre-stellar cloud, its density, density fluctuations, angular momentum, and magnetic field. It is usually assumed that the IMF is invariant in time.

At the present time, no quantitatively satisfactory model of element building in galaxies is available, which is not surprising in view of the complexity of the problem. We may hope that accurate H II region abundances will contribute important clues since these diffuse nebulae can be seen in distant galaxies as well as in nearby systems.

References

Stellar Winds With Application to a Star's Evolutionary History.

In Chapter 10, stellar winds were discussed from the point of view of mechanics and radiative properties of surrounding shells and interstellar bubbles. They also play important roles in stellar evolution. This is a very active field for which we can give only a few references. The fundamental paper is that by Deutsch, A.J.., 1956, Ap. J., **125**, 210. Useful review articles may be found in Ann. Rev. Astron. Astrophys. by Conti, P., 1978, **16**, 371, and by Cassinelli, J.P., 1979, **17**, 275; Lamers, H., 1983, p. 53, Diffuse Matter in Galaxies, ed. J. Audouze et al., Dordrecht, Reidel Publ. Co., and Garmany, C.D., Olson, G.L., Conti, P.S., van Steenberg, M.E., 1981, Ap. J., **250**, 660.

Late Stages of Stellar Evolution and the Origin of Planetary Nebulae.

Important contributions were made by: B. Paczyński, 1970, Acta Astronomica, **20**, 47, 287; 1971, **21**, 417; 1973, **23**, 191, and by Schwarzschild, M., Harm, R., 1967, Ap. J., **150**, 961.

Kaler, J.B. 1981, Ap. J., **250**, L31; 1983, Ap. J., **271**, 188.
See I.A.U. Symposium No. 103, 1983.

Scalo J., and Ulrich, R.K., 1973, Ap. J., **183**, 151 (plume mixing).

Cameron, A.G.W., and Fowler, W.A., 1971, Ap. J., **164**, 11 (Li enhancement).

Becker, S.A., and Iben, I. 1979, Ap. J., **232**, 831; 1980, **237**, 111.

Wood, P.R., and Faulkner, D.J. 1984, I.A.U. Symp. No. 105, Reidel.

References (Continued)

Iben, I. 1982, Ap. J., 260, 821.
Iben, I., Kaler, J.B., Truran, J.W., and Renzini, A. 1983, Ap. J., in press.
Iben, I., and Renzini, A. 1983, Ann. Rev. Astron. Astrophys., 21, 271, and references therein cited.
Renzini, A., and Voli, M. 1981, Astron. Astrophys., 94, 175.
Peimbert, M. 1981, Physical Processes in Red Giants, ed. I. Iben and A. Renzini, Dordrecht, Reidel Publ. Co., p. 409.
Schönberner, D. 1981, Astron. Astrophys., 103, 119.
Kwok, S. 1980, Ap. J., 236, 592.

Early Stages of Planetary Nebula Development

For a general review of observations of giant and supergiant stars relevant to the formation of PN, see:
Zuckerman, B. 1980, Ann. Rev. Astron. Astrophys., 18, 263, and references cited therein.
V1016 Cygni, HM Sge, and HBV 475, which are probably protoplanetary objects, have been studied by many observers. See Nussbaumer, H., and Schild, H., 1981, Astron. Astrophys., 101, 118, and Feibelman, W.A., 1982, Ap. J., 258, 548, and references cited therein.

Variability in the Spectrum of a Planetary Nebula was first established definitely for IC 4997. See Liller, W., and Aller, L.H., 1957, Sky and Telescope, 16, 222; 1966, M.N.R.A.S., 132, 337. Rapid variations in the [O III]/Hγ ratio were observed by A. Purgathofer and M. Stoll, 1981, Astron. Astrophys., 99, 218. See also W. Feibelman, 1982, Ap. J., 258, 562.

Evolution of Chemical Abundances in Massive Stars is discussed by Maeder, A., 1981, Astron. Astrophys., 99, 97; 101, 385; 102, 401; 1982, 105, 149, and references therein quoted.

Ejecta of η Carinae: Davidson, K., Walborn, N., and Gull, T., 1982, Ap. J., 254, L47.

For a discussion of novae see: Gallagher, J.S., and Starrfield, S., 1978, Ann. Rev. Astron. Astrophys., 16, 171.

Nitrogen-Rich Of Stars: Walborn, N.R., 1976, Ap. J., 204, L17; 205, 419; 1982, 256, 452, and references therein cited.

Current views on Wolf-Rayet stars are summarized in Wolf-Rayet Stars: Observations, Physics, Evolution, ed. de Loore, C.W.H., and Willis, A.J., Dordrecht, Reidel Publ. Co., 1982.

A good overall summary of elemental origins is given by: Trimble, V. Reviews of Modern Physics, 54, 1183, 1982, and 55, 511, 1983. Interesting accounts of relations between stellar compositions and element-building scenarios are given in papers by Lambert and his coworkers. See, e.g., Clegg, R., Lambert, D., and Tomkin, J., 1981, Ap. J., 250, 262.

References (Continued)

Supernovae

Some possible characteristics of presupernovae are discussed by Jura and Morris, 1982, Ap. J., 251, 181. The chemical compositions of young supernova remnants (SNR) have been successfully investigated for a number of objects. A number of examples: The Casseopeia A Supernova remnant was first studied by W. Baade and R. Minkowski, 1954, Ap. J., 119, 206. Analysis and discussion may be found in the following papers and references cited therein: Peimbert, M., 1971, Ap. J., 170, 261; Peimbert, M., and van den Bergh, S., 1971, Ap. J., 167, 223; Chavalier, R.A., and Kirshner, R.P., 1977, Ap. J., 218, 142; 1978, 219, 931; 1979, 233, 154.

Crab Nebula Composition: Davidson, K., Walborn, N., and Gull, T., 1982, Ap. J., 254, L47.

For recent X-ray studies of the spectra of a number of young SNR ejecta, see:

Tycho's supernova, Becker, R.H., et al., 1980, Ap. J. Letters, 235, L5;

Kepler's supernova, Becker, R.H., et al., 1980, Ap. J., Letters, 237, L77.

Puppis A, Winkler, P.F., et al., 1981, Ap. J. Letters, 245, 574.

G292.0 + 1.8, Clark, J.H., et al., M.N.R.A.S., 193, 129.

Chemical Composition Variations Within and Chemical Evolution of Galaxies are topics which have received much attention in recent years. The following articles in Annual Reviews of Astronomy and Astrophysics, and references cited therein, are useful: van den Bergh S., 1975, 13, 217; Peimbert, M., 1975, 13, 113; Audouze, J., and Tinsley, B.M., 1976, 14, 43; Pagel, B.E.J., and Edmonds, M.G., 1981, 19, 77.

See also: Pagel, B.E.J., 1979, Stars and Stellar Systems, ed. B.E. Westerlund, Dordrecht, Reidel Publ. Co., and Searle, L., and Sargent, W.L.W., 1972, Ap. J., 173, 25; Edmunds, M.G., 1977, I.A.U. Colloquium No. 45, p. 67; Hunter, D.A., Gallagher, J.S., and Rautenkranz, D., 1982, Ap. J. Suppl., 49, 53; Tinsley, B.M., 1980, Fundamentals of Cosmic Physics, 5, 287; Audouze, J., 1983, Diffuse Nebulae in Galaxies, p. 95, ed. Audouze et al., Reidel Publ Co. The importance of the primordial helium abundance in Big Bang cosmology is discussed by: Yang, J., Schramm, D.N., Steigman, G., and Rood, R.T., 1979, Ap. J., 227, 697, and by Olive, K.A., Schramm, D.N., Steigman, G., Turner, M., and Yang, J., 1981, Ap. J., 246, 557.

List of Illustrations

Fig. 12-1. a) Core structure during the first ascent of the giant branch;

b) Late evolutionary stages during ascent of the asymptotic giant branch.

Fig. 12-2. Evolutionary Track for a Star of Residual Mass 0.6 $M_\odot$ With C-O Core. We plot log L against log T_{eff}, where T_{eff} is the effective temperature. The position of the zero-age main sequence

List of Illustrations (Continued)

is labelled by the line ZAMS; HB denotes the horizontal branch, AGB the asymptotic giant branch. TP indicates the region where thermal pulses occur. The subsequent evolution is indicated by the nearly horizontal line. Times in millenia are indicated by the figures; t = 0 corresponds to the epoch when T(*) = 30,000°K, at which point the PN can be "turned on." At t = 18.6 millenia, the final thermal pulse, indicated by FP, may occur. The star moves quickly along a path sketched by the arrows to R where it is now a cool supergiant. Thereafter, it fades along the path indicated by the descending arrows. From the time when T(*) = 30,000°K again, it takes about 40,000 years for its luminosity to fade a hundredfold. The excitation of the planetary may be rekindled during this phase (adapted from a diagram by Iben, Kaler, Truran, and Renzini 1983).

Fig. 12-3. Comparison of log n(C)/n(O) With the He/H Ratio. The solid curve gives the predicted value for the parameter $\alpha = \ell/H = 2$ (where ℓ is the mixing length and H is the scale height); the dotted curve gives the predicted value for $\alpha = 1.5$ (Renzini and Voli 1981). The numbers in parentheses indicate the initial masses of the progenitor stars ranging from 1.0 to 5.0. Solid dots indicate the more accurate determinations; open circles the less accurate ones. Note the position of the nitrogen-rich object, NGC 6302. Compare Kaler (1983).

Fig. 12-4. Comparison of log n(N)/n(O) With the He/H Ratios. The notation and symbols are similar to those used in Fig. 3. The Renzini-Voli theoretical curve rises too steeply for large He/H ratios. Evidently, they have assumed too efficient a mechanism for converting C to N (compare Fig. 6, Kaler 1983).

Fig. 12-5. Comparison of log n(C)/n(O) With log n(N)/n(O). The notation and symbols are similar to those used in Fig. 3 (compare Peimbert 1981; Kaler 1983). The agreement is not good for large N/O ratios. As noted for Fig. 4, the assumed efficiency for conversion of C to N appears to be too high.

Fig. 12-6. Comparison of log N/O With He/H. Here additional observational data, involving many nebulae of low surface brightness and large N/H ratios, are included. Solid circles: Population I, open circles: Population II. The solid curve gives the mean of the predicted Becker-Iben and Renzini-Voli values through the third dredge-up. The dotted curve gives the Renzini-Voli values with deep envelope burning. The dashed curve gives the Becker-Iben results with the carbon-to-nitrogen conversion assumed to be half the full rate of envelope burning. (Courtesy, J.B. Kaler)

APPENDIX

(Reproduced from IAU Symposium 103, Planetary Nebulae, Courtesy: C. Mendoza)

COMPILATION OF TRANSITION PROBABILITIES, ELECTRON EXCITATION RATE COEFFICIENTS AND PHOTOIONIZATION CROSS SECTIONS

We include in this section a selected and critically evaluated compilation of transition probabilities, electron excitation rate coefficients and photoionization cross sections for use in the study of planetary nebulae. We have attempted to present the data in a clear, practical and self-explanatory manner, but find it necessary to discuss briefly the general arrangements of the tables.

1 INDICES

Table 1 is an index of the ions and transitions for which A-values and electron excitation rate coefficients are included in this compilation. It also contains the numbers of the relevant tables (tables 2 - 9), the methods used, estimates of accuracy and the reference sources (referenced at the end of the Appendix). Similarly, table 10 contains a selected bibliography of photoionization cross sections; for each ion, the initial stages for which cross sections have been obtained, the states of the final ion, the method, accuracy and the reference source are given.

2 METHODS

The following abbreviations are used:
CI: Configuration interaction method.
nCC: n-state close-coupling approximation.
MV: Matrix variational method.
ER: Exact resonance approximation.
DW: Distorted wave approximation.
CBII: Unitarised Coulomb-Born approximation.
MP: Polarization model potential calculations.
CP: Central potential calculations.
MBPT: Many-body perturbation theory.
TDHF: Time-dependent Hartree-Fock method.
SE: Semi-empirical calculations.
TABLES: Values obtained from NBS tables.
INT(EXT): Interpolated (Extrapolated).
COMP: Compilation of several methods.
EXPT: Experiment.

3 ACCURACY

The following scheme is used:
A: Uncertainties within 10%
B: " " 20%

C: Uncertainties within 30%
D: " " 40%

A + sign means "much better than". Accuracy ratings given in this work should only be treated as a rough guide as it is very difficult to critically evaluate these data due to the general lack of experimental results, theoretical comparisons and even error estimates provided by the authors themselves. Furthermore, it is not practical within the present context to give estimates of uncertainties for every transition within a multiplet or configuration, and in some cases they can be significantly different. For instance, for the $ns^2 - nsnp$ transitions of the Be and Mg sequences, it is the transition probability for the intercombination line $^1S_0 - {}^3P^o_1$ that shows the greater uncertainties whereas the A-values for the other transitions are probably correct to 5%; for the forbidden transitions within the np^q configurations the accuracy of the M1 transition probabilities is appreciably greater than that of the E2 type; also the accuracy of the photoionization cross section of the ground state of an ion can differ from that of the excited states, and the error in the cross sections can also vary with the energy region particularly where resonances are present.

4 TRANSITION PROBABILITIES AND EXCITATION RATE COEFFICIENTS

Tables 2 - 9 are arranged in isoelectronic sequences. For each transition they list: the observed energy level separation $\Delta E_{ij}(\mathrm{cm}^{-1})$, the transition probability $A_{ij}(\mathrm{sec}^{-1})$, and the effective collision strength $\Upsilon_{ij}(T_e)$ as a function of electron temperature. The ith level always corresponds to the upper level unless ΔE appears with a negative sign (the A and Υ values are always for the downward transition). The electron de-excitation rate coefficient, q_{ij}, can be obtained from the relation

$$q_{ij}(T_e) = \frac{8.63 \times 10^{-6}\ \Upsilon_{ij}(T_e)}{\omega_i\ T_e^{\frac{1}{2}}} \qquad (\mathrm{cm}^3\,\mathrm{sec}^{-1})$$

where ω_i is the statistical weight of the ith level and T_e is the electron temperature in ^{o}K. The excitation rate coefficient is given by

$$q_{ji} = (\omega_i/\omega_j)\ q_{ij}\ \exp(-\Delta E_{ij}/kT_e) \qquad (i>j)$$

where k is the Boltzmann constant ($1/k = 1.43883$ cm K).

When only one value of Υ is given it is assumed to be temperature independent to within the accuracy of the calculation. A value of the effective collision strength bracketed by arrows ($\leftarrow \Upsilon \rightarrow$) corresponds to the value for the whole multiplet (LS coupling); in this case the Υ's for the fine-structure transitions can be obtained from the ratios of the statistical weights of the multiplet levels. For the $^2P^o_J - {}^4P_{J'}$ transition of the B and Aℓ isoelectronic sequences both the total T for the multiplet and that for each fine-structure transition are given, as in some cases they have been computed by different methods (eg N^{2+}) or, as explained by the author (eg O^{3+}), the sum of the fine-structure components does not add up to the total LS value given.

Finally, the notation $a \pm b$ signifies $a \times 10^{\pm b}$.

NOTE: The author kindly requests users to quote the original sources whenever data from this compilation are referenced.

ION	TRANSITION(S)	TABLE	A_{IJ} METHOD	A_{IJ} ACC	A_{IJ} SOURCE	Υ_{IJ} METHOD	Υ_{IJ} ACC	Υ_{IJ} SOURCE
H^o	n=2	2	TABLES	A+	1	CC	B+	126
He^o	n=2	2	COMP	A	1,2,3,4,5	5CC	C+	6
C^o	$2p^2\ {}^1S,\ {}^1D,\ {}^3P$	6	CI	A	7	CC+CI	C	8
C^o	$2s2p^3\ {}^5S^o_2 - 2s^22p^2\ {}^3P_J$	6	CI	B	39	MV		9
C^+	$2s2p^2\ {}^4P_J - 2s^22p\ {}^2P^o_{J'}$	5	CI	B+	10	8CC+CI	B	11
C^+	$2s2p^2\ {}^4P - 2s^22p\ {}^2P^o$	5				5CC+CI	B+	12
C^{2+}	$2s2p\ {}^1P^o_1,\ {}^3P^o_J,\ 2s^2\ {}^1S_0$	4	CI	A	13	6CC+CI	A	14
C^{3+}	$2p\ {}^2P^o_J - 2s\ {}^2S_{\frac{1}{2}}$	3	TABLES	A+	1	EXPT;5CC	A+	15;16
N^o	$2p^3\ {}^2P^o,\ {}^2D^o,\ {}^4S^o$	7	CI	A	17	8CC+CI	A	18
N^o	${}^2P^o_J - {}^2D^o_{J'}$	7	CI	A	17	CC	C	19
N^+	$2p^2\ {}^1S,\ {}^1D,\ {}^3P$	6	CI	A+	7	CC+CI	A	20
N^+	$2p^2\ {}^3P_J - {}^3P_{J'}$	6	CI	A+	7	ER	C	21
N^+	$2s2p^3\ {}^5S^o_2 - 2s^22p^2\ {}^3P_J$	6	CI	B+	22	7CC+CI	B+	12
N^{2+}	$2s2p^2\ {}^4P_J - 2s^22p\ {}^2P^o_{J'}$	5	CI	B+	23	DW	C	23
N^{2+}	$2s2p^2\ {}^4P - 2s^22p\ {}^2P^o$	5				6CC+CI	A	24
N^{3+}	$2s2p\ {}^1P^o_1,\ {}^3P^o_J,\ 2s^2\ {}^1S_0$	4	CI	A	25	INT	B+	
N^{4+}	$2p\ {}^2P^o_J - 2s\ {}^2S_{\frac{1}{2}}$	3	TABLES	A+	1	2CC	A+	26,27
O^o	$2p^4\ {}^1S,\ {}^1D,\ {}^3P$	8	CI	A	28	6CC+CI	A	18
O^o	$2p^4\ {}^3P_J - {}^3P_{J'}$	8	CI	A	28	MV	B+	29
O^+	$2p^3\ {}^2P^o_J,\ {}^2D^o_{J'},\ {}^4S^o_{\frac{3}{2}}$	7	CI	A+	17	5CC+CI	A	30
O^{2+}	$2s2p^3\ {}^5S^o,\ 2s^22p^2\ {}^1S,\ {}^1D,\ {}^3P$	6	CI	A+	31	12CC+CI	A	32
O^{2+}	$2p^2\ {}^3P_J - {}^3P_{J'}$	6	CI	A+	31	12CC+CI	A	33
O^{3+}	$2s2p^2\ {}^4P_J - 2s^22p\ {}^2P^o_{J'}$	5	CI	B	34	7CC+CI	A	34
O^{4+}	$2s2p\ {}^1P^o_1,\ {}^3P^o_J,\ 2s^2\ {}^1S_0$	4	CI	A	25	6CC+CI	A	14
O^{5+}	$2p\ {}^2P^o_J - 2s\ {}^2S_{\frac{1}{2}}$	3	TABLES	A+	1	CBII	A	26

TABLE 1. Index of transition probabilities, A_{ij}, and effective collision strengths, Υ_{ij}, for transitions in ions of interest in the study of planetary nebulae. For each ion the number of the relevant table (tables 2 - 9), the method used, an estimate of the accuracy and the reference source (referenced at the end of the appendix) is given.

ION	TRANSITION(S)		A_{IJ}			Υ_{IJ}		
		TABLE	METHOD	ACC	SOURCE	METHOD	ACC	SOURCE
Ne^{+}	$2p^5\ ^2P^o_{\frac{1}{2}} - ^2P^o_{\frac{3}{2}}$	9	CI	A+	28	ER(adj.)	A	20
Ne^{2+}	$2p^4\ ^1S,\ ^1D,\ ^3P$	8	CI	A+	28	4CC+CI	A	35
Ne^{2+}	$2p^4\ ^3P_J - ^3P_{J'}$	8	CI	A+	28	ER	C	21
Ne^{3+}	$2p^3\ ^2P^o_J,\ ^2D^o_{J'},\ ^4S^o_{\frac{3}{2}}$	7	CI	A+	17	9CC+CI	A	36
Ne^{4+}	$2p^2\ ^1S,\ ^1D,\ ^3P$	6	CI	A+	7	12CC+CI	A	37
Ne^{4+}	$2s2p^3\ ^5S^o_2 - 2s^22p^2\ ^3P_J$	6	CI	A	39	12CC+CI	A	37
Ne^{4+}	$2p^2\ ^3P_J - ^3P_{J'}$	6	CI	A+	7	ER	C	21
Ne^{5+}	$2s^22p\ ^2P^o_{\frac{3}{2}} - ^2P^o_{\frac{1}{2}}$	5	TABLES	A+	1	ER	C	21
Ne^{6+}	$2s2p\ ^1P^o_1,\ ^3P^o_J,\ 2s^2\ ^1S_0$	4	CI	A	25	6CC+CI	A	38
Na^{2+}	$2p^5\ ^2P^o_{\frac{1}{2}} - ^2P^o_{\frac{3}{2}}$	9	CI	A+	28	ER	C	21
Na^{3+}	$2p^4\ ^1S_0,\ ^1D_2,\ ^3P_J$	8	CI	A+	28	ER	C	21
Na^{4+}	$2p^3\ ^2P^o_J,\ ^2D^o_{J'},\ ^4S^o_{\frac{3}{2}}$	7	CI	A+	17	ER	C	21
Mg^{o}	$3s3p\ ^3P^o_1,\ ^3P^o_J,\ 3s^2\ ^1S_0$	4	CI,MP	B	39,40	DW,CC		41,42
Mg^{+}	$3p\ ^2P^o_J - 3s\ ^2S_{\frac{1}{2}}$	3	MP	A+	43	4CC	A	44
Mg^{3+}	$2p^5\ ^2P^o_{\frac{1}{2}} - ^2P^o_{\frac{3}{2}}$	9	CI	A+	28	ER	C	21
Mg^{4+}	$2p^4\ ^1S_0,\ ^1D_2,\ ^3P_J$	8	CI	A+	28	ER	C	21
Mg^{5+}	$2p^3\ ^2P^o_J,\ ^2D^o_{J'},\ ^4S^o_{\frac{3}{2}}$		CI	A+	17	ER	C	21
Si^{o}	$3p^2\ ^1S_0,\ ^1D_2,\ ^3P_J$	6	CI	A	45	DW		46
Si^{+}	$3s3p^2\ ^4P - 3s^23p\ ^2P^o$	5	CI	B	47	7CC+CI	B	48
Si^{2+}	$3s3p\ ^1P^o_1,\ ^3P^o_J,\ 3s^2\ ^1S_0$	4	CI,MP	B	39,40	12CC+CI	B+	49
Si^{3+}	$3p\ ^2P^o_J - 3s\ ^2S_{\frac{1}{2}}$	3	1CC+MP	A+	28	DW	B	26,50
Si^{5+}	$2p^5\ ^2P^o_{\frac{1}{2}} - ^2P^o_{\frac{3}{2}}$	9	CI	A+	28	ER	C	21
S^{o}	$3p^4\ ^1S_0,\ ^1D_2,\ ^3P_J$	8	CI	A	51			
S^{+}	$3p^3\ ^2P^o_J,\ ^2D^o_{J'},\ ^4S_{\frac{3}{2}}$	7	CI	A+	52	6CC+CI	A	53
S^{2+}	$3p^2\ ^1S,\ ^1D,\ ^3P$	6	CI	A+	45	12CC+CI	B+	53
S^{2+}	$3p^2\ ^3P_J - ^3P_{J'}$	6	CI	A+	45	7CC+CI	B+	53
S^{3+}	$3s3p^2\ ^4P_J - 3s^23p\ ^2P^o_{J'}$	5	CI	B+	54	6CC+CI	B+	54
S^{4+}	$3s3p\ ^1P^o,\ ^3P^o,\ 3s^2\ ^1S$	4	CI,MP	B	39,40	5CC+CI	B+	56
S^{4+}	$3s3p\ ^3P^o_J - ^3P^o_{J'}$	4	CI	A+	39	DW(EXT)		55
S^{5+}	$3p\ ^2P^o_J - 3s\ ^2S_{\frac{1}{2}}$	3	1CC+MP	A+	28	DW	B	26,50

TABLE 1. (continued)

ION	TRANSITION(S)	TABLE	A_{IJ} METHOD	A_{IJ} ACC	A_{IJ} SOURCE	Υ_{IJ} METHOD	Υ_{IJ} ACC	Υ_{IJ} SOURCE
$C\ell^{o}$	$3p^5\ {}^2P^o_{\frac{1}{2}} - {}^2P^o_{\frac{3}{2}}$	9	CI	A+	28			
$C\ell^{+}$	$3p^4\ {}^1S_0,\ {}^1D_2,\ {}^3P_J$	8	CI	A+	51	DW	D	57
$C\ell^{2+}$	$3p^3\ {}^2P^o_J,\ {}^2D^o_{J'},\ {}^4S^o_{\frac{3}{2}}$	7	CI	A+	52	DW	D	57
$C\ell^{3+}$	$3p^2\ {}^1S,\ {}^1D,\ {}^3P$	6	CI	A+	45	12CC+CI	B	53
$C\ell^{3+}$	$3p^2\ {}^3P_J - {}^3P_{J'}$	6	CI	A+	45	DW	D	57
$C\ell^{4+}$	$3s^23p\ {}^2P^o_{\frac{3}{2}} - {}^2P^o_{\frac{1}{2}}$	5	CI	A+	28	DW	D	57
A^{+}	$3p^5\ {}^2P^o_{\frac{1}{2}} - {}^2P^o_{\frac{3}{2}}$	9	CI	A+	28	DW	D	57
A^{2+}	$3p^4\ {}^1S_0,\ {}^1D_2,\ {}^3P_J$	8	CI	A+	51	DW	D	57
A^{3+}	$3p^3\ {}^2P^o_J,\ {}^2D^o_{J'},\ {}^4S^o_{\frac{3}{2}}$	7	CI	A+	52	DW	D	57
A^{4+}	$3p^2\ {}^1S,\ {}^1D,\ {}^3P$	6	CI	A+	45	12CC+CI	B	53
A^{4+}	$3p^2\ {}^3P_J - {}^3P_{J'}$	6	CI	A+	45	DW	D	57
A^{5+}	$3s^23p\ {}^2P^o_{\frac{3}{2}} - {}^2P^o_{\frac{1}{2}}$	5	CI	A+	28	DW	D	57
K^{2+}	$3p^5\ {}^2P^o_{\frac{1}{2}} - {}^2P^o_{\frac{3}{2}}$	9	CI	A+	28	DW	D	57
K^{3+}	$3p^4\ {}^1S_0,\ {}^1D_2,\ {}^3P_J$	8	CI	A+	51	DW	D	57
K^{4+}	$3p^3\ {}^2P^o_J,\ {}^2D^o_{J'},\ {}^4S^o_{\frac{3}{2}}$	7	CI	A+	52	DW	D	57
Ca^{+}	$4p\ {}^2P^o_J - 4s\ {}^2S_{\frac{1}{2}}$	3	MP	A	43	EXPT	A	26,58
Ca^{3+}	$3p^5\ {}^2P^o_{\frac{1}{2}} - {}^2P^o_{\frac{3}{2}}$	9	CI	A+	28	DW	D	57
Ca^{4+}	$3p^4\ {}^1S_0,\ {}^1D_2,\ {}^3P_J$	8	CI	A+	51	DW	D	57
Fe^{o}	Forbidden transitions		CI		59			
Fe^{+}	Forbidden transitions		CI	D	60	4CC+CI	D	60
Fe^{2+}	Forbidden transitions		SE		61	5CC		62
Fe^{3+}	Forbidden transitions		SE		63,64			
Fe^{4+}	Forbidden transitions		SE		61			
Fe^{5+}	Forbidden transitions		CI	C	65	DW	C	65
Fe^{6+}	Forbidden transitions		CI	B	66	DW	C	66

TABLE 1. (continued)

PARM	$T_e(10^4K)$	2s-1s	2p-1s
ΔE		82258.9	82259.2
A			4.699+8
T	0.5	0.244	0.403
	1.0	0.280	0.485
	1.5	0.300	0.563
	2.0	0.311	0.631

(i)

PARM	$T_e(10^4K)$	2^3S-1^1S	2^1S-1^1S	$2^3P^o-1^1S$	$2^1P^o-1^1S$
ΔE		159850.3	166271.7	169081.3	171129.2
A		1.13-4	5.13+1	1.76+2	1.799+9
T	0.5	0.0686	0.0342	0.0126	0.0081
	0.75	0.0719	0.0393	0.0169	0.0122
	1.0	0.0736	0.0435	0.0213	0.0164
	2.0	0.0781	0.0612	0.0382	0.0347
	3.0	0.0817	0.0792	0.0526	0.0543

(ii)

PARM	$T_e(10^4K)$	2^1S-2^3S	$2^3P^o-2^3S$	$2^1P^o-2^3S$	$2^3P^o-2^1S$	$2^1P^o-2^1S$	$2^1P^o-2^3P^o$
ΔE		6421.4	9230.9	11278.8	2809.6	4857.5	2047.9
A		1.51-7	1.022+7	1.29	2.7-2	1.976+6	
T	0.1	1.61	3.08	0.312	0.967	1.07	2.47
	0.2	1.97	5.47	0.464	1.15	2.87	2.83
	0.5	2.39	14.0	0.750	1.41	9.35	3.42
	0.75	2.49	21.9	0.906	1.55	14.5	3.75
	1.0	2.52	29.5	1.03	1.66	19.1	4.05
	2.0	2.45	57.0	1.28	1.89	33.2	4.80
	3.0	2.32	80.0	1.36	1.94	44.1	5.08

(iii)

TABLE 2. Energy level separations ΔE (cm^{-1}), transition probabilities A (sec^{-1}) and effective collision strengths $T(T_e)$ for: (i) 1 → 2 transitions of H^o; (ii) 1 → 2 transitions of He^o; (iii) transitions within n=2 of He^o.

ION	PARM	$T_e(10^4K)$	$^2P^o_{3/2}-^2S_{1/2}$	$^2P^o_{1/2}-^2S_{1/2}$	ION	PARM	$T_e(10^4K)$	$^2P^o_{3/2}-^2S_{1/2}$	$^2P^o_{1/2}-^2S_{1/2}$	ION	PARM	$T_e(10^4K)$	$^2P^o_{3/2}-^2S_{1/2}$	$^2P^o_{1/2}-^2S_{1/2}$
C^{3+}	ΔE		64591.7	64484.0	Mg^+	ΔE		35761.0	35669.4	Ca^+	ΔE		25414.4	25191.5
	A		2.65+8	2.63+8		A		2.55+8	2.54+8		A		1.48+8	1.45+8
	T	0.5				T	0.5	← 15.6 →			T	0.5	← 15.6 →	
		1.0	← 8.88 →				1.0	16.5				1.0	17.5	
		1.5					1.5	17.2				1.5	19.2	
		2.0	8.95				2.0	17.7				2.0	20.8	
N^{4+}	ΔE		80721.9	80463.2	Si^{3+}	ΔE		71748.6	71287.5					
	A		3.38+8	3.36+8		A		9.26+8	9.15+8					
	T	0.5	← 6.61 →			T	0.5	← 16.9 →						
		1.0	6.65				1.0	17.0						
		1.5	6.69				1.5	17.0						
		2.0	6.72				2.0	17.1						
O^{5+}	ΔE		96907.5	96375.0	S^{5+}	ΔE		107137	105874					
	A		4.09+8	4.02+8		A		1.75+9	1.70+9					
	T	0.5	← 4.98 →			T	0.5	← 11.8 →						
		1.0	5.00				1.0	11.9						
		1.5	5.03				1.5	11.9						
		2.0	5.05				2.0	11.9						

TABLE 3. Energy level separations ΔE (cm^{-1}), transition probabilities A (sec^{-1}) and effective collision strengths $T(T_e)$ for the ns - np transition of the Li (n=2), Na (n=3) and K (n=4) isoelectronic sequences.

ION	PARM	$T_e(10^4K)$	$^3P^o_2-{}^1S_0$	$^3P^o_1-{}^1S_0$	$^3P^o_0-{}^1S_0$	$^1P^o_1-{}^1S_0$	$^3P^o_1-{}^3P^o_0$	$^3P^o_2-{}^3P^o_0$	$^3P^o_2-{}^3P^o_1$
C^{2+}	ΔE		52447.1	52390.8	52367.1	102352.0	23.7	80.1	56.4
	A		5.19-3	9.59+1		1.79+9	2.39-7		2.41-6
	T	0.5	←	1.12	→	3.85	0.848	0.579	2.36
		1.0		1.01		4.34	0.911	0.677	2.66
		1.5		0.990		4.56	0.975	0.776	2.97
		2.0		0.996		4.69	1.03	0.867	3.23
N^{3+}	ΔE		67416.3	67272.3	67209.2	130693.9	63.1	207.1	144.0
	A		1.15-2	5.77+2		2.38+9	4.53-6		4.03-5
	T	0.5	←	0.904	→	3.20			
		1.0		0.852		3.46			
		1.5		0.817		3.58			
		2.0		0.798		3.65			
O^{4+}	ΔE		82385.3	82078.6	81942.5	158797.7	136.1	442.8	306.7
	A		2.16-2	2.25+3		2.92+9	4.54-5		3.89-4
	T	0.5	←	0.733	→	2.66			
		1.0		0.721		2.76			
		1.5		0.674		2.82			
		2.0		0.639		2.85			
Ne^{6+}	ΔE		112711.5	111717	111264.9	214951.6	452	1446.6	995
	A		5.78-2	1.98+4		4.08+9	1.69-3		1.32-2
	T	0.5	←	0.129	→	1.39			
		1.0		0.172		1.56			
		1.5		0.205		1.63			
		2.0		0.228		1.66			
Mg^o	ΔE		21911.2	21870.5	21850.4	35051.3	20.1	60.8	40.7
	A		4.13-4	1.80+2		4.93+8	1.45-7	4.08-12	9.10-7
	T								
Si^{2+}	ΔE		53115.0	52853.3	52724.7	82884.4	128.6	390.3	261.7
	A		1.20-2	1.26+4		2.60+9	3.82-5	3.20-9	2.42-4
	T	0.5	←	6.90	→	5.48			
		1.0		5.43		5.82			
		1.5		4.80		6.21			
		2.0		4.41		6.54			
S^{4+}	ΔE		84155.2	83393.5	83024.0	127150.7	369.5	1131.2	761.7
	A		6.59-2	1.26+5		5.13+9	9.07-4	1.63-7	5.96-3
	T	0.5	←	0.911	→	7.30	0.272	0.400	1.24
		1.0		0.910		7.30			
		1.5		0.914		7.29			
		2.0		0.905		7.27			

TABLE 4. Energy level separations ΔE (cm^{-1}), transition probabilities A (sec^{-1}) and effective collision strengths $T(T_e)$ for the ns^2 - nsnp transitions of the Be (n=2) and Mg (n=3) isoelectronic sequences.

ION	PARM	$T_e(10^4K)$	$^4P-^2P^o$	$^2P^o_{3/2}-^2P^o_{1/2}$	$^4P_{1/2}-^2P^o_{1/2}$	$^4P_{1/2}-^2P^o_{3/2}$	$^4P_{3/2}-^2P^o_{1/2}$	$^4P_{3/2}-^2P^o_{3/2}$	$^4P_{3/2}-^4P_{1/2}$	$^4P_{5/2}-^2P^o_{1/2}$	$^4P_{5/2}-^2P^o_{3/2}$	$^4P_{5/2}-^4P_{1/2}$	$^4P_{5/2}-^4P_{3/2}$
C^+	ΔE			63.42	43003.3	42939.9	43025.3	42961.9	22.0	43053.6	42990.2	50.3	28.3
	A			2.29-6	5.53+1	6.55+1	1.71	5.24	2.39-7		4.32+1	3.49-14	3.67-7
	T	0.4	3.25										
		1.0	3.17	1.25									
		1.5	3.09										
		2.0	2.97										
N^{2+}	ΔE			174.4	57187.1	57012.7	57246.8	57072.4	59.7	57327.9	57153.5	140.8	81.1
	A			4.77-5	3.39+2	3.64+2	8.95	5.90+1			2.51+2		
	T	0.4	2.00	0.701	0.0952	0.0616	0.139	0.175	0.695	0.0800	0.390	0.397	1.26
		1.0	2.03										
		1.4	2.07										
		2.0	2.11										
O^{3+}	ΔE			386.3	71440.0	71053.7	71571.4	71185.1	131.4	71755.9	71369.6	315.9	184.5
	A			5.20-4	1.22+3	1.24+3	3.24+1	2.36+2	5.07-5		9.37+2		1.02-4
	T	1.0	1.27	2.33	0.113	0.087	0.174	0.234	0.989	0.122	0.506	0.592	1.82
		1.5	1.39										
		2.0	1.48	2.40	0.128	0.106	0.197	0.274	1.05	0.147	0.568	0.652	1.97
Ne^{5+}	ΔE			1306.6									
	A			2.02-2									
	T			0.433									
Si^+	ΔE			287.3	42824.4	42537.0	42932.7	42645.4	108.3	43108.0	42820.7	283.6	175.3
	A			2.17-4	4.55+3	3.00+3	1.32+1	1.62+3			2.40+3		
	T	0.5	5.28										
		1.0	5.14										
		1.5	5.05										
		2.0	4.97										
S^{3+}	ΔE			950.2	71180	70230	71524	70574	344	72071	71121	891	547
	A			7.73-3	5.50+4	3.39+4	1.40+2	1.95+4			3.95+4		
	T	1.0		6.42	0.51	0.66	0.87	1.47	3.04	0.95	2.53	2.92	7.01
		1.59		6.41	0.47	0.62	0.82	1.38	2.82	0.90	2.39	2.68	6.49
		2.51		6.38	0.44	0.59	0.77	1.30	2.59	0.85	2.26	2.40	5.85
Cl^{4+}	ΔE			1490.8									
	A			2.98-2									
	T			1.052									
A^{5+}	ΔE			2207.5									
	A			9.66-2									
	T			0.798									

TABLE 5. Energy level separations ΔE (cm^{-1}), transition probabilities A (sec^{-1}), and effective collision strengths T(E) for the $ns^2np - nsnp^2$ transitions of the B (n=2) and Aℓ (n=3) isoelectronic sequences.

ION	PARM	$T_e(10^4K)$	$^1D_2-^3P_0$	$^1D_2-^3P_1$	$^1D_2-^3P_2$	$^1S_0-^3P_1$	$^1S_0-^3P_2$	$^1S_0-^1D_2$	$^3P_1-^3P_0$	$^3P_2-^3P_0$	$^3P_2-^3P_1$	$^5S^o_2-^3P_1$	$^5S^o_2-^3P_2$
C^o	ΔE		10192.6	10176.2	10149.2	21631.6	21604.6	11455.4	16.4	43.4	27.0	33718.8	33691.8
	A		7.77-8	8.21-5	2.44-4	2.71-3	2.00-5	5.28-1	7.93-8	1.71-14	2.65-7	6.94	1.56+1
	T	0.05	←	0.0625	→	← 0.0172 →		0.0620				← 0.150 →	
		0.1		0.125		0.0339		0.0877				0.212	
		0.5		0.603		0.149		0.196				0.475	
		1.0		1.14		0.252		0.277				0.671	
		1.5		1.60		0.320		0.340				0.822	
		2.0		1.96		0.365		0.392				0.950	
N^+	ΔE		15316.2	15267.5	15185.4	32640.1	32558.0	17372.6	48.7	130.8	82.1	46735.9	46653.8
	A		5.35-7	1.01-3	2.99-3	3.38-2	1.51-4	1.12	2.08-6	1.16-12	7.46-6	4.8+1	1.07+2
	T	0.5	←	2.64	→	← 0.352 →		0.405	0.401	0.279	1.13	← 1.27 →	
		1.0		2.68		0.352		0.411				1.28	
		1.5		2.72		0.359		0.418				1.29	
		2.0		2.73		0.365		0.425				1.27	
O^{2+}	ΔE		20273.3	20160.1	19967.1	43072.5	42879.5	22912.4	113.2	306.2	193.0	60211.8	60018.8
	A		2.74-6	6.74-3	1.96-2	2.23-1	7.85-4	1.78	2.62-5	3.02-11	9.76-5	2.12+2	5.22+2
	T	0.5	←	2.02	→	← 0.248 →		0.516	0.517	0.257	1.22	← 1.05 →	
		1.0		2.17		0.276		0.617	0.542	0.271	1.29	1.18	
		1.5		2.30		0.299		0.638	0.553	0.281	1.32	1.22	
		2.0		2.39		0.314		0.634	0.556	0.288	1.34	1.24	
Ne^{4+}	ΔE		30291.5	29879.1	29181.4	63501.2	62803.5	33622.1	412.4	1110.1	697.7	87950.7	87253.0
	A		2.37-5	1.31-1	3.65-1	4.21	6.69-3	2.85	1.28-3	5.08-9	4.59-3	2.37+3	6.06+3
	T	0.5	←	1.70	→	← 0.284 →		0.581	0.244	0.122	0.578	← 1.19 →	
		1.0		1.78		0.248		0.518				1.51	
		1.5		1.85		0.240		0.550				1.53	
		2.0		1.92		0.238		0.602				1.51	
Si^o	ΔE		6298.9	6221.7	6075.7	15317.3	15171.2	9095.5	77.1	223.2	146.0	33248.9	33102.9
	A		4.70-7	7.93-4	2.25-3	3.13-2	9.02-4	1.14	8.25-6		4.21-5		
	T												
S^{2+}	ΔE		11320	11023	10488	26866	26331	15843	297.2	832.5	535.3	59401	58866
	A		5.82-6	2.21-2	5.76-2	7.96-1	1.05-2	2.22	4.72-4	4.61-8	2.07-3		
	T	0.5	←	9.07	→	← 1.16 →		1.42	2.64	1.11	5.79		
		1.0		8.39		1.19		1.88	2.59	1.15	5.81		
		1.5		8.29		1.21		2.02	2.38	1.15	5.56		
		2.0		8.20		1.24		2.08	2.20	1.14	5.32		
$C\ell^{3+}$	ΔE		13767.6	13275.6	12425.7	32055.8	31205.9	18780.2	492.0	1341.9	849.9		
	A		1.54-5	7.23-2	1.79-1	2.47	2.62-2	2.80	2.14-3	2.70-7	8.25-3		
	T	0.5	←	5.10	→	← 2.04 →		0.935	0.475	0.400	1.50		
		1.0		5.42		2.27		1.39					
		1.5		5.88		2.32		1.73					
		2.0		6.19		2.30		1.92					
A^{4+}	ΔE		16299.4	15535.5	14270.2	37148.6	35883.3	21613.1	763.9	2029.2	1265.3		
	A		3.50-5	2.04-1	4.76-1	6.55	5.69-2	3.29	7.99-3	1.24-6	2.72-2		
	T	0.5	←	4.37	→	← 1.17 →		1.26	0.257	0.320	1.04		
		1.0		3.72		1.18		1.25					
		1.5		3.52		1.11		1.24					
		2.0		3.42		1.03		1.23					

TABLE 6. Energy level separations ΔE (cm^{-1}), transition probabilities A (sec^{-1}) and effective collision strengths $T(T_e)$ for the ns^2np^2 and $nsnp^3$ transitions of the C (n=2) and Si (n=3) isoelectronic sequences.

ION	PARM	$T_e(10^4K)$	$^2D^o_{5/2}-^4S^o_{3/2}$	$^2D^o_{3/2}-^4S^o_{3/2}$	$^2P^o_{3/2}-^4S^o_{3/2}$	$^2P^o_{1/2}-^4S^o_{3/2}$	$^2D^o_{5/2}-^2D^o_{3/2}$	$^2P^o_{3/2}-^2P^o_{1/2}$	$^2P^o_{3/2}-^2D^o_{5/2}$	$^2P^o_{3/2}-^2D^o_{3/2}$	$^2P^o_{1/2}-^2D^o_{5/2}$	$^2P^o_{1/2}-^2D^o_{3/2}$
N^0	ΔE		19224.5	19233.2	28839.3	28838.9	-8.71	0.386	9614.8	9606.1	9614.5	9605.7
	A		7.27-6	2.02-5	6.58-3	2.71-3	1.27-8		6.14-2	2.76-2	3.45-2	5.29-2
	T	0.05	9.18-3	6.12-3	3.2-3	1.6-3						
		0.1	2.30-2	1.53-2	8.53-3	4.27-3						
		0.5	0.155	0.103	0.0597	0.0298	0.128	0.0329	0.162	0.0856	0.0626	0.0601
		1.0	0.290	0.194	0.113	0.0567	0.269	0.071	0.266	0.147	0.109	0.097
		2.0	0.476	0.318	0.189	0.0947	0.465	0.153	0.438	0.252	0.190	0.157
O^+	ΔE		26810.7	26830.2	40467.5	40468.6	-19.5	-1.1	13656.8	13637.3	13657.9	13638.4
	A		3.82-5	1.65-4	5.64-2	2.32-2	1.20-7	2.08-11	1.17-1	6.14-2	6.15-2	1.02-1
	T	0.5	0.795	0.530	0.265	0.133	1.22	0.280	0.718	0.401	0.290	0.270
		1.0	0.801	0.534	0.270	0.135	1.17	0.287	0.730	0.408	0.295	0.275
		1.4	0.808	0.538	0.274	0.137	1.14	0.292	0.740	0.413	0.299	0.279
		2.0	0.818	0.545	0.280	0.140	1.11	0.300	0.755	0.422	0.305	0.284
Ne^{3+}	ΔE		41234.6	41279.5	62441.3	62434.6	-44.9	6.7	21206.7	21161.8	21200.0	21155.1
	A		4.84-4	5.54-3	1.27	5.21-1	1.48-6	2.68-9	4.00-1	4.37-1	1.15-1	3.93-1
	T	0.6	0.843	0.562	0.308	0.154	1.37	0.323	0.867	0.482	0.347	0.327
		1.0	0.838	0.559	0.313	0.156	1.36	0.343	0.900	0.509	0.368	0.336
		1.4	0.834	0.556	0.312	0.156	1.36	0.355	0.908	0.515	0.373	0.339
		2.0	0.824	0.550	0.309	0.155	1.33	0.370	0.909	0.516	0.374	0.339
Na^{4+}	ΔE		48313.5	48359.3	73236.4	73201.9	-45.8	34.5	24922.9	24877.1	24888.4	24842.6
	A		1.46-3	2.69-2	4.27	1.76	1.55-6	3.64-7	9.18-1	1.29	1.41-1	9.55-1
	T		0.551	0.368	0.239	0.120	0.696	0.438	0.502	0.279	0.201	0.190
S^+	ΔE		14884.8	14853.0	24571.8	24524.9	31.8	46.9	9687.0	9718.8	9640.1	9671.9
	A		2.60-4	8.82-4	0.225	9.06-2	3.35-7	1.03-6	0.179	0.133	7.79-2	1.63-1
	T	0.5	4.38	2.92	1.44	0.722	8.15	2.39	4.74	3.36	2.55	1.50
		1.0	4.19	2.79	1.52	0.759	7.59	2.38	4.79	3.38	2.56	1.52
		1.5	4.00	2.67	1.54	0.772	7.11	2.33	4.74	3.32	2.51	1.51
		2.0	3.84	2.56	1.54	0.769	6.71	2.27	4.62	3.21	2.43	1.48
$C\ell^{2+}$	ΔE		18118.6	18053	29907	29812	66	95	11788	11854	11693	11759
	A		7.04-4	4.83-3	0.754	0.305	3.22-6	7.65-6	0.316	0.323	0.100	0.303
	T		1.88	1.26	1.26	0.627	3.19	1.34	3.33	1.91	1.38	1.24
A^{3+}	ΔE		21219.3	21090.4	35032.6	34855.5	128.9	177.1	13813.3	13942.2	13636.2	13765.1
	A		1.77-3	2.23-2	2.11	0.862	2.30-5	4.94-5	0.598	0.789	0.119	0.603
	T		0.854	0.570	0.423	0.212	1.35	0.601	2.50	1.24	0.865	1.01
K^{4+}	ΔE		24249.6	24012.5	40080.2	39758.1	237.1	322.1	15830.6	16067.7	15508.5	15745.6
	A		4.59-3	8.84-2	5.19	2.14	1.42-4	2.96-4	1.21	1.86	0.141	1.25
	T		0.455	0.303	0.173	0.086	0.971	0.281	2.12	0.974	0.665	0.884

TABLE 7. Energy level separations ΔE (cm^{-1}), transition probabilities A (sec^{-1}) and effective collision strengths $T(T_e)$ for the np^3 transitions of the N (n=2) and P (n=3) isoelectronic sequences.

ION	PARM	$T_e(10^4K)$	$^1D_2-^3P_0$	$^1D_2-^3P_1$	$^1D_2-^3P_2$	$^1S_0-^3P_1$	$^1S_0-^3P_2$	$^1S_0-^1D_2$	$^3P_0-^3P_1$	$^3P_0-^3P_2$	$^3P_1-^3P_2$
O^o	ΔE		15640.9	15709.6	15867.9	33634.3	33792.6	17924.7	68.7	227.0	158.3
	A		7.23-7	2.11-3	6.34-3	7.32-2	2.88-4	1.22	1.74-5		8.92-5
	T	0.05	←	0.0058	→	← 0.00065 →		0.0221	0.0008	0.0006	0.0027
		0.1		0.0151		0.00184		0.0310	0.0018	0.0022	0.0076
		0.5		0.124		0.0153		0.0732	0.0112	0.0148	0.0474
		1.0		0.266		0.0324		0.105	0.0265	0.0292	0.0987
		2.0		0.501		0.0607		0.148	0.0693	0.0536	0.207
Ne^{2+}	ΔE		24920.3	25197.9	25840.8	55107.7	55750.6	29909.8	277.6	920.5	642.9
	A		8.51-6	5.42-2	1.71-1	2.00	3.94-3	2.71	1.15-3	2.18-8	5.97-3
	T	0.5	←	1.35	→	← 0.152 →		0.220	0.185	0.131	0.527
		1.0		1.34		0.151		0.236			
		1.5		1.33		0.152		0.262			
		2.0		1.32		0.157		0.284			
Na^{3+}	ΔE		29264	29734	30840	65390	66496	35656	469.3	1575.6	1106.3
	A		2.24-5	1.86-1	6.10-1	7.10	1.05-2	3.46	5.57-3	1.67-7	3.04-2
	T		←	1.17	→	← 0.163 →		0.157	0.177	0.111	0.471
Mg^{4+}	ΔE		33410	34149	35932	75503	77286	41354	738.7	2521.8	1783.1
	A		5.20-5	5.41-1	1.85	2.14+1	2.45-2	4.23	2.17-2	1.01-6	1.27-1
	T		←	1.02	→	← 0.146 →		0.129	0.156	0.0908	0.400
S^o	ΔE		8665.0	8842.6	9238.6	21783.9	22180.0	12941.3	177.6	573.7	396.1
	A		3.84-6	8.16-3	2.78-2	3.50-1	8.23-3	1.53	3.02-4	6.71-8	1.39-3
	T										
Cl^+	ΔE		10657.1	10957.6	11653.6	27182.0	27878.0	16224.4	300.5	996.5	696.0
	A		9.82-6	2.92-2	1.04-1	1.31	1.97-2	2.06	1.46-3	4.57-7	7.57-3
	T		←	3.86	→	← 0.456 →		1.15	0.933	0.443	2.17
A^{2+}	ΔE		12439.8	12897.9	14010.0	32153.6	33265.7	19255.7	458.1	1570.2	1112.1
	A		2.21-5	8.23-2	3.14-1	3.91	4.17-2	2.59	5.17-3	2.37-6	3.08-2
	T		←	4.74	→	← 0.680 →		0.823	1.18	0.531	2.24
K^{3+}	ΔE		14062.9	14712.7	16384.1	36874.9	38546.3	22162.2	649.8	2321.2	1671.4
	A		4.54-5	1.98-1	8.14-1	1.00+1	8.17-2	3.18	1.48-2	1.01-5	1.04-1
	T		←	1.90	→	← 0.292 →		0.798	0.421	0.290	1.16
Ca^{4+}	ΔE		15554.7	16425.6	18830.3	41431.8	43836.5	25006.2	870.9	3275.6	2404.7
	A		8.42-5	4.26-1	1.90	2.31+1	1.45-1	3.73	3.54-2	3.67-5	3.10-1
	T		←	0.904	→	← 0.116 →		0.793	0.202	0.224	0.760

TABLE 8. Energy level separations ΔE (cm^{-1}), transition probabilities A (sec^{-1}) and effective collision strengths T(T_e for the np^4 transitions of the O (n=2) and S (n=3) isoelectronic sequences.

ION	PARM	$T_e(10^4K)$	$^2P^o_{\frac{1}{2}}-{}^2P^o_{\frac{3}{2}}$	ION	PARM	$T_e(10^4K)$	$^2P^o_{\frac{1}{2}}-{}^2P^o_{\frac{3}{2}}$
Ne^+	ΔE		780.4	$C\ell^o$	ΔE		882.4
	A		8.55-3		A		1.24-2
	T	0.5	0.362		T		
		1.0	0.368				
		1.5	0.375	A^+	ΔE		1431.6
		2.0	0.381		A		5.27-2
					T		0.635
Na^{2+}	ΔE		1366.7				
	A		4.59-2	K^{2+}	ΔE		2166.1
	T		0.300		A		1.83-1
					T		1.78
Mg^{3+}	ΔE		2229.5				
	A		1.99-1	Ca^{3+}	ΔE		3117.9
	T		0.300		A		5.45-1
					T		1.06
Si^{5+}	ΔE		5094.1				
	A		2.38				
	T		0.242				

TABLE 9. Energy level separations ΔE (cm^{-1}), transition probabilities A (sec^{-} and effective collision strengths $T(T_e)$ for the np^5 transitions of the F (n=2) and Cℓ (n=3) isoelectronic sequences.

PARENT(STATES)	FINAL ION(STATES)	METHOD	ACC	SOURCE
$H^o(n\ell)$	H^+	Hydrogenic	A+	67
$He^o(1\,^1S, 2\,^1S, ^3S, ^1P^o, ^3P^o)$	$He^+(1s, 2s, 2p)$	COMP;EXPT	A	68-71;72-74
$He^+(n\ell)$	He^{2+}	Hydrogenic	A+	67
$C^o(^3P, ^1D, ^1S)$	$C^+(^2P^o, ^4P, ^2D, ^2S, ^2P)$	8CC+CI,5CC+CI;EXPT	B+;C	75,105;76
$C^+(^2P^o, ^4P)$	C^{2+}	6CC+CI;CC	;D	77;85
$C^{2+}(^1S, ^3P^o)$	C^{3+}	5CC; TDHF	A;	78;79
$C^{3+}(^2S)$	C^{4+}	5CC	A	78,80
$N^o(^4S^o, ^2D^o, ^2P^o)$	$N^+(^3P, ^1D, ^1S, ^5S^o, ^3D^o, ^3P^o, ^1D^o, ^3S^o, ^1P^o)$	5CC+CI,8CC+CI;EXPT	B+	81,82;83
$N^+(^3P, ^1D, ^1S)$	$N^{2+}(^2P^o)$	6CC+CI;CC	;D	84;85
N^{2+}(ground+excited)	N^{3+}	6CC+CI	A	86
$N^{3+}(^1S, ^3P^o)$	N^{4+}	5CC; TDHF	A;	87;79
$N^{4+}(^2S)$	N^{5+}	5CC	A	80
$O^o(^3P, ^1D, ^1S)$	$O^+(^4S^o, ^2D^o, ^2P^o)$	CC+CI;CC;EXPT	B+;D;A	88,125;85;89
$O^+(^4S^o, ^2D^o, ^2P^o)$	$O^{2+}(^3P, ^1D, ^1S)$	CC	D	85
$O^{2+}(^3P, ^1D, ^1S, ^5S^o)$	$O^{3+}(^2P^o, ^4P)$	8CC+CI;CC	A;D	90;85
O^{3+}(ground+excited)	O^{4+}	6CC+CI	A	91
$O^{4+}(^1S)$	O^{5+}	CP;TDHF	B;	92;79
$O^{5+}(^2S)$	O^{6+}	5CC	A	80
$Ne^o(^1S)$	Ne^+	2CC+CI;MBPT;EXPT	A	93;94;73,95
$Ne^+(^2P^o)$	$Ne^{2+}(^3P, ^1D, ^1S, ^3P^o, ^1P^o)$	5CC+CI	B+	96
$Ne^{2+}(^3P, ^1D, ^1S)$	$Ne^{3+}(^4S^o, ^2D^o, ^2P^o, ^4P, ^2D, ^2S, ^2P)$	7CC+CI;CC	B+;D	96;85
$Ne^{3+}(^4S^o, ^2D^o, ^2P^o)$	$Ne^{4+}(^3P, ^1D, ^1S, ^5S^o, ^3D^o, ^3P^o, ^1D^o, ^3S^o, ^1P^0)$	9CC+CI;CC	B+;D	96;85
$Ne^{4+}(^3P, ^1D, ^1S)$	$Ne^{5+}(^2P^o)$	CC	D	85
$Ne^{5+}(^2P^o)$	$Ne^{6+}(^1S)$	CC	D	85
Na^o(ground+excited)	Na^+	MP;EXPT	D	97-99;100,101
$Na^+(^1S)$	Na^{2+}	MBPT	B+	102,103
$Na^{2+}-Na^{4+}$	$Na^{3+}-Na^{5+}$	CP		92
$Mg^o(^1S, ^3P^o, ^1P^o, ^3S)$	Mg^+	4CC	A	104
$Mg^+(^2S)$	Mg^{2+}	MP	A	43
$Mg^{2+}(^1S)$	Mg^{3+}	MBPT	B+	102,103
$Mg^{3+}-Mg^{4+}$	$Mg^{4+}-Mg^{5+}$	CP		92

TABLE 10. Selected bibliography of photoionization cross sections for ions of interest in planetary nebulae studies.

PARENT(STATES)	FINAL ION(STATES)	METHOD	ACC	SOURCE
$Si^{o}(^{3}P, ^{1}D, ^{1}S)$	$Si^{+}(^{2}P^{o})$	7CC+CI;CC	B+;D	104;106
$Si^{+}(^{2}P^{o})$	Si^{2+}	12CC+CI;MBPT;CC		107;108;106
$Si^{2+}(^{1}S)$	Si^{3+}	5CC;CP	;B	109;92
$Si^{3+}(^{2}S)$	Si^{4+}	1CC+MP;MP	;B	109;110
$Si^{4+}(^{1}S)$	Si^{5+}	MBPT	B+	103
$S^{o}(^{3}P, ^{1}D, ^{1}S)$	$S^{+}(^{4}S^{o}, ^{2}D^{o}, ^{2}P^{o})$	6CC+CI;CC	B+; D	104;111
$S^{+}(^{4}S^{o}, ^{2}D^{o}, ^{2}P^{o})$	$S^{2+}(^{3}P, ^{1}D, ^{1}S)$	10CC+CI;CC	B+; D	104;111
$S^{2+}(^{3}P, ^{1}D, ^{1}S)$	$S^{3+}(^{2}P^{o})$	7CC+CI;CC	B+; D	104;111
$S^{3+}(^{2}P^{o})$	$S^{4+}(^{1}S)$	CC	D	111
$S^{4+}(^{1}S)$	S^{5+}	5CC;CP	;B	109;92
$S^{5+}(^{2}S)$	S^{6+}	1CC+MP;CP	;B	109;92
$C\ell^{o}(^{2}P^{o})$	$C\ell^{+}$	MBPT	B+	112
$C\ell^{+} - C\ell^{4+}$	$C\ell^{2+} - C\ell^{5+}$	CP		92
$A^{o}(^{1}S)$	A^{+}	COMP;EXPT	A	93,94,113;73,114
$A^{+}(^{2}P^{o})$	$A^{2+}(^{3}P, ^{1}D, ^{1}S)$	CC	D	106
$A^{2+}(^{3}P, ^{1}D, ^{1}S)$	$A^{3+}(^{4}S^{o}, ^{2}D^{o}, ^{2}P^{o})$	CC	D	106
$A^{3+}(^{4}S^{o}, ^{2}D^{o}, ^{2}P^{o})$	$A^{4+}(^{3}P, ^{1}D, ^{1}S)$	CC	D	106
$A^{4+}(^{3}P, ^{1}D, ^{1}S)$	$A^{5+}(^{2}P^{o})$	CC	D	106
$A^{5+}(^{2}P^{o})$	$A^{6+}(^{1}S)$	CC	D	106
$K^{o}(^{2}S)$	K^{+}	MP;EXPT	D	115;116,117
$K^{+}(^{1}S)$	K^{2+}	MBPT	B+	102,118
$K^{2+}(^{2}P^{o})$	K^{3+}	CC	B+	119
$K^{3+} - K^{4+}$	$K^{4+} - K^{5+}$	CP		92
$Ca^{o}(^{1}S)$	Ca^{+}	4CC;EXPT	C	104;120,121
$Ca^{+}(^{2}S)$	Ca^{2+}	MP	B	43
$Ca^{2+}(^{1}S)$	Ca^{3+}	MBPT	B+	102,118
$Ca^{3+} - Ca^{4+}$	$Ca^{4+} - Ca^{5+}$	CP		92
Fe^{o}	Fe^{+}	MBPT;CP;EXPT		122;123;124
$Fe^{+} - Fe^{5+}$	$Fe^{2+} - Fe^{6+}$	CP		123

TABLE 10. (continued)

REFERENCES

1. Wiese WL, Smith MW and Glennon BM, 1966, "Atomic Transition Probabilities", Vol.1 (NSRDS-NBS4).
2. Hata J and Grant IP, 1981, J.Phys.B, 14, 2111.
3. Drake GWF, 1979, Phys.Rev.A, 19, 1387.
4. Drake GWF and Dalgarno A, 1969, Astrophys.J., 157, 459.
5. Lin CD, Johnson WR and Dalgarno A, 1977, Phys.Rev.A, 15, 154.
6. Berrington KA, Fon WC and Kingston AE, 1982, Mon.Not.R.Astr.Soc., 200, 347.
7. Nussbaumer H and Rusca C, 1979, Astron. Astrophys., 72, 129.
8. Péquignot D and Aldrovandi SMV, 1976, Astron. Astrophys., 50, 141.
9. Thomas LD and Nesbet RK, 1975, Phys.Rev.A., 12, 2378.
10. Nussbaumer H and Storey PJ, 1981, Astron. Astrophys., 96, 91.
11. Tambe BR, 1977, J.Phys.B, 10, L249.
12. Jackson ARG, 1973, Mon.Not.R.Astr.Soc., 165, 53.
13. Nussbaumer H and Storey PJ, 1978, Astron. Astrophys., 64, 139.
14. Dufton PL, Berrington KA, Burke PG and Kingston AE, 1978, Astron. Astrophys., 62, 111; + Daresbury Laboratory Data Bank.
15. Taylor PO, Gregory D, Dunn GH, Phaneuf RA and Crandall DH, 1977, Phys.Rev.Lett., 39, 1256.
16. Gau JN and Henry RJW, 1977, Phys.Rev.A, 16, 986.
17. Zeippen CJ, 1982, Mon.Not.R.Astr.Soc., 198, 111.
18. Berrington KA and Burke PG, 1981, Planet. Space Sci., 29, 377.
19. Dopita MA, Mason DJ and Robb WD, 1976, Astrophys.J., 207, 102.
20. Seaton MJ, 1975, Mon.Not.R.Astr.Soc., 170, 475.
21. Saraph HE, Seaton MJ and Shemming J, 1969, Phil.Trans.R.Soc.A, 264, 77.
22. Hibbert A and Bates DR, 1981, Planet. Space Sci., 29, 263.
23. Nussbaumer H and Storey PJ, 1979, Astron. Astrophys., 71 L5.
24. Nussbaumer H and Storey PJ, 1982, Astron. Astrophys., 109, 271.
25. Nussbaumer H and Storey PJ, 1979, Astron. Astrophys., 74, 244.
26. Osterbrock DE and Wallace RK, 1977, Astrophys.Lett., 19, 11.
27. van Wyngaarden WL and Henry RJW, 1976, J.Phys.B, 9, 1461.
28. Mendoza C and Zeippen CJ, 1982, preliminary results.
29. Le Dourneuf M and Nesbet RK, 1976, J.Phys.B, 9, L241.
30. Pradhan AK, 1976, Mon.Not.R.Astr.Soc., 177, 31.
31. Nussbaumer H and Storey PJ, 1981, Astron. Astrophys., 99, 177.
32. Baluja KL, Burke PG and Kingston AE, 1980, J.Phys.B, 13, 829; 1981, 14, 119.
33. Aggarwal KM, Baluja KL and Tully JA, 1982, Mon.Not.R.Astr.Soc., in press.
34. Hayes MA, 1982, Mon.Not.R.Astr.Soc., 199, 49P; + private communication.
35. Pradhan AK, 1974, J.Phys.B, 7, L503.
36. Giles K, 1981, Mon.Not.R.Astr.Soc., 195, 63P.
37. Baluja KL, Burke PG and Kingston AE, 1980, J.Phys.B, 13, 4675.
38. Dufton PL, Doyle JG and Kingston AE, 1979, Astron. Astrophys., 78, 318; + Daresbury Laboratory Data Bank.
39. Butler K and Mendoza C, 1982, preliminary results.
40. Lin CD, Laughlin C and Victor GA, 1978, Astrophys.J., 220, 734.
41. Clark REH, Magee NH, Mann JB and Merts AL, 1982, Astrophys.J., 254, 412.
42. Fabrikant II, 1974, J.Phys.B, 7, 91.

43. Black JH, Weisheit JC and Laviana E, 1972, Astrophys.J., 177, 567.
44. Mendoza C, 1981, J.Phys.B, 14, 2465.
45. Mendoza C and Zeippen CJ, 1982, Mon.Not.R.Astr.Soc., 199, 1025.
46. Pindzola MS, Bhatia AK and Temkin A, 1977, Phys.Rev.A, 15, 35.
47. Nussbaumer H, 1977, Astron. Astrophys., 58, 291.
48. Mendoza C, 1982, preliminary results.
49. Baluja KL, Burke PG and Kingston AE, 1980, J.Phys.B, 13, L543; 1981, 14, 1333.
50. Flower DR and Nussbaumer H, 1975, Astron. Astrophys., 42, 265.
51. Mendoza C and Zeippen CJ, 1982, Mon.Not.R.Astr.Soc., in press.
52. Mendoza C and Zeippen CJ, 1982, Mon.Not.R.Astr.Soc., 198, 127.
53. Mendoza C, 1982, J.Phys.B, to be published.
54. Dufton PL, Hibbert A, Kingston AE and Doschek GA, 1982, Astrophys.J., 257, 338.
55. Feldman U, Doschek GA and Bhatia AK, 1981, Astrophys.J., 250, 799.
56. Giles K, 1980, Ph.D thesis, University of London.
57. Krueger TK and Czyzak SJ, 1970, Proc.R.Soc.Lond. A, 318, 531.
58. Taylor PO and Dunn GH, 1973, Phys.Rev.A, 8, 2304.
59. Grevesse N, Nussbaumer H and Swings JP, 1971, Mon.Not.R.Astr.Soc., 151, 239.
60. Nussbaumer H and Storey PJ, 1980, Astron. Astrophys., 89, 308.
61. Garstang RH, 1957, Mon.Not.R.Astr.Soc., 117, 393.
62. Garstang RH, Robb WD and Rountree SP, 1978, Astrophys.J., 222, 384.
63. Ekberg JO and Edlén B, 1978, Physica Scripta, 18, 107.
64. Garstang RH, 1958, Mon.Not.R.Astr.Soc., 118, 572.
65. Nussbaumer H and Storey PJ, 1978, Astron. Astrophys., 70, 37.
66. Nussbaumer H and Storey PJ, 1982, Astron. Astrophys., in press.
67. Burgess A, 1964, Mem.R.Astr.Soc., 69, 1.
68. Bell KL and Kingston AE, 1970, J.Phys.B, 3, 1433.
69. Jacobs VL, 1971, Phys.Rev.A, 3, 289; 1974, 9, 1938.
70. Jacobs VL and Burke PG, 1972, J.Phys.B, 5, L67; 1972, 5, 2272.
71. Stewart AL, 1978, J.Phys.B, 11, L431; 1978, 11, 2449; 1979, 12, 401.
72. Marr GV, 1978, J.Phys.B, 11, L121.
73. West JB and Marr GV, 1976, Proc.R.Soc.Lond.A, 349, 397.
74. Woodruff PR and Samson JAR, 1980, Phys.Rev.Lett., 45, 110; 1982, Phys.Rev.A, 25, 848.
75. Burke PG and Taylor KT, 1979, J.Phys.B, 12, 2971.
76. Cantú AM, Mazzoni M, Pettini M and Tozzi GP, 1981, Phys.Rev.A, 23, 1223.
77. Drew JE and Storey PJ, 1982, in progress.
78. Drew JE and Storey PJ, 1982, J.Phys.B, 15, 2357.
79. Watson DK, Dalgarno A and Stewart RF, 1978, Phys.Rev.A, 17, 1928.
80. Pradhan AK, 1982, Phys.Rev.A, 25, 592.
81. Le Dourneuf M, Vo Ky Lan and Zeippen CJ, 1979, J.Phys.B, 12, 2449.
82. Zeippen CJ, Le Dourneuf M and Vo Ky Lan, 1980, J.Phys.B, 13, 3763.
83. Samson JAR and Cairns RB, 1965, J.Opt.Soc.Am., 55, 1035.
84. Nussbaumer H and Storey PJ, 1982, in progress.
85. Henry RJW, 1970, Astrophys.J., 161, 1153.
86. Butler K, J.Phys.B, to be published.
87. Butler K, Lugo L and Mendoza C, 1982, J.Phys.B, to be submitted.
88. Taylor KT and Burke PG, 1976, J.Phys.B, 9, L353.

89. Kohl JL, Lafyatis GP, Palenius HP and Parkinson WH, 1978, Phys.Rev.A, 18, 571.
90. Saraph HE, J.Phys.B, 1982, to be published.
91. Saraph HE, 1980, J.Phys.B, 13, 3129.
92. Reilman RF and Manson ST, 1979, Astrophys.J.Suppl.Ser., 40, 815; 1981, 46, 115.
93. Burke PG and Taylor KT, 1975, J.Phys.B, 8, 2620.
94. Chang TN, 1977, Phys.Rev.A, 15, 2392.
95. Wuilleumier F and Krause MO, 1979, J.Electron Spectrosc.Relat.Phenom., 15, 15.
96. Pradhan AK, 1979, J.Phys.B, 12, 3317; 1980,Mon.Not.R.Astr.Soc., 190, 5P.
97. Laughlin C, 1978, J.Phys.B, 11, 1399.
98. Aymar M, Luc-Koenig E and Combet-Farnoux F, 1976, J.Phys.B, 9, 1279.
99. Aymar M, 1978, J.Phys.B, 11, 1413.
100. Hudson RD and Carter VL, 1967, J.Opt.Soc.Am., 57, 651.
101. Rothe DE, 1969, J.Quant.Spectrosc.Radiat.Transfer, 9, 49.
102. Chang TN, 1977, Phys.Rev.A, 16, 1171.
103. Chang TN and Olsen T, 1981, Phys.Rev.A, 24, 1091.
104. Mendoza C, 1982, to be published.
105. Hofmann H and Trefftz E, 1980, Astron. Astrophys., 82, 256.
106. Chapman RD and Henry RJW, 1972, Astrophys.J., 173, 243.
107. Le Dourneuf M and Zeippen CJ, 1982, in progress.
108. Daum GR and Kelly HP, 1976, Phys.Rev.A, 13, 715.
109. Mendoza C and Zeippen CJ, 1982, in progress.
110. Shevelco VP, 1974, Opt.Spektrcsk.,36, 14.
111. Chapman RD and Henry RJW, 1971, Astrophys.J., 168, 169.
112. Brown ER, Carter SL and Kelly HP, 1980, Phys.Rev.A, 21, 1237.
113. Kelly HP and Simons RL, 1973, Phys.Rev.Lett., 30, 529.
114. Madden RP, Ederer DL and Codling K, 1969, Phys.Rev., 177, 136.
115. Weisheit JC, 1972, Phys.Rev.A, 5, 1621
116. Sandner W, Gallagher TF, Safinya KA and Gounand F, 1981, Phys.Rev.A, 23, 2732.
117. Hudson RD and Carter VL, 1965, Phys.Rev., 139, A1426.
118. Chang TN, 1979, Phys.Rev.A, 20, 291.
119. Combet-Farnoux F, Lamoureux M and Taylor KT, 1978, J.Phys.B, 11, 2855.
120. Carter VL, Hudson RD and Breig EL, 1971, Phys.Rev.A, 4, 821.
121. McIlrath TJ and Sandeman RJ, 1972, J.Phys.B, 5, L217.
122. Kelly HP and Ron A, 1972, Phys.Rev.A, 5, 168.
123. Reilman RF and Manson ST, 1978, Phys.Rev.A, 18, 2124.
124. Lombardi GG, Smith PL and Parkinson WH, 1978, Phys.Rev.A, 18, 2131.
125. Pradhan AK and Saraph HE, 1977, J.Phys.B, 10, 3365.
126. Aggarwal KM, 1982, Mon.Not.R.Astr.Soc., submitted

References to Individual Nebulae

Planetaries

Planetaries (Continued)

H II Regions and Supernova Remnants (SNR)

H II Regions and Supernova Remnants (SNR) (Continued)

Index of Principlal Symbols Used

Note: Symbols used on only one page or two consecutive pages usually are not indexed.

A_m	Magnetic dipole A-value (Eq. 7, P. 119).
$A_{nn'}$	Einstain coefficient (Eq. 2, P. 39).
A_n	Eq. 3, P. 74; Eq. 19, P. 79.
A_q	Electric quadrupole A-value (Eq. 14, P. 76).
A_λ	Extinction coefficient (P. 188).
$A_{\lambda i}$	Eq. 40, P. 129.
a_ν	Atomic absorption coefficient (Eq. 1, P. 39).
$a_n(\nu)$	Continuous atomic absorption coefficient (Eq. 17, P. 41); primes indicate correction for negative absorption.
$B_{n'n}$, $B_{nn'}$	Einstein coefficients of normal and induced absorption (Eq. 2, P. 39).
B_ν	Planck function (Eq. 34, P. 43).
b	Eq. 39, P. 129 *et seq*.
b_n	Factor measuring deviation from local thermodynamic equlibrium (Eq. 19, P. 41).
C	Extinction parameter (Eq. 1, P. 187).
$C_{n''n}$	Eq. 4, P. 74; Eq. 21, P. 79.
c	Velocity of light.
D	Eq. 29, P. 42 (Chap. 3 only).
d	Eq. 39, P. 129 (Chap. 5).
$E_{nn'}$	Eq. 25, P. 42.
$E_{n\kappa}$	Eq. 30, P. 43.
$E_{\kappa n}$	Eq. 59, P. 50.
$E_{\kappa\kappa'}$	Eq. 61, P. 50.
$\underset{o}{E}1\ (X_n)$	Eq. 41, P. 44.
$E_{4,2}(T_\varepsilon)$	Eq. 15, P. 76.
F_n	Eq. 5, P. 74.
$F_{nn'}$	Eq. 24, P. 42.
$F_{n'n}$	Eq. 21, P. 41.
$F_{n\kappa}$	Eq. 27, 28, P. 42.
$F_{\kappa n}$	Eq. 38, P. 44.
$F_{\kappa\kappa'}$	Eq. 1, P. 153.
$F^*(H\beta)$, $F(H\beta)$	Eq. 100, P. 64.
$\pi F^*_\nu(o)$	Emergent energy flux from stellar surface, P. 47.
$\mathcal{F}_{1n}$, $\mathcal{F}_{1\kappa}$	Eq. 1, 2, P. 115.
$\mathcal{F}_{J'J}$	Eq. 23, P. 121.
$f_{n'n}$	Eq. 1, P. 39.
$f(\lambda)$	Extinction function Eq. 4, P. 181, 191 (Seaton).
$f(v, T_\varepsilon)$	Maxwellian velocity distribution function Eq. 32, P. 44.
$G_A(T_\varepsilon)$, $G_B(T_\varepsilon)$	Eq. 48 *et seq*., P. 45.
$g_{n'n}$	Gaunt factor, Eq. 5, P. 39.
g_{II}	Gaunt factor for bound-free transitions, Eq. 61, P. 98.

g_{III}	Gaunt factor for free-free transitions, Eq. 31, P. 90.
h	Planck constant.
$\mathcal{H}$	Eq. 5, 6, P. 116.
I_ν	Specific intensity.
IRE	Infrared excess, Eq. 11, P. 197.
J_ν	Intensity average over 4π solid angle.
K	Eq. 39, P. 44.
k	Boltzmann constant.
$k_\nu^{'ff}$	Corrected mass absorption coefficient for free-free transitions Eq. 28, P. 89.
L_i	Auxiliary function for calculation of ionic concentrations in p^3 configuration, P. 132.
$L^*(\nu)$	Eq. 26, P. 63.
M	Mach number, P. 211.
$M_s(t)$	P. 233.
$N_n(X^i)$	Number of atoms of element X in i^{th} ionization, stage and level n.
$N(H^+)$	Number of protons/cm^3.
$N(n,\ell)$	Eq. 17, P. 76.
$N^*(ion)$	Eq. 38, P. 129.
N_ε	Number of electrons/cm^3.
n	Principal quantum numbers.
P	Pressure.
$P_{nn'}$	Eq. 13, P. 75; Eq. 20, P. 79.
$P_{uv}(*)$	Total number of Lyman quanta emitted by star, P. 25; Eq. 56, P. 48.
Q	Radiated energy per unit mass, P. 209.
Q_{BA}, $Q_{Ly\alpha}$, $Q(2q)$	Eq. 96, P. 63.
$Q(H\beta)$	Eq. 98, P. 64.
$q(J',J)$ or $q_{J'J}$	Activation coefficient, Eq. 24, P. 122.
$q(J,J')$ or $q_{J,J'}$	Deactivation coefficient, Eq. 26, P. 122.
q_ε	Rate coefficient for electron collisions. } Eq. 71,
q_p	Rate coefficient for proton collisions. } P. 100
R	Rydberg constant, Eq. 7, P. 40.
R_c	Rydberg constant for infinite mass, P. 89.
$R_s(t)$	Eq. 7, P. 234.
r_{gal}	Distance from galactic center, Chap. 11.
S_n	Eq. 49, P. 45; Eq. 5, P. 154
S_n, S_q	Forbidden line strengths, Eqs. 8 and 14, Chap. 5.
S_o	Stromgren sphere radius, Eq. 57., P. 48.
T_ε	Electron temperature.
t	$T_\varepsilon/10{,}000$, Eq. 33, P. 128.
U	Excitation parameter, Eq. 58, P. 48.
V	Velocity (cm/sec).
W_g	Geometrical dilution factor (P. 18).
X_n	Eq. 20, P. 41.
$X_{\kappa'}$	Eq. 60, P. 50.
x	$10^{-2} N_\varepsilon\sqrt{T_\varepsilon}$, Eq. 33, P. 128.
Y	Helium content by mass, P. 310.
Z	Nuclear charge on ion.

$\alpha(T_\varepsilon)$	Total direct recombination coefficient, Eq. 68, P. 52.
α_d	Dielectronic recombination coefficient, P. 60.
α_T	Total recombination rate for all processes, Eq. 95, P. 63.
$\alpha_A(T_\varepsilon)$	Total recombination coefficient for all hydrogenic levels, Eq. 42, P. 44.
$\alpha_B(T_\varepsilon)$	Total recombination coefficient on second and higher hydrogenic levels, Eq. 44, P. 45.
$\alpha_n(T_\varepsilon)$	Recombination coefficient on nth level, Eq. 40, P. 44.
$\alpha(H\beta)$	Eq. 14, P. 42.
β	Escape probability for a photon, Eq. 75, P. 108.
β_{CE}	Charge exchange coefficient, P. 63.
$\beta_{n'n}$	Eq. 48, P. 94.
β_{2q}	Eq. 71, P. 100.
$\Gamma(I_v)$	Eqs. 67, 68, P. 52.
$\gamma_{J'J}$	Eq. 30, P. 127.
γ_T	Eq. 67, P. 99.
$\gamma(2q)$	Eq. 72, P. 100.
$\gamma_{n\ell}$	Eq. 74, P. 106.
Δ_{grad}	Galactic abundance gradient, P. 266.
$\Delta\nu_o$	Eq. 5, P. 178.
ε	Charge on electron.
η_ν	Eq. 58, P. 96.
κ	Quantum number for continuum, Eq. 8, P. 40.
κ_c'	Continuum absorption coefficient per unit volume, Eq. 45, P. 94.
κ_L'	Line absorption coefficient per unit volume, Eq. 45, P. 94; primes indicate corrections for negative absorptions.
λ	Wavelength.
λ_μ	Wavelength in micrometers.
ν	Frequency (Hz).
ν_{ff}	Eq. 62, P. 50.
ν_n	P. 44.
ξ	Nonkinetic energy in shock phenomena, P. 209.
ρ	Density (gm cm^{-3}).
σ_v	Target area for collisional excitation by an electron of velocity v, Eq. 22, P. 121.
$\sigma_{\kappa n}$	Target area for recapture on level n from state κ of continuum, Eq. 33, P. 3-5; for H, Eq. 19, P. 161.
τ_ν	Optical depth at frequeny ν.
τ_c	Optical depth in continuum, Eq. 33, P. 90.
ϕ_ν	Line profile function, Eq. 40, P. 93.
χ	Excitation potential.
χ	[In Chap. 5(iii) only: a measure of deviation from LS coupling.]
$\chi_{J'J}$	Excitation potential difference between levels J' and J.
Ψ	Quantum mechanical wave function, Eq. 11, P. 119.
Ω	Collision strength, Eq. 22, P. 121.
ω_n	Statistical weight, P. 39.

Subject Index